APPLIED URBAN ECOLOGY

APPLIED URBAN ECOLOGY: A GLOBAL FRAMEWORK

Edited by

Matthias Richter

Environmental Scientist, Publicist and University Lecturer, Germany

Ulrike Weiland

Institute for Geography, University of Leipzig, Leipzig, Germany

WILEY-BLACKWELL

A John Wiley & Sons, Ltd., Publication

This edition first published 2012 © 2012 by Blackwell Publishing Ltd.

Blackwell Publishing was acquired by John Wiley & Sons in February 2007. Blackwell's publishing program has been merged with Wiley's global Scientific, Technical and Medical business to form Wiley-Blackwell.

Registered office: John Wiley & Sons, Ltd, The Atrium, Southern Gate, Chichester, West Sussex, PO19 8SQ, UK
Editorial offices: 9600 Garsington Road, Oxford, OX4 2DQ, UK
The Atrium, Southern Gate, Chichester, West Sussex, PO19 8SQ, UK
111 River Street, Hoboken, NJ 07030-5774, USA

For details of our global editorial offices, for customer services and for information about how to apply for permission to reuse the copyright material in this book please see our website at www.wiley.com/wiley-blackwell.

Library of Congress Cataloging-in-Publication Data
Applied urban ecology : a global framework / edited by Matthias Richter and Ulrike Weiland.
p. cm.
Includes bibliographical references and index.
ISBN 978-1-4443-3339-8 (hardcover : alk. paper) – ISBN 978-1-4443-3340-4 (pbk. : alk. paper) 1. Urban ecology (Biology) 2. Applied ecology. I. Richter, Matthias, 1964- II. Weiland, Ulrike.
QH541.5.C6A67 2011
577.5′6 – dc23

2011019097

A catalogue record for this book is available from the British Library.

This book is published in the following electronic formats: ePDF [9781444344998]; Wiley Online Library [9781444345025]; ePub [9781444345001]; Mobi [9781444345018]

Set in 9/11pt Photina by Laserwords Private Limited, Chennai, India
Printed and bound in Malaysia by Vivar Printing Sdn Bhd

1 2012

Contents

List of contributors

Surinder Aggarwal
Department of Geography, University of Delhi, Delhi, India

Dinand Alkema
Faculty of Geo-Information Science and Earth Observation (ITC), University of Twente, Department of Earth Systems Analysis, Enschede, The Netherlands

Peter Baccini
ETH Zürich, Zurich, Kriesbühl, CH-6027 Römerswil, Switzerland

Ellen Banzhaf
Working Group Geomatics, Department of Urban and Environmental Sociology, Helmholtz Centre for Environmental Research – UFZ, Leipzig, Germany

Carsten Butsch
Institute for Geography, University of Cologne, Cologne, Germany

Alexander Buyantuyev
Sino-US Center for Conservation, Energy, and Sustainability Science (SUCCESS), Inner Mongolia University, Hohhot, China

Sarel Cilliers
School of Environmental Sciences and Development, North-West University, Potchefstroom, South Africa

Elandrie Davoren
School of Environmental Sciences and Development, North-West University, Potchefstroom, South Africa

Christina von Haaren
Institute of Environmental Planning, Gottfried Wilhelm Leibniz University of Hannover, Hanover, Germany

Harriet Herbst
Freelance translator, Leisniger Chaussee 10, 04769 Ablass, Germany

Maria Ignatieva
Department of Urban and Rural Development, Division of Landscape Architecture, Swedish University of Agricultural Sciences, Uppsala, Sweden

G. Darrel Jenerette
Department of Botany and Plant Sciences, University of California, Riverside, California, USA

Norman Kerle
Faculty of Geo-Information Science and Earth Observation (ITC), University of Twente, Department of Earth Systems Analysis, Enschede, The Netherlands

Jennifer Litteral
School of Life Sciences, Arizona State University, Tempe, Arizona, USA

Alex Loftus
Department of Geography, Royal Holloway, University of London, Egham, Surrey, United Kingdom

Rina Lubbe
School of Environmental Sciences and Development, North-West University, Potchefstroom, South Africa

Kaesha Neil
School of Life Sciences and Global Institute of Sustainability, Arizona State University, Tempe, Arizona, USA

Maik Netzband
Geomatics, Geography Department, Ruhr-University Bochum, Bochum, Germany

Marialena Nikolopoulou
Kent School of Architecture, University of Kent, Canterbury, United Kingdom

Matthias Richter
Environmental Scientist, Publicist and University Lecturer, Germany
Contact information: http://www.Dr-Matthias-Richter.de

Dieter Rink
Department of Urban and Environmental Sociology, Helmholtz Centre for Environmental Research – UFZ, Leipzig, Germany

Weijun Shen
South China Botanical Garden, Chinese Academy of Sciences, Guangzhou, China

Stefan Siebert
School of Environmental Sciences and Development, North-West University, Potchefstroom, South Africa

Antje Stokman
Institute for Landscape Planning and Ecology, University of Stuttgart, Stuttgart, Germany

Ulrike Weiland
Institute for Geography, University of Leipzig, Leipzig, Germany

Jianguo Wu
School of Life Sciences and Global Institute of Sustainability, Arizona State University, Tempe, Arizona, USA; Sino-US Center for Conservation, Energy, and Sustainability Science (SUCCESS), Inner Mongolia University, Hohhot, China

Kongjian Yu
Graduate School of Landscape Architecture, Peking University, Beijing, China

Foreword

Jürgen Breuste

Urban and Landscape Ecology, Department of Geography and Geology, Paris Lodron University Salzburg, Salzburg, Austria

The majority of people worldwide live in cities and their number continues to increase. It seems that our common future has an urban perspective. Worldwide, cities offer an attractive living environment for the majority of people. This growth process of the cities was observed and described decades ago, but only in the last few decades has the drama of this development been recognized. Not only do highly effective socioeconomic urban systems arise, which compete with one another, but also new living areas for people, whose current and future main living space is the city.

In the old cities, which can already look back upon several centuries of development, sometimes hard-won experiences already exist for the continuing improvement of this living area for people and to technically modify that which exists to the changing demands of humans and the economy. Shrinkage, multiethnicity and the development of the city into the surrounding countryside – the last being a process which endangers nature and is energetically questionable – provide new challenges.

In the dynamically growing cities, in particular in Asia and parts of Africa and Latin America these experiences first have to be made. Here, it is often difficult to obtain an overview of the growth processes, let alone to be able to control these optimally or even at all. Thus cities arise, which not only grow more quickly than their economic base would necessitate, but also environmental conditions arise that are unhealthy and risky for the people living in and moving to the cities. This does not only apply to the mega-cities, but also to the many "small" cities with a population of much less than 1 million. Unprepared city administrations are confronted with what are often unsolvable problems. A few targeted measures of environmental hygiene, traffic development or the control of residential development could already hugely improve the situation.

The global perspective on the development of the cities is thus heterogeneous and is dependant on growth dynamics, socioeconomic conditions, cultural relationships, and last but not least on the ecological framework conditions of the regions. Cities cannot be meaningfully compared to one another, but their problematic situations can be. These require structuring interventions and controls, often very quickly, in order to avoid deep-rooted, long-lasting and encumbering effects on people's health, the environmental conditions and the reduction in efficiency of the future urban development.

It is no longer possible to support this through one or a few research fields. The problem conditions are too complex, the connections between sociocultural, economic and environmental systems too strong. Technical solutions alone are no longer sufficient and their arbitrary transfer to culturally and economically different urban systems has often even caused further problems, which had not been thought of before. The singular optimization of urban development aspects is almost ruled out in favor of a multi-criteria approach.

Since the 1970s this point of view has gained more and more notice from academics of various disciplines in Central Europe. There is and was one common element: to provide a contribution to the improvement in management through the academic research of the complex urban systems or their parts. It was, and is, obvious that the ecosystem city (created, used and more or less controlled by people) has to be the object of research and design. This open and highly complex ecosystem city became the object of specialists, but even more so of concerted and cooperative research from many different specialist perspectives. Here the traditional, limited outlook first had to be overcome, which did not want to interpret the city in terms of ecological systems.

Although it was first biologists, climatologists and urban planners who occupied themselves with the urban ecosystems, hydrologists, pedologists, geographers, economists, political scientists, and planners from various fields quickly came as well, and many others too. A cross-sectional discipline – urban ecology – was formed. This development process of the discipline of urban ecology, which has already existed for 40 years, has, however, not yet been finalized. The theoretical–methodical basis is naturally heterogeneous and strongly determined by the participating disciplines and their approaches. What urban ecology is or can be depends less on the relationship to a fixed theory and methodical structure, but more on the perspective of the participating academics and disciplines. The terminus ecology itself, from an urban context, has developed from relating to the life–environment–relations in the biosciences, for example, into the normative aspects of planning and politics. There is no ordering institutionalization here. It is, at least at first, also not completely necessary.

Landscape ecology is also a complex cross-sectional discipline integrated into urban ecology early on, first as an exotic section. Now with research topics, research institutions and user requirements worldwide urban ecology has proved itself to be an important part of landscape ecology. In 2009 it established itself for the first time independently and self confidently in a worldwide Society of Urban Ecology (SURE).

Our knowledge of the urban ecosystems, their spatial–functional process pattern and their interaction are still incomplete and limited and thus urgently need to be extended. On the other hand the requirements of the design process aimed at the urban ecological viewpoint are so pressing and in need of being served, that these cannot remain unfulfilled with reference to the not yet complete knowledge about the urban ecosystems. For this reason urban ecologists must work in a particularly applied manner. Their field of operation lies here, it is here that they can gain increasing acceptance and help to solve manifold problems of the cities.

The merit of this book is to provide a contribution for this from a global viewpoint. The variety of approaches to urban ecology and its challenges are consciously made the focus here. The approaches of the participating disciplines will be portrayed, although this is only possible in exemplary form, and the wide field of uses will be dealt with to a notable extent. Here it is mainly the huge challenges such as the retention or production of healthy living conditions in cities, dynamic changes of use, adaptation to climate change and the preservation of biodiversity, which are dealt with as focal points. The synthesis, to continue to further develop a multicriteria, methodical and manifold urban environmental management, without ignoring the extension of the academic-theoretical basis of urban ecology, is obvious.

With this book Ulrike Weiland and Matthias Richter have produced an excellent and suitable text book of applied urban ecology, which takes on the current challenges in a global perspective and uses examples to demonstrate approaches and solutions.

PART I

INTRODUCTION

CHAPTER 1

Urban ecology – brief history and present challenges

Ulrike Weiland[1] and Matthias Richter[2]

[1] *Institute for Geography, University of Leipzig, Leipzig, Germany*
[2] *Environmental Scientist, Publicist and University Lecturer, Germany**

1.1 INTRODUCTION

Urban ecology plays an important role in understanding urban systems. In order to analyze and apprehend, for instance, urban land use changes and their impact on the regional water balance, the role of urban green spaces for the local climate, conditions for the coexistence of species in an urban setting, or resource fluxes and opportunities to reduce and optimize them, it is necessary to know how urban systems function and how and to what extent they both impact and are affected by global or regional processes. Urban ecology is characterized by a variety of approaches. It is an interdisciplinary research field at the interface of natural sciences, social sciences and humanities as well as engineering. As an interdisciplinary research field urban ecology investigates the interrelations between environmental compartments and human activities such as construction, production, housing, and transport. As an applied scientifically based approach, urban ecology gives hints about where and how the urban environment may be protected from further harm, and how environmental quality and thus physical human living conditions may be improved.

Today's urban ecology differs widely from its beginnings. Its traditional lines have a close connection to the scientific and social context of their time as well as to the respective urban structures. Five traditional strands of urban ecology can be differentiated by analyzing their preferred research aims, their scientific approach, and their research motives; they occur partly in parallel but at the same time. Knowledge of the history of urban ecology is helpful in order to understand the recent approaches presented in this book; therefore general lines of tradition of urban ecology will be discussed first in the following prior to depicting present challenges (Richter and Weiland 2008, Weiland and Richter 2009).

1.2 BRIEF HISTORY

1.2.1 Initials in urban natural history

The beginning of urban ecological research is rooted in 16th century observations of nature in cities when spontaneously growing species in the cities drew the attention of botanists to stone walls, castles, and ruins, which were identified as the first "habitats" (Sukopp 1994, 2002). Since the notion of "urban ecology" did not yet exist at that time, this approach has been called the "line of tradition rooted in natural history" (Weiland and Richter 2009, p.50). At that time the European city was characterized by a very high building density within the city walls. Extensions of the built-up area set up the preconditions for further urban development (Mumford 1963, Lichtenberger 2002). Knowledge of the medical use of herbs was widespread. Thus, it is not surprising that interest in acquiring knowledge about nature was applied to cities at these

*Contact information: http://www.Dr-Matthias-Richter.de

Applied Urban Ecology: A Global Framework, First Edition. Edited by Matthias Richter and Ulrike Weiland.

early times. Today, this traditional line continues as a partial aspect of bioecological (e.g., Penev *et al.* 2005) and biogeographical research approaches (e.g., Wania *et al.* 2006).

1.2.2 Socioecological tradition

During the industrialization phase, the sociology branch of the Chicago School had a major influence on the development of urban ecology. In the 1920s Chicago was in the heyday of industrialization, and was a typical example of the rapidly growing, unsanitary industrial cities of the 19th and early 20th centuries in North America and Europe, with extremely high-density tenements, deficits in water supply, sewage and waste disposal, poor air quality, and poor lighting conditions. These conditions and their consequences provoked Robert E. Park to commence his socioecological studies, which were strongly influenced by Georg Simmel and Max Weber, amongst others (Park *et al.* 1925). The Chicago School investigated the interrelations between city and society, in particular the living conditions of the industrial workers. Robert E. Park and Ernest W. Burgess tried to explain the urban development processes of Chicago and their impacts on social groups by means of a human–ecological or "quasi biological" research approach (Feagin 1998, p.2) using theoretical concepts of animal and plant ecology: e.g., succession, symbiosis, competition, and adaptation (Kurtz 1984, p.21). They explained phenomena such as migration and segregation phases of different population classes and social minorities with the help of "invasion-succession cycles" and city structure models. This approach has been profoundly criticized because of its biological basis. Later on, the socioecological tradition of urban ecology was superseded by a collateral human ecological perspective (e.g., Winter and Mack 1988, Fellenberg 1991).

1.2.3 Complex bioecological tradition

The perception of the finite nature and instability of the supply of fossil resources increased after the climax of the economic boom in the United States and Europe that followed the Second World War. As a consequence, urban ecology received much more attention than before. In the 1970s Herbert Sukopp and a group of colleagues developed a complex bioecological approach to urban ecology (e.g., Sukopp 1973, 2005). In its early years, the Berlin School of Urban Ecology carried out mainly ecological site analyses and field botany research on wasteland that existed in great quantities in Berlin in the years after the Second World War. Its approach, which maintains urban flora, fauna, and habitats as its core, can be considered a more versatile strand of the line of tradition rooted in natural history. In this approach humans influence and superimpose natural habitat conditions, especially in the form of land use and land use changes. Research is centered on organisms, species, and their habitats; additionally urban climate, soil, and water bodies are investigated, mainly as habitat conditions for urban flora and fauna. Humans play a role as a source of disturbance and as users of urban nature, above all for recreational purposes. A further central and application-oriented motive for research is to transfer nature conservation to cities and urban areas in order to protect urban nature for the human inhabitants. Further research approaches that can be considered to be examples of this line of thought are presented, for example, by Wittig (1991) and Gilbert (1989).

1.2.4 Ecosystem-related tradition

During the same time period as the Berlin School of Urban Ecology, an (eco-)system related tradition of urban ecology evolved on an international level. This approach is highly influenced by American and German landscape ecology (Tansley 1935, Troll 1939, 1968, Schmithüsen 1942, Neef 1967) and systems theory (von Bertalanffy 1953), the systemic approach linking both research directions; later influences can be characterized by the keywords "patterns and processes." Major international research programs such as UNESCO's Man and the Biosphere Program (MAB) (Spooner 1986) and the International Biological Program (IBP) initiated large research projects in this area. The ecosystem-related tradition is heterogeneous; two main directions can be distinguished.

The first direction: ecological analyses of urban landscapes

Landscape ecological studies were assigned to cities and city sectors with the aim of identifying ecological patterns and processes. Within this theoretical framework, a long history of approaches focuses on the analysis of urban–rural gradients (McDonnell and Pickett 1990, Kinzig and Grove 2001). Studies of metapopulation theory, which also have been carried out in cities since the 1990s, often show overlaps between organism approaches and landscape ecology approaches (Niemelä *et al.* 2002). Further research projects focus on the interrelations between urban structures and compartments of the urban natural environment (e.g., Breuste *et al.* 1998).

The second direction: analyses of urban material and energy flows

The *Fundamentals of Ecology* (Odum 1953), serving as the scientific basis of this research approach, explains physical and chemical processes of aquatic, terrestrial, and anthropogenic ecosystems using a systems approach. Research is not focused on organisms, but on substances and material flows. From the 1970s onward, energy flows were also included, induced by the oil crisis that promoted awareness of the impermanent character of natural resources. Independently from each other, the ecologist Howard T. Odum (1953) and the urbanist Lewis Mumford (1963) influenced architects as well as urban and regional planners in the subsequent ecological movements – triggered by the Club of Rome's publication *Limits to Growth* (Meadows *et al.* 1972) – to take into consideration both the cultural–historical and the ecosystems approach in their plans and concepts. Material and energy flow studies of, for example, Brussels (Duvigneaud 1974) and Hong Kong (Boyden *et al.* 1981) were conducted (see also Baccini 1996). Since then, cities have been considered "importers" and "intermediate stores" (Baccini and Bader 1996) of large masses and of a variety of resources (Chambers *et al.* 2001). Quantitatively the most important fluxes are those of energy, water, food, and building materials. Since recycling processes barely exist, warmth, waste water, garbage, and waste air are deposited, pass through the urban environmental systems and cause local, regional, and sometimes global environmental problems (Davíla and Atkinson 1999). The most important merits of this approach include its contribution to an increased understanding of how (and which) substances accumulate in different ecosystem compartments (e.g., urban soil, ground floor vegetation, trees) and how they can become dangerous for plants, animals, and humans via food webs. Furthermore, the identification and quantification of regional to global material and energy fluxes has increased the understanding of global interconnectedness of the single city, not only in economic aspects, but also with respect to resource flows and environmental pollution.

Long-term ecological research (LTER) sites

A considerable step forward in international recognition of urban ecology can be attributed to the interdisciplinary research teams at the Long-Term Ecological Research (LTER) sites in Baltimore and Phoenix in the United States and to cooperating research groups (Alberti 2008, Grimm *et al.* 2008, Marzluff *et al.* 2008). The research on urban LTER sites also can be traced back to the ecosystem-related tradition and at the same time it deepens the urban ecological knowledge and enriches it by using up-to-date techniques.

The urban ecology research program in Baltimore, Maryland, seeks to understand an urban region as an ecological system and investigates

1 the relations between spatial structures of socioeconomic, ecological, and physical features and their changes over time,
2 the fluxes of energy and matter as well as human influences on these, and
3 options to improve the quality of the urban environment.

The urban ecological research program in Phoenix, Arizona investigates the structures and functions of an urban ecosystem in an arid environment – the Sonoran Desert – and aims at understanding the distribution and perception of ecosystem services in the metropolitan region of Phoenix (see Chapter 4). Urban ecology, according to this research approach, considers urban agglomerations as complex systems with integrated social, economic, ecological and technical subsystems. It analyses, for example, socioecological drivers of land management and ecosystem responses, nitrogen fluxes as well as social vulnerability, environmental inequity and health (ASU 2011).

1.3 RECENT AND PRESENT CHALLENGES

Several main steps forward in urban ecology can be observed in recent times: on the one hand, urban ecological research is increasingly carried out in international settings, parallel to economic globalization, while most research activities are still based in North America and Europe (Deeter 2003). On the other hand, a tendency of convergence of research themes and methods can be observed, although the orientation of urban ecological research towards ecosystem research is stronger in North America than in Europe (Zipperer *et al.* 2000).

Furthermore a complex systems approach to cities and their ecology has become common within the majority of the scientific community. Today, cities are considered to be coupled complex human and ecological systems. They are characterized by non-linear development over time with unforeseen changes and leaps forward, and new emerging properties. Multiple agents, feedback mechanisms and their variability in time and space imply complexity, heterogeneity, and surprise as inherent characteristics of urban ecosystems. As a consequence, their behavior is not exactly predictable and prognoses are uncertain (Alberti 2008, p.225 ff.). Urban risks as well as vulnerability and resilience of urban systems represent recent research issues. These advancements in urban ecological research correspond with the growing challenges to urban ecology posed by the larger urban impacts on the environment.

Today, urbanization, i.e., the change of lifestyles, and urban growth, metropolization and mega-urbanization are dominant urban development processes in most parts of the world. Since 2008, the majority of the world population is urban[1], and in 2050, this share is expected to be 70%. The 21st century is addressed as the "urban century" (UN ESA 2008).

Urban areas are subject to large-scale influencing factors such as economic, sociodemographic, technological, and environmental change, referred to as global, demographic and climate change. They imply chances and risks for the cities and urban agglomerations.

In regard to this, urban development processes are characterized by a large variability; dynamically growing agglomerations contrast with economically weak and stagnating cities, whereby growth and shrinkage processes also occur at the same time within cities. In (former) industrialized cities large brownfield areas remain as relicts of the Industrial Age comprising socioenvironmental risks as well as opportunities for new beginnings.

From an environmental point of view, urban sprawl (urban expansion up to the "networking of cities"), causes not only an increase of resource use, but also a fragmentation of landscapes and the loss of natural areas. Economic activities and transport lead to environmental pollution far beyond the physical and administrative urban borders (Hall and Pfeiffer 2000). Urban growth and urbanization are changing a growing number of natural or seminatural habitats and former agriculturally productive areas, for example, by producing heat islands, causing water pollution, and by contributing to desertification. Urban systems are "flow-through systems"; by far the largest part of energy and material fluxes of the human economy returns in an altered form as pollution and waste to the ecosphere. A number of scientists, prominent among them Saskia Sassen, consider that "this makes cities a source of most of the environmental damage, and some of the most intractable conditions feeding the damage" (Sassen 2009, p.46). However, it is not urbanization *per se* that necessarily causes negative impacts on the environment, but rather the particular types of urban systems and the way of industrial production as well as deficits in urban governance. Since cities are not only polluters, but also sites for innovation, it is "within the complexity of the city that we must find the solutions to much environmental damage and the formulas for reconfiguring the socioecological system that is urbanization. [. . .] Cities make the multi-scalar property of ecological systems present and recognizable" to decision makers and urban inhabitants. As a consequence, improving the urban environment and reducing the environmental impacts of urbanization on regional, national, and the global environment entails a multiscalar approach (Sassen 2009, p.46).

Global urbanization and its unintended negative effects, not only on the urban environment and the urban dwellers, but also on the entire environment, necessitate sustainable urban development on various scales. Habitat Agenda and Agenda 21 contributed to establishing a common understanding that cities rebound to global and local environmental problems and are a potential arena in which to address sustainability. Many cities try to implement sustainable urban development (c.f. extensive database in IISD 2010); it is about – according to Agenda 21 (UNEP 1992) – integrating ecological, economic, social, and cultural aspects of urban development in a long-term perspective, including good human health conditions. Sustainable urban development requires the cooperation of a variety of authorities, stakeholders, and social groups on different political levels, including the heads

[1] The definition of "urban" differs between countries; here it depicts the percentage of the total population living in areas termed "urban" by the respective country. The definitions range from population centers of 100 or more dwellings to only the population living in national and provincial capitals.

of international organizations and consortia. It applies to the local level while considering regional, national and global interrelationships, because the ecological regulation of cities can no longer be separated from wider questions of regional, national, and global governance. Sustainable urban development is a goal of political negotiation with no foreseeable end point (Grunwald and Kopfmüller 2006, Atkinson *et al.* 2007, Girardet 2007).

Considering the differences between cities and the variety of urban development processes, it becomes obvious that the general meaning of sustainable development has to be transferred to the prevailing local conditions, and that every city has to find its own way of striving for sustainability. As a consequence, differing understandings of sustainable urban development exist; the complexity of the integrated model and its need for interpretation are both its characteristic and its problem. In Europe important objectives are, for example, mixed urban development and the "city of short distances" (*Stadt der kurzen Wege*) in order to avoid further urban sprawl (Baccini and Oswald 1998, Oswald and Baccini 2003). The large amount of public awareness that the sustainability discussion enjoyed in the 1990s has diminished to date due to changes in the relevance of issues on the political agenda. Nevertheless, sustainable urban development has never disappeared from the political agenda; international organizations work on its implementation, and the scientific discussion has never broken off (Girard *et al.* 2005, Elliott 2006, Atkinson *et al.* 2007, Girardet 2007).

Parallel to the sustainability discourse, but only partly linked to it, and only indirectly referring to urban ecology, the discourse on "new urbanity" has spread among architects and planners in North America and Europe since the 1980s (Häußermann and Siebel 1995, Swyngedow *et al.* 2002, Oswald 2003). "New urbanity" is understood as "the way of living of the majority of people in developed countries" that is "based on modern technical, social and organisational preconditions" (Oswald and Baccini 2003, p.291). It is an alternative draft to the decline of heavy industry, river- and seaports, to suburbanization and urban sprawl. The concept of "new urbanity" is multifaceted: urban reconstruction, revitalization of urban brownfields, rebuilding of historic townscapes, and the rediscovery of urban waterfronts characterize the concept. Some examples are the Docklands in London and Dublin, Kop van Zuid in Rotterdam, or Harbour City in Hamburg.

Both concepts "sustainable urban development" and "new urbanity," have an influence on urban development that is difficult to measure and mostly concentrated in scattered projects. In the majority of cases "new urbanity" specifically addresses urban ecology only indirectly. Nevertheless, urban environmental research has been carried out and discussed more frequently under the aspects of resource protection and diminution of environmental pollution aiming at equal opportunities for future generations, which are considered constituent parts of the integrative concept of sustainable (urban) development (IWM EB 2002, Marchettini *et al.* 2004, Mander *et al.* 2006, Weiland 2006, Kennedy *et al.* 2007).

At the International Conference of Urban Ecology in 1997 in Leipzig, a broad scope of issues related to urban ecology, the perception of urban ecological issues by urban citizens, and policy approaches were discussed (Breuste *et al.* 1998). Furthermore, "redesigning the urban metabolism in view of sustainability goals" is considered a relevant research question for urban ecology (Brunner 2007). These examples show that urban ecological research has both thematically expanded compared to previous decades and has also shifted towards investigating the applicability of research findings in urban decision making (Baccini 1996, Alberti *et al.* 2003, Pickett *et al.* 2004, Müller *et al.* 2008).

1.4 PURPOSE AND STRUCTURE OF THE BOOK

1.4.1 Purpose of the book

This book *Applied Urban Ecology – A Global Framework* bridges the gap between theory and practice and presents a broad spectrum of urban ecology approaches from systems research to environmentally sound urban design, exemplified by selected case studies from different continents. This claim is met by engaging experts from geographically different parts of the world (in alphabetical order): from China, Germany, India, the Netherlands, New Zealand, South Africa, Sweden, Switzerland, United Kingdom, and the United States, and by including widely acknowledged scientists on a keynote speaker level accompanied by up-and-coming scientists. The conception of the book derives from a 5-year study "Urban Ecology – an International Comparison" and an investigation of the preferred up-to-date research questions at large research institutions.

Based on this study, the book portrays a range of recent approaches to urban ecology and focuses on providing knowledge and expertise for the application of urban ecological findings. *Applied Urban Ecology – A Global Framework* is a contribution to support the improvement of environmental quality in urban settings by mutually learning from other attempts.

The purpose of this book is to provide urban ecological knowledge in a nutshell tailored to supporting environmentally sound urban development and based on a sound theoretical and contextual framework, including case studies. It is conceived as a textbook targeted for an international readership, especially for advanced students and PhD students, researchers, and experienced practitioners in the fields of urban ecology and landscape ecology, urban environmental research, environmental geography, urban planning and landscape architecture, and sustainable urban development.

1.4.2 Structure of the book

The structure of the book reflects the variety of approaches to modern urban ecology.

Parts I–II

In Part I "Introduction", Ulrike Weiland and Matthias Richter lead into the subject matter of the book by portraying urban ecological research approaches briefly within the context of their respective societal context, and by depicting recent and present challenges to urban ecology.

In Part II "Urban Ecology: Related Disciplines and Methods", Matthias Richter and Ulrike Weiland give an overview of disciplines and methods related to urban ecology. Ellen Banzhaf and Maik Netzband discuss opportunities of monitoring urban land use changes with remote sensing techniques.

Part III

In Part III "Selected Fields of Applied Urban Ecology" case studies on various topics from different parts of the world are presented. It becomes obvious that in different countries different problems are on the agenda of urban ecologists. It is of the utmost importance to be aware of the ways in which urban ecological topics are determined and constructed, as well as the respective goals of research. The following issues are addressed:

Pathways of the ecosystem approach

The ecosystem approach has been one of the most influencing research pathways in the field of urban ecology during the last 25 years. Pathways of the ecosystem approach are discussed from two different perspectives. Jianguo Wu and his co-authors report about the research experiences in long-term research areas that are in the framework of the LTER network in the United States: Baltimore, Maryland and Phoenix, Arizona. Another prominent pathway of the ecosystem approach is represented by the work of Peter Baccini. He shows how sustainable use of substances and energy in the urban area should be combined with urban planning, thus leading to a resource-sparing urban development. This approach uses the "Schweitzer Mittelland" as a case study.

Socioenvironmental threats

The focus in this section is on changes in urban areas resulting in socioenvironmental consequences and it explores the question of how everyday life for humans is influenced. The chapters cover a wide range of urban environments in industrialized countries as well as developing countries and sheds light on different types of socioenvironmental and health threats. Surinder Aggarwal and Carsten Butsch show how the growth of Indian mega-cities has been and is being accompanied by severe environmental problems and health risks, especially related to air and water pollution and poor sanitary conditions. Dieter Rink and Harriet Herbst highlight abandoned green open spaces from different perspectives including socioecological aspects. The different meanings of urban wilderness for citizens are contextualized in their contribution.

Flooding and climate adaptation

In recent times global change has been a more intensively discussed topic in general and this also influences its relevance for urban ecology. Therefore, this up-to-date topic is included. Dinand Alkema and Norman Kerle report about their experiences with flood risk assessment in Southeast Asian cities, such as Naga, Philippines. They use different methods (e.g., GIS and scenario techniques) for warning urban citizens about different types of flooding. Marialena Nikolopoulou demonstrates how urban open spaces can be adapted to climate change using her case studies from different parts of the world.

Urban biodiversity

Urban biodiversity is one of the key topics in urban ecology. Especially in Central Europe, there is a long research tradition in urban biodiversity. More recently, this topic is often connected with socioeconomic activities. Sarel Cilliers and his coauthors write about the connection between biodiversity in selected urban areas of South Africa and local land use management practice, which has been shown to be class- and income-dependent. Maria Ignatieva demonstrates that globalization, combined with the dispersion of plant material, results in problems for locally adapted native species. Despite these negative influences she names techniques and examples from Russia, the United States, and New Zealand showing how to foster native biodiversity.

Environmental urban design

Design and planning of urban green and open spaces are the focus of this section. Kongjian Yu demonstrates that planning for green infrastructure can guide city planning and describes how this is implemented in growing Chinese cities. He combines and applies the principles of landscape urbanism with Feng Shui for his planning attempt. Landscape planning and landscape design for urban areas are mostly separated procedures originating from different disciplinary backgrounds. Antje Stokman and Christina von Haaren are building a bridge between these two disciplines and show how they should be linked for win–win situations and advantages for users.

Environmental urban politics

The availability of water, its quality and quantity, is one of the major environmental challenges for the future. Many of the conflicts are combined with the distribution of political power and how this is negotiated. Originating from a strongly coined "natural science background", urban ecologists are in the process of integrating governance aspects into their research. Alex Loftus compares different attempts of solving water conflicts in selected urban areas in South Africa and South America. Here the suitability of public, private or intermediate water supply is addressed. This is discussed by taking into account the differing political background of capitalism and socialism.

Part IV-synthesis

The synthesis in Part IV drawn by Matthias Richter and Ulrike Weiland provides a cross-cutting assessment of the approaches presented before. The findings from different parts of the world are reflected against recent challenges of urban ecology. Conclusions are drawn regarding how to improve urban environment taking into consideration the findings of urban ecology research. Hence a framework for "urban green governance" and for supporting an ecological urban development is presented, which is focused on relevant and up-to-date knowledge.

REFERENCES

Alberti, M. (2008) *Advances in Urban Ecology. Integrating Humans and Ecological Processes in Urban Ecosystems.* Springer, New York.

Alberti, M., Marzluff, J.M., Shulenberger, E., Bradley, G., Ryan, C., Zumbrunnen, C. (2003) Integrating Humans into Ecology: Opportunities and Challenges for Studying Urban Ecosystems. *BioScience* 53(12), 1169–1179.

ASU (Arizona State University, Global Institute of Sustainability) (2011) Central Arizona- Phoenix Long-Term Ecological Research (LTER) – Making New Discoveries. http://caplterasu.edu/research/ (accessed 18 April 2011).

Atkinson, G., Dietz, S., Neumayer, E. (eds) (2007) *Handbook of Sustainable Development*. Edward Elgar, Northampton.

Baccini, P. (1996) Understanding regional metabolism for a sustainable development of urban systems. *Environmental Science and Pollution Research* 3(2), 108–111.

Baccini, P., Bader, H.P. (1996) *Regionaler Stoffhaushalt: Erfassung, Bewertung und Steuerung*. Spektrum, Heidelberg.

Baccini, P., Oswald, F. (eds) (1998) *Netzstadt. Transdisziplinäre Methoden zum Umbau urbaner Systeme.* vdf Hochschulverlag, Zurich.

von Bertalanffy, K.L. (1953) *Biophysik des Fließgleichgewichtes. Einführung in die Physik offener Systeme und ihre Anwendung in der Biologie*. Vieweg, Braunschweig.

Boyden, S., Millar, S., Newcombe, K., O'Neill, B. (1981) *The Ecology of a City and its People: The Case of Hong Kong.* Australian National University Press, Canberra.

Breuste, J., Feldmann, H., Uhlmann, O. (eds) (1998) *Urban Ecology*. Springer, Berlin.

Brunner, P.H. (2007) Reshaping urban metabolism. *Journal of Industrial Ecology* 11(2), 11–13.

Chambers, H., Simmons, C., Wackernagel, M. (2001) *Sharing Nature's Interest: Ecological Footprints as an Indicator of Sustainability*. Earthscan, London.

Dávila, J.D., Atkinson, A. (1999) Organisation and politics in urban environmental management. In: Atkinson, A., Dávila, J.D., Fernandes, E., Mattingly, M. (eds) *The Challenge of Environmental Management in Urban Areas*. Ashgate, Aldershot, pp. 193–202.

Deeter, T. (2003) International Compendium of Urban Ecology Organizations. http://www.douglas.bc.ca/__shared/assets/Compendium33050.pdf (accessed 22 March 2011).

Duvigneaud, P. (1974) L'écosystème "Urbs". *Mémoires de la Société Royale de Botanique de Belgique* 6, 5–35.

Elliott, J.A. (2006) *An Introduction to Sustainable Development*, third edition. Taylor & Francis, London.

Feagin, J.R. (1998) *The New Urban Paradigm*. Rowman & Littlefield, Oxford, UK.

Fellenberg, G. (1991) *Lebensraum Stadt*. Verlag der Fachvereine, Zurich.

Gilbert, O.L. (1989) *The Ecology of Urban Habitats*. Chapman & Hall, London.

Girard, L.F., Forte, B., Cerreta, M., de Toro, P., Forte, F. (eds) (2005) *The Human Sustainable City – Challenges and Perspectives from the Habitat Agenda*. Ashgate, Aldershot.

Girardet, H. (2007) *Creating Sustainable Cities*. Schumacher Briefing No. 2. Reprint. Green Books, Bristol.

Grimm, N.B., Faeth, S.H., Golubiewski, N.E., *et al.* (2008) Global change and the ecology of cities. *Science* 319(5864), 756–760.

Grunwald, A., Kopfmüller, J. (2006) *Nachhaltigkeit*. Campus Verlag, Frankfurt/Main.

Häußermann, H., Siebel, W. (1995) *Neue Urbanität: Bauen und Gestalten für eine menschengerechte Stadt*, fifth edition. Suhrkamp, Frankfurt/Main.

Hall, P., Pfeiffer, U. (2000) *Urban 21. Der Expertenbericht zur Zukunft der Städte*. Deutsche Verlagsanstalt, Stuttgart.

IISD (International Institute for Sustainable Development) (2010) *Measurement and Assessment*. http://www.iisd.org/measure/compendium/searchinitiatives.aspx (accessed 22 March 2011).

IWM EB (Chartered Institutions of Wastes Management Environmental Body) (2002) *City Limits. A Resource Flow and Ecological Footprint Analysis of Greater London*. London.

Kennedy, C., Cuddihy, J., Joshua, E-Y. (2007) The changing metabolism of cities. *Journal of Industrial Ecology* 11(2), 43–59.

Kinzig, A.P., Grove, J.M. (2001) Urban–suburban ecology. In: Levin, S.A. (ed.) *The Encyclopedia of Biodiversity*. Academic Press, San Diego, pp. 733–746.

Kurtz, L.R. (1984) *Evaluating Chicago Sociology. A Guide to the Literature, with an Annotated Bibliography*. University of Chicago Press, Chicago.

Lichtenberger, E. (2002) *Die Stadt – Von der Polis zur Metropolis*. Wissenschaftliche Buchgesellschaft, Darmstadt.

Mander, U., Brebbia, C.A., Tiezzi, E. (eds) (2006) *The Sustainable City IV: Urban Regeneration and Sustainability*. WIT Press, Southampton.

Marchettini, N., Brebbia, C.A., Tiezzi, E., Wadhwa, L.C. (eds) (2004) *The Sustainable City III: Urban Regeneration and Sustainability*. WIT Press, Southampton.

Marzluff, J.M., Shulenberger, E., Endlicher, W., *et al.* (eds) (2008) *Urban Ecology. An International Perspective on the Interaction between Humans and Nature*. Springer, New York.

McDonnell, M.J., Pickett, S.T.A. (1990) Ecosystem structure and function along urban-rural gradients: An unexploited opportunity for ecology. *Ecology* 71, 1232–1237.

Meadows, D.H., Meadows, D.L., Randers, J., Behrens III, W.W. (1972) *The Limits to Growth*. Earth Island, London.

Müller, N., Knight, D., Werner, P. (eds) (2008) *Urban Biodiversity & Design. Implementing the Convention on Biological Diversity in Towns and Cities*. BfN-Skripten 229/1. Federal Agency for Nature Conservation (BfN), Bonn.

Mumford, L. (1963) *Die Stadt – Geschichte und Ausblick*. Kiepenheuer & Witsch, Cologne.

Neef, E. (1967) *Die theoretischen Grundlagen der Landschaftslehre*. Haack, Gotha.

Niemelä, J., Kotze, D.J., Venn, S., *et al.* (2002) Carabid beetle assemblages (Coleoptera, Carabidae) across urban–rural gradients: An international comparison. *Landscape Ecology* 17, 387–401.

Odum, E.P. (1953) *Fundamentals of Ecology*. Saunders, Philadelphia.

Oswald, F. (ed.) (2003) *Neue Urbanität – Das Verschmelzen von Stadt und Landschaft*. gta Verlag, ETH Zürich, Zurich.

Oswald, F, Baccini, P. (2003) *Netzstadt. Einführung in das Stadtentwerfen*. Birkhäuser, Basel.

Park, R.E., Burgess, E.W., McKenzie, R.D. (1925) *The City*. University of Chicago Press, Chicago.

Penev, L., Niemelä, J., Kotze, D.J., Chipev, N. (eds) (2005) *Ecology of the City of Sofia*. Pensoft, Sofia.

Pickett, S.T.A., Cadenasso, M.L., Grove, J.M. (2004) Resilient cities: Meaning, models, and metaphor for integrating the ecological, socio-economic, and planning realms. *Landscape and Urban Planning* 69, 369–384.

Richter, M., Weiland, U. (2008) Stadtökologische Traditionslinien. In: Rehberg, K.-S. (ed.) Die Natur der Gesellschaft. Bd. 1. Campus, Frankfurt/Main, pp. 475–488.

Sassen, S. (2009) *Bridging the Ecologies of Cities and of Nature*. The 4th International Conference of the International Forum on Urbanism (IFoU): The New Urban Question – Urbanism beyond Neo-Liberalism. Delft, pp. 45–52.

Schmithüsen, J. (1942) Vegetationsforschung und ökologische Standortslehre in ihrer Bedeutung für die Geographie der Kulturlandschaft. *Zeitschrift der Gesellschaft für Erdkunde zu Berlin*, 113–157.

Spooner, B. (1986) *MAB Urban and Human Ecology Digest.* UNESCO, Paris.

Sukopp, H. (1973) Die Großstadt als Gegenstand ökologischer Forschung. *Schriften des Vereines zur Verbreitung naturwissenschaftlicher Kenntnisse in Wien* 113, 90–140.

Sukopp, H. (1994) Stadtforschung und Stadtökologie in Vergangenheit und Gegenwart. *Geobotanische Kolloquien* 11, 3–16.

Sukopp, H. (2002) On the early history of urban ecology in Europe. *Preslia* 74, 373–393.

Sukopp, H. (2005) History of urban ecology. In: Müller, N. (ed.) *Biodiversität im besiedelten Bereich.* Conturec Schriftenreihe, Volume 1, Darmstadt, pp. 93–99.

Swyngedouw, E., Moulaert, F., Rodriguez, A. (2002) Neoliberal urbanization in Europe: Large-scale urban development projects and the new urban policy. *Antipode* 34, 542–577.

Tansley, A.G. (1935) The use and abuse of vegetational concepts and terms. *Ecology* 16, 284–307.

Troll, C. (1939) Luftbildplan und ökologische Bodenforschung. *Zeitschrift der Gesellschaft für Erdkunde zu Berlin* 7/8, 241–298.

Troll, C. (1968) Landschaftsökologie. In: Tuexen, R. (ed.) *Pflanzensoziologie und Landschaftsökologie.* Junk, Den Haag, pp. 1–21.

UNEP (United Nations Environment Programme) (ed.) (1992) *Agenda 21 – Environment and Development Agenda.* Nairobi.

UN ESA (UN Department of Economic and Social Affairs, Population Division) (2008) *World Urbanization Prospects: The 2007 Revision.* http://www.un.org/esa/population/publications/wup2007/2007_urban_agglomerations_chart.pdf (accessed 22 March 2011).

Wania, A., Kühn, I., Klotz, S. (2006) Plant richness patterns of plants in agricultural and urban landscapes in Central Germany. *Landscape and Urban Planning* 75, 97–110.

Weiland, U. (2006) Sustainability indicators and sustainable development. In: Wuyi, W., Krafft, T., Kraas, F. (eds) *Global Change, Urbanization and Health.* China Meteorological Press, Beijing, pp. 241–250.

Weiland, U., Richter, M. (2009): Lines of tradition and recent approaches to urban ecology, focussing on Germany and the USA. *GAIA* 1, 49–57.

Winter, J., Mack, J. (1988) *Herausforderung Stadt – Aspekte einer Humanökologie.* Ullstein, Frankfurt/Main.

Wittig, R. (1991) *Ökologie der Großstadtflora.* Fischer, Stuttgart.

Zipperer, W.C., Wu, J., Pouyat, R.V., Pickett, S.T.A. (2000) The application of ecological principles to urban and urbanizing landscapes. *Ecological Applications* 10, 685–688.

PART II

URBAN ECOLOGY: RELATED DISCIPLINES AND METHODS

CHAPTER 2

Thematic–methodical approaches to applied urban ecology

Matthias Richter[1] and Ulrike Weiland[2]

[1] *Environmental Scientist, Publicist and University Lecturer, Germany**
[2] *Institute for Geography, University of Leipzig, Leipzig, Germany*

The aim of this chapter is to give a short interdisciplinary overview of applied urban ecology and leading the way to its transdisciplinary future. This requires a basic understanding of the concept "applied urban ecology", which is laid out as follows.

Sometimes it is claimed that urban ecology does not have its own research areas, since urban ecological principles and ecosystem functions in urban areas are not different from those in the "traditional cultural landscape."[1] However, the urban land use and land cover pattern differ greatly from those of other landscapes, and the urban habitat results in other environmental conditions and other combinations of species. Last but not least the actors relevant to the implementation of concrete projects are different. Thus "applied urban ecology" is created and defined as a separate research area.

One of the main concerns of "applied urban ecology" is – while building on a profound understanding of interrelations in urban ecological natural sciences – to produce knowledge on environmental and health quality in cities and urban agglomerations and on how to improve it. Relevant objectives can be, for example: habitat and species protection, climate change mitigation and adaptation, sustainable management of water resources or the protection of human health with respect to diverse environmental impacts, just to mention a few. Thus a thematic–methodical overview of the topic has to consider both, basic research as well as applied research and their respective "research subjects."

The traditional research core of urban ecology includes urban vegetation sciences, urban zoology, urban soil science and urban climatology (see Figure 2.1, C1 (core one)).

The traditional core of urban ecology was extended as part of the historical development of the knowledge area through the following research areas: urban land use patterns and their changes, energy and material flows and, last but not least, human beings as receivers and sources of various environmental effects and their consequences, such as noise and its health effects (see C2 (core two) in Figure 2.1).[2]

If one looks at the disciplines[3] from which people come (those who identify themselves as urban ecologists), various natural sciences as well as landscape ecology and several planning "disciplines" stand out,

*Contact information: http://www.Dr-Matthias-Richter.de.

[1] Niemelä has argued this in a similar way (see Niemelä, J. (1999) Is there a need for a theory of urban ecology? *Urban Ecosystems* 3, 57–65).

[2] Both of the core areas are thus identified as such here since they play a central role in the tradition of urban ecology (see Chapter 1).

[3] Under "a discipline" in this context we understand "a field of research activity which has achieved a professional level and has become institutionalized." We do not distinguish subdisciplines because this difference is strongly dependent on the angle of view. Explaining it more in detail belongs to philosophy of science.

Applied Urban Ecology: A Global Framework, First Edition. Edited by Matthias Richter and Ulrike Weiland.

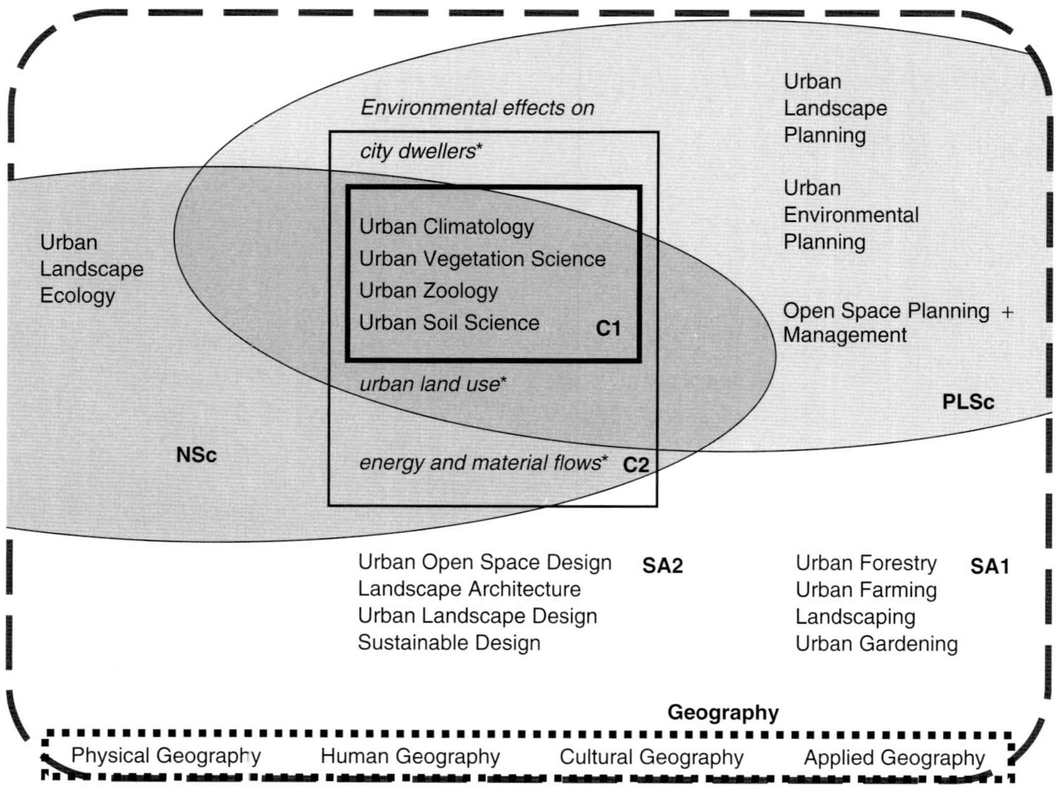

Figure 2.1 Overview of the thematic–methodical approaches to applied urban ecology.
Abbreviations:

C1: core one, established and more traditional central research areas of urban ecology
C2: core two, includes all established and central research areas of urban ecology
NSc: natural sciences oriented
PLSc: oriented towards urban development and planning
SA1: supplementary angle one, agronomically oriented
SA2: supplementary angle two, design-oriented.

Further explanations are found in the text.

*Note: The terms "environmental effects on city dwellers", "urban land use", and "energy and matter flows" are not independent disciplines. They represent, however, central research areas in the context of urban ecological research.

for example, urban landscape planning, urban environmental planning and open space planning.

The area of geography is added at the bottom of Figure 2.1. Due to its different research areas and interests it contributes to applied urban ecology in several ways. Physical geography with a focus on urbanity overlaps with urban landscape ecology, and human geography as well as applied geography can be closely linked to urban development and planning, whereas cultural geography can pick out the differences of urban landscapes in different cultures and the related reasons for these differences. However, the links between applied urban ecology and geography are only optional. This means that within the geographical strands mentioned above urban ecological topics can, but not "must", be chosen.

An overview of and knowledge about different approaches to applied urban ecology is important in order to maintain an orientation framework for judging management suggestions and measures motivated by urban ecology. The objectives, which are emphasized in the different urban ecology sections, can be contrary to each other. Different approaches also determine not only different methods but also differently constituted spheres of knowledge and partly also different research objects.

Here landscape ecologists are usually more strongly oriented towards the basic ecological relationships (natural sciences oriented, NSc) and planners to applied aspects (oriented towards urban development and planning, PLSc). Both the urban ecology oriented planner and the landscape ecologist with the focus on urban ecology usually refer not only to one of the central disciplines shown above (for instance only urban soil science or only urban climatology), but rather they focus on different cross-sections according to the focus of interest. However there are also "pure" urban climatologists, urban soil scientists or urban vegetation scientists.

In this book we are dealing primarily with the portrayal of applied urban ecology as a inter- and transdisciplinary[4] area of knowledge and as an orientation framework for differing management approaches in an international social context.

Although the research area of urban ecology shows a relatively stable "research core" the emphasis of the focal points of research shifts according to social and sociopolitical interests. The current (research) focal points portrayed in this book will be taken up in a synthesis in Chapter 16 in order to readjust the awareness for urban ecological facts, correlations, connections, and for different normative perspectives.

The two established central areas of urban ecology (see C1 and C2 in Figure 2.1), together with the cross-section oriented perspectives by landscape ecologists and planners are complemented by "designer disciplines", such as urban landscape design and sustainable design (see SA2 in Figure 2.1) and landscape agronomic areas, such as urban forestry, urban farming or urban gardening (SA1). "SA" stands for "supplementary angle."

[4]"Transdisciplinarity" here is understood as an integrative approach for the integration of knowledge, methods, and applications from different disciplines including the research areas listed in Figure 2.1 that are considered relevant for applied urban ecology.

These fields referred to under SA1 and SA2 have, up to now, been insufficiently recognized in their overlap with urban ecology, although they address applied urban ecology through graphical drafts with an ecological basis – mostly project related – as under SA2 or they are established in "everyday land use practice," as under SA1. This means that the working fields particularly in SA1 are not only disciplines (e.g., taught at university) but they are everyday land use practice as well. And it is very important to integrate the latter into reflections on applied urban ecology.

Landscape architecture or landscaping are traditionally anchored in their own professional organizations and are less bound to normative-ecological goals but instead are oriented primarily in an esthetic–artistic or pragmatic direction with respect to economic management.

However, in SA1 and SA2, sections are shown – such as landscape urbanism, sustainable design or urban forestry – that increasingly reveal overlaps with what constitutes applied urban ecology. The reason for this is that, in the meantime, criteria that are specific to city-related sustainability are more frequently taken up in landscaping, horticulture, forestry, and landscape architecture than they were 30 to 50 years ago.

A further section of applied urban ecology reveals itself to be increasingly important: the sociopolitically relevant field of empirically based protagonists and stakeholder research. This involves the investigation of those persons, groups, and organizations (and their interaction with municipal or state organizational units) that have chosen practical urban ecology as their management goal.

These "interest groups" are involved in, for instance, the improvement of the potential for experiencing nature in "their" city, in bird conservation, habitat protection, or just in ecologically oriented approaches in the design and management of their gardens or municipal parks. Ecopolitical and environmental psychological approaches also belong explicitly to the thematic–methodically important aspects of applied urban ecology. These two aspects will be taken into consideration appropriately in Chapters 6, 7, 9, 10, 15, and 16 providing manifold examples.

In this book the sociopolitical and sociopractical dimension of urban ecology in particular will be emphasized systematically with reference to case studies against the backdrop of increasing globalization and urbanization.

CHAPTER 3

Monitoring urban land use changes with remote sensing techniques

Ellen Banzhaf[1] and Maik Netzband[2]

[1] *Working Group Geomatics, Department of Urban and Environmental Sociology, Helmholtz Centre for Environmental Research – UFZ, Leipzig, Germany*
[2] *Geomatics, Geography Department, Ruhr-University Bochum, Bochum, Germany*

3.1 LAND USE CHANGES AND THEIR CONSEQUENCES FOR URBAN ECOLOGY

The rapid and increasing process of urbanization is leading to recognition of the increasingly vital role of the city. One half of the world's population lives in cities, while the other half depends on cities in a co-existence. Population growth, globalization and democratization are the three factors that have contributed to strengthen the role of the cities as centers of production, consumption and change, whether it be political or social (Hall 2003), thus helping to modify our vision of sustainable development (Girard *et al.* 2007). Whilst this makes challenges environmental, economic, and social sustainability immense, it also provides many promising opportunities.

The cities of the 21st century are spreading out and sucking food, energy, water and resources from the environment, without taking into due account the social, economic and environmental consequences generated at the global level. This shows that urban landscapes are changing faster than the forces that impel the changes can be understood. Changes, and the forces behind them, seem to be operating in opposing directions and at different spatiotemporal scales. Migration patterns allow urban regions to grow and leave rural areas depopulated. Additionally, cross-border migration is largely a reaction to inequalities in world economies, transforming the social and ethnical proportion of cities, thus evoking massive social and political challenges.

The urban environment itself is profoundly changing the entire global ecosystem. In order to understand the environmental role played by cities, the damage caused and further damage averted, it is necessary to see the city as part of the environment and beyond this to investigate the multiple-dimension of cities and their dynamics on various scales. Environmental changes are also expressed in land use changes. Social, economic or political trends are conveyed spatially. In recent decades, the strongest per capita growth shifted to the more rural areas of the urban fringe (Bugliarello 2003). Open spaces are increasingly included between cities, villages and traffic axes. An urbanizing landscape, the accompanying technical infrastructure, and uncontrolled dynamics of urban growth patterns are the results. The conversion from natural land cover to land being used for anthropogenic progresses (i.e., predominantly agricultural surfaces are transformed into settlement and traffic surfaces resulting in decreased settlement density in the suburban areas) increased traffic and costly infrastructure development. Sociospatial patterns are expressed in different building activities for single family houses of different strata, with different amounts of green spaces and shopping facilities,

Applied Urban Ecology: A Global Framework, First Edition. Edited by Matthias Richter and Ulrike Weiland.

to name a few. These activities have driven settlement areas to expand further. The settlement density and thus the inner urban densification process continue to decrease.

Data available from Earth observation systems are geo-coded and present an opportunity to collect information relevant to urban and peri-urban environments at various spatial, temporal, and spectral scales (see Table 3.1). "Urban" is understood as an area of high population concentration with high building density and infrastructure, and "peri-urban" signifies adjacent agricultural regions with lower population concentration (Alberti 2008, Banzhaf *et al.* 2009, Eurostat 2010). Adjacent urban areas are conurbations, and the development of such conurbations inevitably changes the structure and extent of open spaces. On the outskirts of the urban agglomerations, soil sealing reduces open spaces through conversion of formerly agrarian to urban land use and thus from pervious to impervious surfaces. In the central part of urban areas, new green spaces may result from the derelict land of former industrial sites, either revitalizing them into trade and housing locations, or redeveloping them into new green spaces. Observation and evaluation instruments are critically necessary in order to structure and plan for a sustainable future in dynamic cities (Simmons 2007). Land cover and land use changes influence the urban climate and atmospheric deposition of pollutants. Especially the increase of imperviousness at the expense of the decrease of green spaces must be documented from the local to the global scale, and the knowledge integrated into climate change investigations and further global change issues.

Exclusively characterizing and monitoring land cover and land use change is of limited use in understanding the development pathways of cities and their resilience to outside stresses (Longley 2002). Geological, ecological, climatic, social and political data are also necessary to describe the developmental history of an urban center and understand its ecological functioning (Grimm *et al.* 2000).

It is the process of urbanization that must be monitored, analyzed, and even simulated on different scales. Dependent on the issue to be investigated upon, the relevant scale must be selected (see Figure 3.1). Local and regional environmental effects must be documented, analyzed, evaluated, and, if possible, predicted. Without researchers and stakeholders exchanging and collaborating, the goal cannot be achieved.

3.2 URBAN REMOTE SENSING (URS) AND GEOGRAPHICAL INFORMATION SYSTEMS (GIS) FOR RESEARCH IN URBAN ECOLOGY

In recent years URS has proved to be a useful tool for cross-scale urban planning and urban ecological research. Remote sensing in urban areas is, by nature, defined as the measurement of surface radiance and properties connected to the land cover and land use in cities.

Urban remote sensing is applied to create practical models and products for tracking farmland conversion, land use and land cover changes (LULC), floodplain analysis, urban heat island modeling, and vegetation monitoring – from crop types to urban green spaces or forest ecological changes, for example. LULC classifications are incorporated into local and regional ecosystem models to assess the effects of urban change on carbon cycling and source/sink relationships or on atmospheric processes. The urban pattern causes deterioration in air quality, the urban ecosystem processes and biodiversity. In this context URS is a necessary prerequisite to examine how urban forms modify the landscape as a complex system. It can help to detect and evaluate the distribution of impervious or sealed surfaces, a key parameter of urban ecology (surface and groundwater availability and runoff, vegetation dynamics) and planning (storm water runoff, flooding hazards, landslides).

Kühn (2003) explains the development of urban landscapes being shaped by the penetration of settlement and open-space structures. Remotely sensed data will be used to detect and evaluate the physical structure and composition of urban areas, such as the structure of residential, commercial or mixed neighborhoods, green spaces or other open spaces.

The growth of Spatial Data Infrastructures, geoportals and private sector initiatives (e.g., Google Earth, Bing Maps Platform, etc.) produced an increase in geographical data availability at any scale and worldwide. This growth has not been fully coupled by an increase of knowledge to support spatial decisions. Spatial analytical techniques and geographical analysis and modeling methods are therefore required in order to analyze data and to facilitate the decision process at all levels. Since cities can be described as a concentration of people, it is most striking to find coherence between urban land use and sociodemographic as well as socioeconomic parameters. The statistical analysis

Table 3.1 Selected earth observation satellite sensors for quantitative information on land surface characteristics to put parameters on and validate urban land use processes.

Satellite sensor system	Applications	Scale	Spatial extent and ground resolution	Temporal extent and resolution	Price*
MODIS (Terra) Moderate Resolution Image Spectrometer)	Monitoring land cover, urban dynamics, canopy structure, biological productivity, structure and spatial heterogeneity, surface temperature.	Global to regional scale	2330 × 10 km^2 (Pixel size 250–2000 m)	Since 2000 (1–2 days)	Free
Landsat (E)TM (Enhanced) Thematic Mapper	Long-term LULC (monitoring and mapping, esp. rapid urbanization processes. Spectral capacity to map built-up areas; leaf area index; land surface temperature.	Regional scale	185 × 185 km^2 (Pixel size 30 and 120 m)	Since 1984 (16 days)	Free (from USGS)
ASTER (Advanced Spaceborne Thermal Emission and Reflection Radiometer)	LULC monitoring and mapping; identifying the "fingerprint" of human activities; quantitative investigations on the physical processes affecting environmental changes, such as urban land use, patchiness and connectivity of green spaces, urban heat islands.	Regional to local scale	60 × 60 km^2 (Pixel size: 15, 30 and 90 m)	Since 2000 (16 days)	$80
Spot 4/5 (HRVIR: High Resolution Visible and Infrared Sensor)	Urban planning; LULC monitoring and mapping; vegetation pattern; spatial extent of natural hazards. Monitoring landscape changes with respect to imperviousness through urban, peri-urban and rural inventories. Analysis of spatial structure.	Regional to local scale	60 × 60 km^2 (Pixel size 5 to 20 m)	Since 1998/ 2002 (26 days)	€1900 for 20 m color; €2700, for 10 m color; €5400, for 5 m color
Geoeye 1	3-D city scapes (urban landscapes).	Local scale	15 × 15 km^2 single area. (Pixel size 0.41 m/1.65 m) Stereopair	Since 2008 (revisit 2–2 days)	$12.50–$40,-per km^2 (dependent on product precision)
QuickBird	Mapping of urban pattern, mining, infrastructure. Delineation of urban structure types and spatial metrics for local vulnerability analyses.	Local scale	16.5 × 16.5 km^2 single area. Strip 16.5 × 115 km (0.61 cm & 2.4 m)	Since 2001 (2–2 days)	$350–425,- (dependent on product) per 25 km^2 minimum order size for archives
TerraSAR-X Radar Satellite with 3 Sensors	Weather-independent radar sensors; synergistic exploitation with optical data: multiscale, multitemporal and multipolarized observations: Detecting properties for building densities and texture differentiations at local scale; monitoring changes in urban environments.	Regional to local scale – simultaneously	5 × 10 km^2– 100 × 50 km^2 (Pixel size 1–26 m)	Since 2007 (11 days)	€200 (COFUR prices for scientific use)

*Prices as of 8 April 2011.
LULC, land use/land cover.

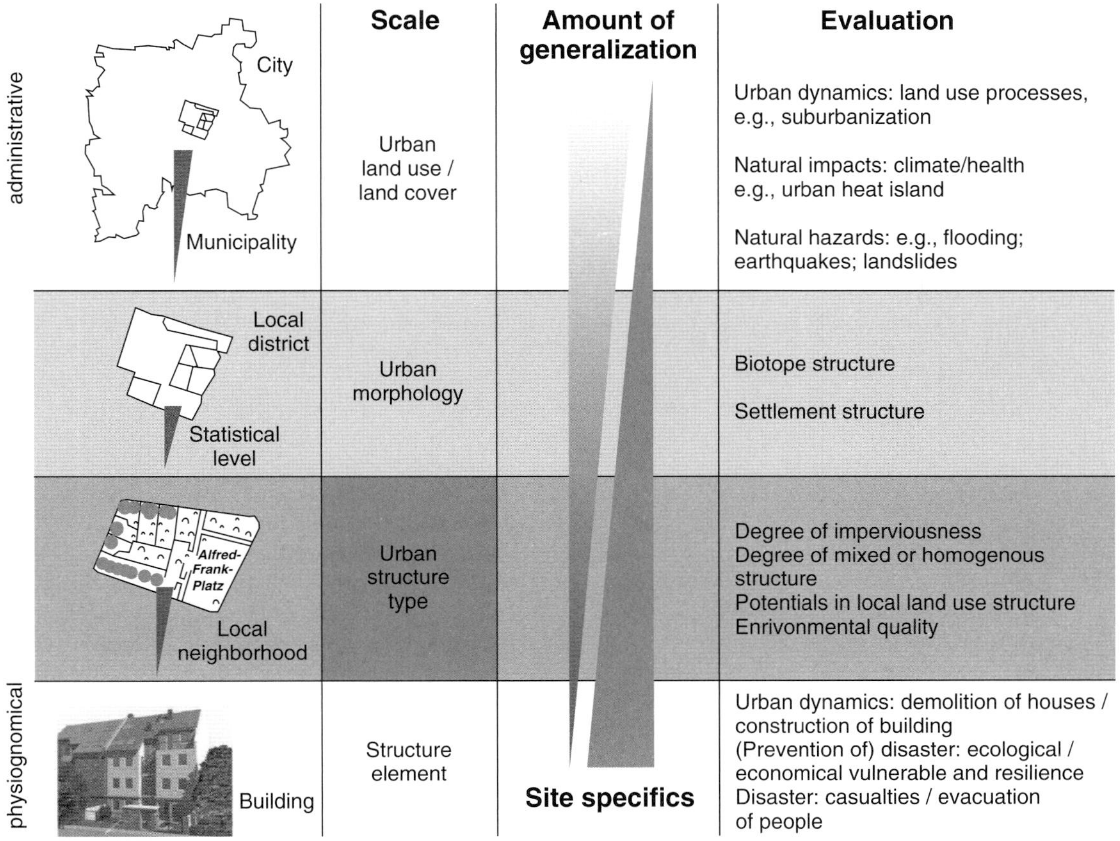

Figure 3.1 Scale-dependent urban analysis (modified from Wickop *et al.* 1998).

of census data infers information on the human usage of the land, the human exposure to potential hazards in the city, and the configuration of each neighborhood indicating the urban quality of life. For example, overlaying choropleth maps of sociodemographic features with land use maps gives information on gender and age distribution connected with proximity to urban green spaces, income and building density, or water consumption and level of provision of infrastructure. In this context URS helps by providing spatial information that is linked to social indicators to explain the interrelations between ecological conditions and sociospatial development (Banzhaf *et al.* 2009).

Table 3.1 lists some of the most important sensors currently still in use and their characteristics and potential applications, especially for urban areas. The table is far from complete, but rather represents a selection of satellite-based instruments that offer two advantages essential for characterizing and monitoring urban and peri-urban regions: first, a large volume of multispectral data obtained at relatively low cost, and second, data for the same region repeatedly acquired over short periods of time (from days to weeks). The selection also depicts satellite sensors at various scales to present the wide scope of quantitative data allowing global, regional and local investigations (also see Section 3.3 and 3.4).

3.3 MEASURING PHYSICAL CHARACTERISTICS OF URBAN AREAS WITH REMOTE SENSING TECHNOLOGY

3.3.1 Effects of urban form on natural and man-made hazards

A problem common to cities around the world is the formation and intensification of urban heat islands (UHI)

(Voogt and Oke 2003). Honjo *et al.* (2003) and Eliasson (1999) have emphasized the role that vegetation and imperviousness play in terms of the climatic performance of the cities. Both factors not only control the spatial distribution of temperatures but also explain microscale differences between urban zones of similar density or similar land uses.

Phoenix, Arizona represents a classic example of the UHI effect, but heat islands are present in almost every city; they increase energy and water use, biodiversity change, and human discomfort (Brazel *et al.* 2000; Stefanov and Brazel 2007; Zehnder 2002). The cost of dealing with rising urban temperatures may also intensify social and environmental injustices in cities (Harlan *et al.* 2006, Jenerette *et al.* 2007, Stefanov *et al.* 2004). The aggregate effects of UHI on regional and global climate are poorly understood. Day and night thermal infrared data acquired by ASTER, MODIS, and Landsat can be used to model the UHI effect and quantify the contributions of different materials to the thermal budget, thus proving to be an effective tool to measure the spatial extent and intensity of the effect of urban development on local temperatures (Hartz *et al.* 2006).

In the case of Santiago de Chile, up to 49% of vegetation cover explains relevant heat reductions in urban spaces (Romero *et al.* 2007). Both imperviousness rates and vegetation cover depend on policy decisions and must be considered as land use allocations in terms of local climate regulations, quality of life and social equity. Parameters for such allocations are urban densities, biodiversity and location of urban parks, green belts, and other ecological corridors (Peña 2008).

Urban and peri-urban regions in both developed and developing countries often have poor air quality due to industrial processes, automobile use, residential wood and coal burning, agricultural activities, and disruption of soil surfaces due to construction or informal settlements (Krzyzanowsk and Schwela 1999, Williams 1999). Although many cities in developed regions use networks of in-place, ground-level sensors to measure air quality on a real-time basis, this capability does not exist in much of the developing world. The characterization of urban atmospheres, and especially of the aerosols, is a major issue. In fact, the atmospheric correction of remote sensing data requires the knowledge of their abundance and of their optical properties (Lachérade *et al.* 2008). Using the spectral behavior of the aerosols from remote sensing images of very high spatial resolution, it seems possible to estimate these parameters through a new retrieval procedure of the atmospheric particles. Based on the observation of sun/shadow transitions, it allows the retrieval of an aerosol model and of its spectral optical thickness.

The surrounding and underlying geology of an urban center determines the types of structures that can be built and the susceptibility of the city to various geological hazards (Valentine 2003). In China's coal mining industrial region surface subsidence is a major issue. Since the 1980s, with the development of industry and agriculture, the amplitude of groundwater withdrawal has increased year by year, which leads to serious surface subsidence and ground fissure hazards in many of the coal mining cities (Ferretti *et al.* 2000, Yang *et al.* 2009, Guang *et al.* 2008). Multiband SAR data and interferometric pairs of D-InSAR are taken to measure the areas of land subsidence, which are greatly correlated with the groundwater withdrawal, seasonal variation, and the New Economic and Technological Development Zone. Spectral analysis is also used to map areas of potential soil contamination, e.g., noting a certain clay species that can absorb heavy metals (Ben-Dor *et al.* 1999).

Urban geo-hazards and environmental geology are growing subfields within the discipline of geological sciences. An important synergy of these fields combines social data with geo-hazard knowledge for predictive purposes, resulting in greater knowledge of where geological disasters might occur (e.g., areas of earthquake rupture, tsunamis). This can help guide future expansion, hazard response planning and the development of appropriate engineering guidelines for buildings and infrastructure. In this context, TerraSAR-X imagery has been proven to be particularly useful (Kuntz *et al.* 2009). With such interpreted and evaluated data at hand, on-site support can be given, thus ensuring a rapid identification of damages. The identification of exposures to natural and man-made hazards is a fundamental element helping towards risk reduction in the complex urban environment, where structural processes might easily lead to an increase in exposure and vulnerabilities of human and physical assets. Such results can immediately be made available online, ensuring accessibility for national and international, governmental and non-governmental organizations worldwide. An archive of pre-disaster images proves to be most valuable for post-disaster image interpretation (Yonezawa 2001). This archive should consist of images from various sensor types and should be updated on a regular basis.

3.3.2 Urban dynamics and ecosystem function

From an ecological perspective, land development is one of the most disturbing processes since it dramatically alters the natural energy and material cycles of ecosystems (Berry 1990, McDonnell *et al.* 1997, Oke 1989, Pielke *et al.* 1999). Many cities are developing, or have developed, open space models, concepts or strategies to prioritize areas for preservation (Cook 2002). For a long time, urban planning authorities monitored green spaces, their quality and quantity through fieldwork. Vegetation indices and land cover information is now more often derived from remotely sensed data, which enhance monitoring capabilities. Small urban gardens and urban agricultural areas are particularly important for subsistence, as well as for recreational purposes, in both developed and less developed countries. Such areas also feed back into the urban ecological system (i.e., improving the green structure and urban climate).

The carbon cycle is altered due to the removal of developed land from the photosynthetic process and the increase in CO_2 emissions from fossil energy use in urban areas. While not all of the land in urban areas is sealed by impervious surfaces, it has been shown that at least in the less resource limited regions of the United States (eastern and southeastern), urbanization lowers the photosynthetic activity of the landscape (Imhoff *et al.* 2000). This observation is particularly relevant when considering that urbanization in the United States occurs preferentially where the soils are most productive (Imhoff *et al.* 1997), thereby causing a loss of prime farmland. Milesi (2003) proves the impact of land cover changes on ecosystem resources using MODerate resolution Imaging Spectroradiometer (MODIS) sensor on-board the TERRA (EOS-AM1) platform functioning at regional to global scales. She incorporates night time data gained from the Defense Meteorological Satellite Program's Operational Linescan System (DMSP/OLS) to track changes in human settlements. This methodology is suited to understand regional effects of urbanization on primary productivity.

Looking at urban watersheds it is challenging to assess, in a spatially representative way, the combined effects of pollutants, petrochemicals and bacteria that may constitute the runoff from an area of urbanized land. An ecological indicator that integrates such cumulative resource impacts, without regard to specific factors, is impervious surface area. As urbanization progresses over a watershed any subsequent runoff is altered in both time and magnitude, though interestingly, not to the same extent as the degree of urbanization. Nevertheless, increases in runoff volume and discharge rates cause physical changes to stream channels. Streambeds are scoured due to higher velocities and stream channels become deepened and/or widened (Booth 1990); this results in habitat alterations to both in-stream and riparian zones and the loss of critical aquatic habitats (e.g., riffles). Furthermore, a highly problematic aspect of ever-increasing impervious surface area in watersheds is the amount of non-point source pollution delivered to water systems. Gillies *et al.* (2003) found that empirical evidence, in terms of aquatic species richness, was available to assess the relative impact of changes in impervious surface area using multi-temporal Landsat imageries.

Alberti (2008) states that the role of diversity in the functioning and stability of ecosystems becomes even more relevant in the context of increasing human-induced transformation and urbanization. Remotely sensed data are of particular use for monitoring biodiversity in the urban mosaic of artificial and natural surfaces due to the relatively high temporal resolution (compared to ground survey campaigns) for change detection of ongoing and fast land use changes. To focus analysis on the monitoring of the actual LULC and its current and potential changes, products such as derived from different vegetation indices (e.g., normalized difference vegetation index (NDVI), soil-adjusted vegetation index (SAVI), or enhanced vegetation index (EVI)) and day/night surface kinetic temperature maps using the ASTER database are generated. Fragmentation of natural patches is one of the best-known impacts of human activities on the diversity, structure and distribution of vegetation. Connectivity is the critical property of landscapes facilitating or limiting the movement of resources and organisms among natural patches. Urban growth affects connectivity directly by modifying the landscape and indirectly by changing biophysical patterns and processes (Tischendorf *et al.* 2005). Satellite remote sensing data provide universal coverage but typically provide little biotic resolution with respect to the scale of individual flora and fauna. This means that remote sensing information can only be linked to meaningful biodiversity indicators through spatial parameters such as land use and land cover and the related habitat structure. Perhaps the highest contribution of remote sensing methodology to urban ecological remote sensing is in monitoring the highly heterogeneous and fragmented land use and land cover, as well as changes in both, including information on biophysical attributes, vegetation structure, and habitat fragmentation, which are essential determinants of species distributions. Although the effects of alternative urbanization patterns on plants are still not fully understood, it is known that converting natural or rural landscape into an urbanized landscape reduces the diversity of native plant species in the urbanized region. A biodiversity gradient with species richness declines from the urban fringe towards the urban core based on the physical changes and observed on the urban-to-rural gradient, (Pickett *et al.* 2001). Insight needs to be provided into the vulnerability and resilience of various ecosystem types to human impact, which will help address the question of consequences of change for human civilization from the perspective of biodiversity and ecosystem services (Folke 2006, Turner *et al.* 2003). Land use/land cover classifications contribute, together with spatial analysis, to document landscape structure change, and also help to derive spatial metrics from remotely sensed data.

3.4 GLOBAL INITIATIVES TO MEASURE URBAN EXPANSION AND LAND USE CHANGE

Global environmental monitoring, analyses and assessments have become an important element in international, national and local policy, and decision making. The global nature of such efforts emphasizes the role in addressing problems that require cooperation among different countries, between scientists and policy makers, and across different scales. Remote sensing provides the most suitable systematic approach for collecting spatial information on human settlements. Earth observation has been focused on mapping, monitoring and understanding urban phenomena for many years, but with more emphasis on local to regional scales. Global mapping of human settlements is particularly challenging because of the spatial and spectral heterogeneity of urban environments, as well as their small and fragmented spatial configuration. Due to this heterogeneity the key issue is to combine Earth observation indicators for characteristics and change in human settlements (Herold 2009).

3.4.1 Global Urban Observatory of UN-HABITAT

The UN-HABITAT Global Urban Observatory (GUO) (2003) was established after the conference Habitat II in 1997 in response to a decision of the United Nations Commission on Human Settlements, which called for a set mechanism to monitor global progress in implementing the Habitat Agenda and to monitor and evaluate global urban conditions and trends. The goals of GUO are to help all partners monitor and evaluate urban conditions and trends as measures of progress in implementing the Habitat Agenda and Millennium Development Goals (MDG) to support partners using urban data in participatory decision-making processes at all levels. The Global Urban Observatory (GUO) addresses the urgent need to improve the worldwide base of urban knowledge by helping governments, local authorities and organizations of the civil society develop and apply policy-oriented urban indicators, statistics and other urban information. The major products of the GUO are to install a network and databases.

The Global Urban Observatory Network (GUONet) is a worldwide information and capacity-building network established by the United Nations Human Settlement Programme (UN-HABITAT) to help implement the Habitat Agenda at the national and local levels. In the Global Urban Observatory databases a multitude of information is stored on urban indicators, statistics, and city profiles. It is now essential to work further on UN-HABITAT agenda goals such as to provide security of tenure (indicators: durable structures, overcrowding), or to promote access to basic services (indicators: access to safe water, access to improved sanitation). UN-HABITAT states cities and climate change as two of the greatest challenges currently facing humanity in the 21st century for which information is continuously gathered and monitoring needed (UN-HABITAT 2011).

3.4.2 "The Dynamics of Global Urban Expansion" – a contribution by the World Bank

The study of Angel *et al.* (2005), in the World Bank's Urban Development Division, examines the dynamics of global urban expansion by defining a new universe of 3943 cities with populations in excess of 100 000 and draws a stratified global sample of 120 cities from this universe. Population data and satellite images for two time periods – a decade apart – are obtained and analyzed, and several measures of urban extent and expansion are calculated, among them the built-up area of cities and the average density of the built-up area. Data for 90 cities out of the global sample of 120 are presented and analyzed in this report. Weighted averages of the built-up area and the average density, as well as compactness and contiguity measures – and their change over time – are presented for nine regions, four income groups and four city size groups covering the entire globe. Densities in developing country cities were found to be three times higher than densities in cities in industrialized countries, and densities in all regions were found to be decreasing over time. If average densities continue to decline at the annual rate of 1.7% – as they have during the past decade – the built-up area of developing country cities will increase from 200 000 km^2 in 2000 to more than 600 000 km^2 by 2030, while their population doubles. Ten econometric models, which sought to explain the variation in urban extent and expansion in the universe of cities, were constructed, and several hypotheses postulated by neoclassical theories of urban spatial structure were tested. All tests yielded R^2 values in excess of 0.80. The policy implications of the analysis are presented and discussed (Angel *et al.* 2005). The central message of this study is quite clear: cities in developing countries should be making realistic – yet minimal – plans for urban expansion, designating adequate areas for accommodating the projected expansion, investing wisely in basic trunk infrastructure to serve this expansion, and protecting sensitive land from incursion by new urban development.

3.4.3 Socioeconomic data and applications Center (SEDAC) at the Center for International Earth Science Information Network (CIESIN), Columbia University, New York, USA

Given that impervious surfaces and soils often have similar compositions and therefore indistinguishable reflectances, the spectral resolution of broadband sensors may not be sufficient to distinguish subtle differences between impervious surfaces and soils. Thus the spatial resolution of Landsat and SPOT is, in many cases, not sufficient to resolve individual components of the urban mosaic. Thus most pixels represent mixtures of pervious and impervious surfaces. In his investigations and analyses of satellite imagery, Small (2003) proves that spectral mixture analysis (SMA) provides a general, physically consistent representation of mixed pixels and allows partially pervious areas to be mapped more accurately than would be possible with hard classification.

A comparative SMA of Landsat 7 Enhanced Thematic Mapper (ETM+) imagery, for example, being collected for 28 urban areas worldwide provides a physical basis for a spectral characterization of urban reflectance properties (Small 2003). These urban areas have similar mixing space topologies and can be represented by three-component linear mixture models in both scene-specific and global composite mixing spaces. The results of the analysis indicate that the reflectance of these cities can be accurately described as linear combinations of high albedo, dark and vegetation spectral endmembers within a two-dimensional mixing space containing over 90% of the variance in the observed reflectance.

It has been found that the relative proportions of these endmembers vary considerably among different cities and within individual cities, but in all cases the reflectance of the urban core lies near the dark end of a mixing line between the high albedo and dark endmembers. The most consistent spectral characteristic of the urban mosaic is spectral heterogeneity at scales of 10 to 20 m.

In spite of their heterogeneity, built-up areas do occupy distinct regions of the spectral mixing space. This localization in mixing space allows spectrally mixed pixels in built-up areas to be discriminated from undeveloped land cover types.

This analysis also highlights the diffuse nature of urban areas and the complexity of the urban-to-rural transition. Representing reflectance as endmember fractions makes it possible to classify urban areas and their transitional surroundings on the basis of a continuous field variable, such as vegetation fraction or albedo. Continuous field variables can more accurately represent the true "fuzzy" nature of complex entities like cities. SMA provides a physically based analog to fuzzy classification. At the same time it avoids the unrealistic thresholds required by traditional "hard" classification schemes in which each pixel receives a single binary class assignment. In this sense, continuous fields better represent the true nature of urban areas characterized by scale-dependent transitions between different combinations of land use. By accommodating heterogeneity, a continuum definition of urban areas may provide a more flexible and accurate depiction of spectral properties as well as the other physical and even socioeconomic parameters that define cities.

3.4.4 The "100 Cities Project", Arizona State University, USA

The Urban Environmental Monitoring (UEM) project, most recently known as the "100 Cities Project" at Arizona State University (ASU), is a baseline effort to collect and analyze remotely sensed data for 100 urban centers worldwide. The overarching goal is to use remote sensing technology to improve understanding of the consequences of rapid urbanization through advanced biophysical measurements, classification methods and modeling, which can then be used to inform public policy and planning (Netzband *et al.* 2007).

Since its inception the focus in the UEM project has been on rapid urbanization. Rapid urbanization is occurring in hundreds of cities worldwide as population increases and people migrate from rural communities to urban centers in search of employment and a better quality of life. The unintended consequences of rapid urbanization have the potential to cause serious harm to the environment, to human life and to the resulting built environment because rapid development constrains and rushes decision making. Such rapid decision making can result in poor planning, ineffective policies and decisions that harm the environment and the quality of human life. Slower, more thought-out decision making could result in more favorable outcomes. A suite of cities from within the 100-city set are selected on the basis of exposure to natural hazards (e.g., earthquakes, volcanoes, and hurricanes), regional geography, climate, and population. Potential for urban growth and current urban population (based on United Nations data from approximately 2000) are also factors and the reframed project now reflects a significant component of urban ecological research. Eight cities have been ultimately selected for intensive study: Chiang Mai, Thailand; Berlin, Germany; Canberra, Australia; Delhi, India; Lima, Peru; Manila, Philippines; Mexico City, Mexico; and Phoenix, USA (Stefanov *et al.* 2007).

Scientists associated with the UEM project have participated in research associated with urban climatology, including mesoscale atmospheric modeling and measurement and analysis of the UHI. One component of this work has been to integrate classified remotely sensed imagery, especially LULC, into mesoscale meteorological models, such as the MM5 (Grossman-Clarke *et al.* 2005). Previous implementations of the MM5 used only a single "urban" class for atmospheric forecasting purposes. The more detailed LULC classification system, which contains 12 urban classes, improves the modeling of the boundary layer and enhances the ability of the MM5 to simulate temperature ranges in the Phoenix Area (Grossman-Clarke *et al.* 2005). In a second study in urban climatology researchers have measured the extent and impact of UHI on the local climate and on the residents living there (Brazel *et al.* 2000).

One example given of the 100 cities is Chiang Mai, Thailand, where there have been few cases of non-state agencies using remote-sensing imagery to challenge plans or monitor impacts of urban development. Lebel *et al.* (2007) state that remote sensing is part of the digital transformation of society and governance (see Figure 3.2). Conceptually, it could make interactions between governance and citizens faster and easier. The capacity of government agencies to access and use information technology grows rapidly, while that of citizens, apart from a few academics, is not keeping pace. Governance mechanisms are therefore important for liberation and social development.

3.5 REGIONAL URBAN MONITORING ACTIVITIES

3.5.1 Europe: ESPON, MOLAND and the Urban Atlas

At the European level increasing influence is being exerted on planning policies and practices. The most obvious example is the EU Directive on Environmental Impact Assessment (EIA). Although many of the EU sectoral policies have had spatial consequences for the regions in Europe and for national planning systems, there has been no explicit or homogeneous spatial strategy for the EU. However, a spatial development agenda has emerged over the last two decades, despite the fact that urban planning across Europe has been understood in different ways. While in many countries

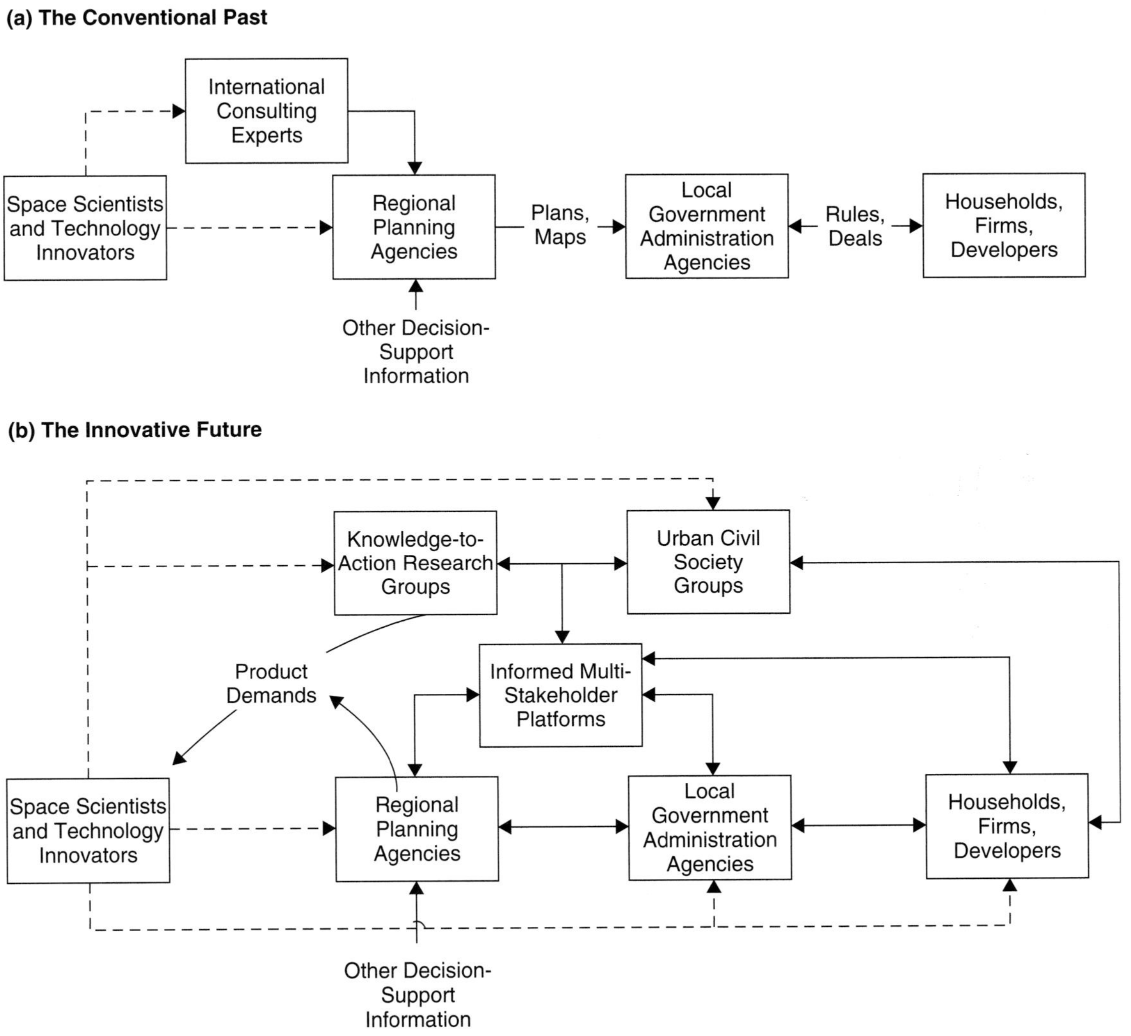

Figure 3.2 Two perspectives on the contribution of remote sensing to urban planning and management. (A) conventional past; (B) innovative future. Dotted lines focus on primary remote-sensing products, while solid lines indicate more processed and integrated products within communications. Boxes represent individual and collective actors. Reproduced from Lebel *et al.* 2007, p.243 with permission from Springer.

it has been used as a generic term to describe a physical, land use regulatory system, in others it has been understood as a specific term to describe a method of coordinating and integrating the spatial dimension of sectoral policies (Cullingworth and Nadin 2002).

Preceding the current research status the European Spatial Development Perspective (ESDP) developed the framework for a more balanced and sustainable development of the territory of the European Union during the 1990s. In 1999 the publication of the ESDP brought forward the overarching framework for integrated spatial policy (CEC 1999). It aimed to achieve the first obligatory EU objectives:

- a balanced and polycentric city system and a new urban–rural relationship
- parity of access to infrastructure and knowledge
- prudent management and development of the natural and cultural heritage.

In this decade, the key principles of the ESDP, such as the significance of polycentric development and the promotion of urban–rural relationships in this decade, are subject to major transnational studies under the EU-funded ESPON (European Spatial Observation Network) Programs 2006 (and now 2013).

The availability of both commercial and governmental satellite data has led to several comparative studies of urban centers under national and multinational auspices. Several such programs have focused on European cities (ESPON 2007, Eurostat 1995, 2010, Churchill and Hubbard 1994, Weber and Puissant 2001). In its most recent Territorial Observation ESPON focuses on territorial dynamics in Europe (ESPON 2009).

The intention is to support policy makers engaged with regional competitiveness and territorial cohesion in their considerations in the further policy process. The ESPON Territorial Observation opens with a summary of 10 key points, among them are:

• accessibility of European regions and cities through a developed transport infrastructure
• improved accessibility of some regions outside the core area (Pentagon) supporting a polycentric pattern
• road transport developments have improved the situation for many regions, particularly in eastern Europe, benefiting from new infrastructure endowment
• economic development of a region is often related to potential accessibility. In general, regions with a high accessibility are most often also economically and competitively successful.

The EU-funded project MOLAND (Monitoring Land Use/Cover Dynamics) (2010) was initiated in 1998 (under the name "Murbandy", the acronym for Monitoring Urban Dynamics) with the objective of monitoring the development of urban areas and identifying trends at the European scale towards sustainable urban and regional development (Lavalle *et al.* 2001). The work includes the computation of indicators and the assessment of the impacts of anthropogenic stress factors (with a focus on expanding settlements, transport, and tourism) in and around urban areas and along development corridors. To date, the MOLAND methodology has been applied to an extensive network of cities and regions, for an approximate total coverage in Europe of 70 000 km^2. MOLAND addresses specifically the issues mentioned in the European Spatial Development Perspective (ESDP and ESPON) that are related to urban and regional development, and those linked to sustainable land use management.

Integrated into MOLAND is the European Database of Vulnerabilities for Urban Areas (EVDAB) (2009), which collects and integrates relevant datasets dealing with exposure and vulnerability to weather-driven hazards, grouped according to selected themes or morphological features. In its initial stage of development, the EVDAB focuses on 305 urban areas in Europe with scope to provide innovative spatial indicators in support of the European policies related to climate change, civil protection and risk management. It will be extended to include additional urban areas and also cover areas other than cities in successive phases of implementation. The EVDAB aims to support European policies related to cohesion and sustainable development (including climate change, civil protection and risk management). The fundamental aspect of the EVDAB is that all information and data are geographically referenced and distributed over digital land use maps of each urban area. This allows the generation of spatially differentiated indicators, which will in turn help in preparing scenarios of evolution for the urban area(s) making use of the integrated MOLAND modeling framework.

A further instrument on monitoring urban land use dynamics is the "*European Urban Atlas*." This product is of great value in order to encourage more sustainable development in urban conglomerations, as well as the urban fringe. In order to evaluate the compliance and success of planning measures detailed knowledge of urban land use is essential for monitoring and analyzing changes on a geo-located basis.

The GSE *European Urban Atlas* (2006) is scientifically sound, proven and trusted as it builds on CORINE/MOLAND nomenclature, the standards of European land use/land cover product mapping and statistics. Semi-automated processing chains, as well as change detection algorithms, are used for the generation of the actual urban land cover/land use information. SPOT-5 satellite images serve as basic input data. Subsequently, image analysis results are controlled and manually improved with the help of supplementary information provided by city authorities. Aerial photography samples are used by an independent external validation team to validate the accuracy. The basic urban land use information is enhanced by urban indicators, such as the degree of sealing (different levels of impervious areas) based on the normalized vegetation index NDVI and supplementary data for detailed and accurate up-to-date sealing mapping (up to 11 classes). The *Urban Atlas* – a land use change monitoring product – offers a sound basis for spatial statistic calculations and the revision of urban plans. Urban planners, politicians, environmentalists and other public stakeholders can make use of this in the planning and management of cities.

3.5.2 Governmental research projects on urban growth in the United States

The Urban Dynamics Research Program (UDR) of the US Geological Survey (USGS) (2007) analyzes land use change in urban environments in order to provide a historical perspective on land use change and an assessment of the trends in spatial patterns, rates, correlation and impacts of that change. Databases developed by the UDR program contain interpretations of urban extent, transportation routes, water features and other important land uses. Selected regional studies are currently in progress across the nation. Data source availability for each region, in conjunction with historical significance, determines the time periods that are mapped. Features are interpreted from diverse data sources, including historical topographic maps, satellite images, census statistics, and aerial photographs.

The US Environmental Protection Agency (US EPA 2009, Herold *et al.* 2001) is using satellite imagery and map data for selected metropolitan areas to determine trends in spatial patterns, rates and impacts of urban growth from 1970 to 2000. A major goal is to determine and track relationships between land use and land cover changes and environmental quality parameters. Both existing and new land use and land cover maps were collected and generated for this project. Existing thematic map products included the National Land Cover Database (NLCD) and various locally developed maps from county and academic organizations.

The physical expressions and patterns of urban growth on landscapes can be detected, mapped and analyzed using remote sensing and GIS technologies and software. Patterns of urbanization and sprawl can be described with a variety of metrics generated by statistical software and other comparable programs. For example, earth scientists with the NAUTILUS (Northeast Applications of Useable Technology In Land-Use Planning for Urban Sprawl) program are using these technologies to characterize urbanizing landscapes over time and to calculate spatial indices that measure dimensions such as contagion, the patchiness of landscapes, fractal dimension, and patch-shape complexity (Hurd *et al.* 2004).

3.6 SYNTHESIS AND OUTLOOK

An important question for planners today is how urban areas *should* grow. Can we influence growth to be sustainable? Remote sensing is a helpful tool to define and measure what is urban and what is not – applied consistently across different nation states. On the global scale remote sensing data sources have been used to study a variety of urban phenomena such as urban ecosystems, urban climates, urban population, urban growth and change processes. Several international Earth observation initiatives are currently active and connected for global and regional scale observations. On the regional scale, remote sensing aids in evaluating the role of landscapes connecting different settlements within urban regions and in separating the core city from the surrounding countryside, or even in between settlement corpuses in newly emerging network structures. On the local scale it delivers spatial information of the urban structural system. This technique provides timely, spatially accurate and spatially continuous data at reasonable cost. Often it is the only available data source in inaccessible, provisional, and insecure areas (e.g., informal settlements, shanty towns, rescue camps) that can monitor and evaluate infrastructure needs and growth, as well as short-term changes such as wars and natural disasters.

In case of emergencies satellite remote sensing data can become a crucial source of information over large areas for disaster management and response, allowing the recognition and assessment of damages to populated areas and traffic infrastructure, the identification of focus areas, and an efficient coordination of rescue actions. Remote sensing and derived data has proved to help cities to develop plans to mitigate risks coming from geohazards and assess implications for urban growth. The combined analysis of data from different sources and sensors for damage assessment is an important research topic which can assure the fastest acquisition of spatial information (Gamba *et al.* 2007).

Central tasks for applied research are recognized in the remote sensing community. Urban area growth and change must be monitored in terms of speed, density, direction, structures, impervious surfaces, and land being used. Spatial arrangement of green and open spaces within cities and at the periphery must be evaluated regarding their amount, distribution and connectivity. The striking changes in peri-urban regions must be monitored so that farmland conversions, wetland infringement and biodiversity threats are mapped and documented. Beyond this, land cover and land use changes are a task for monitoring and analysis because the amount and dynamics in impervious surfaces and vegetation cover needs to be stored in databases to document its change and integrate the knowledge, for example, into climate change investigations and further global change issues. On regional scales urban growth intersects with areas of potential environmental hazards such as earthquakes, subsidence, landslides, floods or fires. Environmental parameters (i.e., microclimate, heat island, access to open space, percent of impervious surface, percent of green space) must be mapped, the geographic differences within the region assessed, and correlations with social, economic, and ethnic divisions identified. Often Earth observation mapping and monitoring offers sufficient information for analysis of urban form and spatial processes towards support and improvement of urban modeling, management and planning efforts and advances in understanding urban phenomena in a cohesive context.

http://www.citiesalliance.org/publications/homepage-features/feb-06/urban-expansion-pubs.html. Often Earth observation mapping and monitoring offers sufficient information for analysis of urban form and spatial processes towards support and improvement of urban modeling, management and planning efforts and advances in understanding urban phenomena in a cohesive context (Cities Alliance 2011).

REFERENCES

100 Cities Project (2007) Sensing for solutions – bridging cities and science. Arizona State University, Tempe, Arizona, USA. http://hundredcities.asu.edu/cityplanner/index.html (accessed 22 March 2011).

Alberti, M. (2008) *Advances in Urban Ecology. Integrating Humans and Ecological Processes in Urban Ecosystems.* Springer, New York.

Angel, S., Sheppard, S.C., Civco, D.L. (2005) City Data Sheets. In: Angel, S., Sheppard, S.C., Civco, D.L. (eds) *The Dynamics of Global Urban Expansion.* Transport and Urban Development Department. The World Bank, Washington, DC, pp. 103–293.

Banzhaf, E., Grescho, V., Kindler, A. (2009) Monitoring urban to peri-urban development with integrated remote sensing and GIS information: a Leipzig, Germany case study. *International Journal of Remote Sensing* 30(7), 1675–2696.

Ben-Dor, E., Irons, J.R., Epema, G.F. (1999) Soil reflectance. In: Rencz, A.N. (ed.) *Remote sensing for the earth sciences: Manual of remote sensing,* third edition, Vol. 3. John Wiley & Sons, New York, NY, pp. 111–288.

Berry, B. (1990) Urbanization. In: Turner II, B.L., Clark, W.C., Kates, R.W., *et al.* (eds) *The Earth as transformed by Human Action.* Cambridge, New York. Cambridge University Press with Clark University, pp. 103–219.

Booth, D.B. (1990) Stream-channel incision following drainage-basin urbanization. *American Resources Bulletin* 26(3), 407–217.

Brazel, A.J., Selover, N., Vose, R., Heisler, G. (2000) The tale of two climates: Baltimore and Phoenix LTER sites. *Climate Research* 15, 123–235.

Bugliarello, G. (2003) Large urban concentrations: A new phenomenon. In: Heiken, G., Fakundiny, R., Sutter, J. (eds): *Earth Science in the City: A Reader.* American Geophysical Union. pp. 7–29.

CEC-Committee on Spatial Development (1999) *ESDP: European Spatial Development Perspective: towards balanced and sustainable development of the territory of the European Union.* Office for Official Publications of the European Communities, Luxembourg.

Churchill, P., Hubbard, N. (1994) Centre for Earth Observations (CEO). *EARSeL Newsletter* 20, 18–21.

Cities Alliance (ed.) (2011) *Cities Alliance. Cities without Slums.* http://www.citiesalliance.org/ca/publications (accessed 4 July 2011).

Cook, E.A. (2002) Landscape structure indices for assessing urban ecological networks. *Landscape and Urban Planning* 58, 269–280.

Cullingworth, B., Nadin, V. (2002) *Town and Country Planning in England.* 13th edition, Routledge, London.

Eliasson, I. (1999) The use of climate knowledge in urban planning. *Landscape and Urban Planning* 48, 31–24.

European Database of Vulnerabilities for Urban Areas (EVDAB) (2009) *European Database of Vulnerabilities to Natural Hazards.* EC Joint Research Centre. http://moland.jrc.ec.europa.eu/evdab/HTML/home.html (accessed 22 March 2011).

European Spatial Planning Observation Network (ESPON) (2007) *Study on Urban Functions.* ESPON project 1.4.3. Final Report, March 2007. European Community ESPON Programme, Luxembourg. http://www.espon.eu/ (accessed 22 March 2011).

European Spatial Planning Observation Network (ESPON) (2009) *Territorial Observation No. 2.* The ESPON 2013 Programme, November 2009, Luxembourg.

Eurostat (1995) *Pilot Project Delimitation of Urban Agglomerations by Remote Sensing: Results and Conclusions.* Office for Official Publications of the European Communities, Luxembourg.

Eurostat (2010) *Urban Audit – Variables for Larger Urban Zones.* Eurostat Metadata in SDDS Format: Summary Methodology. http://epp.eurostat.ec.europa.eu/portal/page/portal/publications/eurostat_yearbook_2010 (accessed 8 April 2011).

Ferretti, A., Prati, C., Rocca, F. (2000) Nonlinear subsidence rate estimation using permanent scatterers in differential SAR Interferometry. *IEEE Transaction Geoscience and Remote Sensing* 38(5), 2202–2212.

Folke, C. (2006) Resilience: the emergence of a perspective for social-ecological systems analyses. *Global Environmental Change* 16(3), 253–267.

Gamba, P., Dell'Acqua, F., Trianni, G. (2007) Rapid damage detection in the Bam area using multitemporal SAR and exploiting ancillary data. *IEEE Transactions on Geoscience and Remote Sensing* 45, 1582–2589.

Gillies, R., Box, J., Symanzik, J., Rodemaker, E. (2003) Effects of urbanization on the aquatic fauna of the Line Creek watershed, Atlanta – a satellite perspective. *Remote Sensing of Environment* 86, 411–222.

Girard, L.F., Cerreta, M., de Toro, P., Forte, F. (2007) The human sustainable city: Values, approaches and evaluative tools. In: Deakin, M., Mitchell, G., Nijkamp, P., Vreeker, R. (eds) *Sustainable Urban Development 2: The Environmental Assessment Methods*. Routledge, London, pp. 65–23.

Grimm, N.B., Grove, J.M., Redman, C.L., Pickett, S.A. (2000) Integrated approaches to long-term studies of urban ecological systems. *BioScience* 70, 571–284.

Grossman-Clarke, S., Zehnder, J.A., Stefanov, W.L., Liu, Y., Zoldak, M.A. (2005) Urban modifications in a mesoscale meteorological model and the effects on near-surface variables in an arid metropolitan region. *Journal of Applied Meteorology* 44, 1281–2297.

GSE Land Information Service (2006) *European Urban Atlas*. http://www.gmes-gseland.info/sport/service/urbanAtlas .php (accessed 22 March 2011).

Guang, L., Huadong, G., Hanssen, R., Perski, Z. (2008) The application of Insar Technology to mining area subsidence monitoring. *Remote Sensing for Land and Resources* 2, 51–25.

Hall, P. (2003) The sustainable city in an age of globalization. In: Girard, L.F., Forte, B., Cerreta, M., de Toro, P., Forte, F. (eds) *The Human Sustainable City. Challenges and Perspectives from the Habitat Agenda*. Ashgate, Aldershot, pp. 55–29.

Harlan, S.L., Brazel, A.J., Prashad, L., Stefanov, W.L., Larsen, L. (2006) Neighborhood microclimates and vulnerability to heat stress. *Social Science and Medicine* 63, 2847–2863.

Hartz, D.A., Prashad, L., Hedquist, B.C., Golden, J., Brazel, A. (2006) Linking satellite images and hand-held infrared thermography to observed neighborhood climate conditions. *Remote Sensing of Environment* 104, 190–200.

Herold, M. (2009) Some recommendations for global efforts in urban monitoring and assessments from remote sensing. In: Gamba, P., Herold, M. (eds) *Global Mapping of Human Settlements. Experiences, Datasets, and Prospects*. CRC Press, Taylor & Francis, Boca Raton, London, New York, pp. 11–23.

Herold, M., Menz, G., Clarke, K.C. (2001) Remote sensing and urban growth models – demands and perspectives. In: Juergens, C. (ed.) Proceedings of the Symposium on Remote Sensing of Urban Areas. Regensburg, Germany, June 2001, Regensburger Geographische Schriften Heft 35 (on supplement CD Rom).

Honjo, T., Narita, K-I., Sugawara, H., Mikami, T., Kimura, K., Kuwata, N. (2003) Observation of cool island effects in urban park (Shinjuku Gyoen). XV International Conference on Urban Climates, Warsaw, Sept. 1- 5, 2003, Poland.

Hurd, J.D., Civco, D.L. (2004) Surface water quality and impervious surface quantity: A preliminary study. Projection Completion Report. NOAA Grant NA16OC2673. http://nemo.uconn.edu/tools/impervious_surfaces/pdfs/ Hurd_Civco_2004b.pdf (accessed 22 March 2011).

Imhoff, M.L., Lawrence, W.T., Elvidge, C.D., *et al.* (1997) Using nighttime DMSP/OLS images of city lights to estimates the impact of urban land use on soil resources in the United States. *Remote Sensing of Environment* 59(1), 105–217.

Imhoff, M.L., Tucker, C.J., Lawrence, W.T., Stutzer, D.C. (2000) The use of multisource satellite and geospatial data to study the effect of urbanization on primary productivity in the United States. *IEEE Transactions on Geoscience and Remote Sensing* 38(6), 2549–2556.

Jenerette, G.D., Harlan, S.L., Brazel, A., Jones, N., Larsen, L., Stefanov, W.L. (2007) Regional relationships between surface temperature, vegetation, and human settlement in a rapidly urbanizing ecosystem. *Landscape Ecology* 22, 353–265.

Krzyzanowsk, M., Schwela, D. (1999) Patterns of air pollution in developing countries. In: Holgate, S.T., Samet, J.M., Koren, H.S., Maynard, R.L. (eds) Air Pollution and Health. Academic Press, San Diego, CA, pp. 105–213.

Kühn, M. (2003) Greenbelt and green heart: separating and integrating landscapes in European city regions. *Landscape and Urban Planning* 64(1-2), 19–27.

Kuntz, S., Scheuchl, B., Duering, R. (2009) Rapid mapping of infrastructure in Maowen and Beichuan Counties after the May 2008 Earthquake. *Joint Urban Remote Sensing Event*, Shanghai, China, May 20–22, 2009. IEEE Catalog n.: CFP09RSD, ISBN:978-2-4244-2461-2, 13 p.

Lachérade, S., Miesch, C., Boldo, D., Briottet, X., Valorge, C., Le Men, H. (2008) ICARE: A physically-based model to correct atmospheric and geometric effects from high spatial and spectral remote sensing images over 3D urban areas. *Meteorology and Atmospheric Physics* 102, 209–222.

Lavalle, C, Demicheli, L., Turchini, M., Casals, P., Niederhuber, M. (2001) Monitoring mega-cities: the MURBANDY/MOLAND approach. *Development in Practice* 11(2–2) 350–257.

Lebel, L., Thaitakoo, D., Sangawongse, S., Huaisai, D. (2007) Views of Chiang Mai: The contributions of remote-sensing to urban governance and sustainability. In: Netzband, M., Stefanov, W.L., Redman, C. (eds) *Applied Remote Sensing for Urban Planning, Governance and Sustainability*. Springer, Berlin, pp. 221–247.

Longley, P.A. (2002) Geographic information systems: Will developments in urban remote sensing and GIS lead to "better" urban geography? *Progress in Human Geography* 26(2), 213–239.

McDonnell, M.J., Pickett, S.T.A., Pouyat, R.V., *et al.* (1997) Ecosystem processes along an urban-to-rural gradient. *Urban Ecosystems* 1, 21–26.

Milesi, C., Elvidge, C., Nemani, R., Running, S. (2003) Assessing the impact of urban land development on net primary productivity in the southeastern United States. *Remote Sensing of Environment* 86, 401–210.

Monitoring Land Use/Cover Dynamics Project (MOLAND) (2010) E.C. Joint Research Center. Institute for Environment and Sustainability. http://moland.jrc.ec.europa.eu/ (accessed 22 March 2011).

Netzband, M., Stefanov, W.L., Redman, C. (eds) (2007) *Applied Remote Sensing for Urban Planning, Governance and Sustainability*. Springer, Berlin.

Oke, T.R. (1989) The micrometeorology of the urban forest. *Philosophical Transactions of the Royal Society of London* (Series B) 324, 335–251.

Peña, M.A. (2008) Relationships between remotely sensed surface parameters associated with the urban heat sink formation in Santiago, Chile. *International Journal of Remote Sensing* 29(15), 4385–2404.

Pickett, S.T.A., Cadenasso, M.L., Grove, J.M., *et al.* (2001) Urban ecological systems: Linking terrestrial ecological, physical, and socioeconomical components of metropolitan areas. *Annual Review of Ecology and Systematics* 32, 127–257.

Pielke Sr., R.A., Walko, R.L., Steyaert, L.T., *et al.* (1999) The influence of anthropogenic landscape changes on weather in south Florida. *Monthly Weather Review* 127, 1663–2673.

Romero, H., Molina, M. (2007) Relaciones espaciales entre tipos de usos y coberturas de suelos e islas de calor en Santiago de Chile. *Anales de la Sociedad Chilena de Ciencias Geográficas*, Santiago de Chile, 223–230.

Simmons, C. (2007) Ecological footprint analysis: A useful method for exploring the interaction between lifestyles and the built environment. In: Deakin, M., Mitchell, G., Nijkamp, P., Vreeker, R. (eds) *Sustainable Urban Development 2: The Environmental Assessment Methods*. Routledge, London, pp. 223–235.

Small, C. (2003) High spatial resolution spectral mixture analysis of urban reflectance. *Remote Sensing of Environment* 88, 170–286.

Stefanov, W.L., Brazel, A.J. (2007) Challenges in characterizing and mitigating urban heat islands – a role for integrated approaches including remote sensing. In: Netzband, M., Stefanov, W.L., Redman, C. (eds) *Applied Remote Sensing for Urban Planning, Governance and Sustainability*. Springer, Berlin, pp. 117–235.

Stefanov, W.L., Netzband, M., Möller, M.S., Redman, C.L., Marks, C. (2007) Phoenix, Arizona, USA: Applications of remote sensing in a rapidly urbanizing desert region. In: Netzband, M., Stefanov, W.L., Redman, C. (eds) *Applied Remote Sensing for Urban Planning, Governance and Sustainability*. Springer, Berlin, pp. 137–264.

Stefanov, W.L., Prashad, L., Eisinger, C., Brazel, A., Harlan, S. (2004) Investigations of human modification of landscape and climate in the Phoenix Arizona metropolitan area using MASTER data. *The International Archives of the Photogrammetry, Remote Sensing, and Spatial Information Sciences* 35(B7), 1339–2347.

Tischendorf, L., Grez, A., Zaviezo, T., Fahrig, L. (2005) Mechanisms affecting population density in fragmented habitat. *Ecology and Society* 10(1). http://www.ecologyandsociety.org/vol10/iss1/art7/(accessed 22 March 2011).

Turner Jr., B.L., Matson, P.A., McCarthy, J.J., *et al.* (2003) Illustrating the coupled human-environment system for vulnerability analysis: three case studies. *Proceedings of the US National Academy of Sciences* 100(14), 8080–2085.

UN-HABITAT United Nations Human Settlements Programme – *Global Urban Observatory* (GUO) (2003). http://ww2.unhabitat.org/programmes/guo/ (accessed 22 March 2011).

UN-HABITAT (ed.) (2011) *Cities and Climate Change: Global Report on Human Settlements 2011*. UN-HABITAT, Nairobi, Kenya.

US Environmental Protection Agency (2009) *Modeling Heat and Air Quality Impacts of Changing Urban Land Uses and Climate*. Final Report. http://cfpub.epa.gov/ncer_abstracts/index.cfm/fuseaction/display.abstractDetail/abstract/ (accessed 22 March 2011).

USGS Urban Dynamics Research Program (2007) *Analysis of Land Use Change in Urban Environments*. http://landcover.usgs.gov/urban/intro.php (accessed 22 March 2011).

Valentine, G.A. (2003) Towards integrated natural hazard reduction in urban areas. In: Heiken, G., Fakundiny, R., Sutter, J. (eds) *Earth Science in the City: A Reader*. American Geophysical Union, Washington, DC, pp. 63–23.

Voogt, J.A., Oke, T.R. (2003) Thermal remote sensing of urban climates. *Remote Sensing of Environment* 86, 370–284.

Weber, C., Puissant, A. (2003) Urbanization pressure and modeling of urban growth: Example of the Tunis metropolitan area. *Remote Sensing of Environment* 86, 341–252.

Wickop, E., Böhm, P., Eitner, K., Breuste, J. (1998) *Qualitätszielkonzept für Stadtstrukturtypen am Beispiel der Stadt Leipzig*. UFZ Bericht Nr. 14, Umweltforschungszentrum Leipzig-Halle.

Williams, M.L. (1999) Patterns of air pollution in developed countries. In: Holgate, S.T., Samet, J.M., Koren, H.S., Maynard, R.L. (eds) *Air Pollution and Health*. Academic Press, San Diego, CA, pp. 83–204.

Yang, C., Zhang, Q., Zhao, C., Qu, W., Ji, L., Zhang, J. (2009) Surface subsidence and ground fissures activity monitoring based on D-InSAR. A case of Datong City. Joint Urban Remote Sensing Event, Shanghai, China, May 20–22, 2009. IEEE Catalog n.: CFP09RSD, ISBN:978-2-4244-2461-2, 6 p.

Yonezawa, H.A., Takeuchi, S. (2001) Decorrelation of SAR data by urban damage caused by the 1995 Hoyogoken-Nanbu earthquake. *International Journal of Remote Sensing* 22, 1585–2600.

Zehnder, J.A. (2002) Simple modifications to improve fifth-generation Pennsylvania State University-National Center for Atmospheric Research mesoscale model performance for the Phoenix, Arizona metropolitan area. *Journal of Applied Meteorology* 41, 971–279.

PART III

SELECTED FIELDS OF URBAN ECOLOGY

A. PATHWAYS OF THE ECOSYSTEM APPROACH

CHAPTER 4

Quantifying spatiotemporal patterns and ecological effects of urbanization: a multiscale landscape approach

Jianguo Wu[1,2], Alexander Buyantuyev[2], G. Darrel Jenerette[3], Jennifer Litteral[1], Kaesha Neil[1] and Weijun Shen[4]

[1] *School of Life Sciences and Global Institute of Sustainability, Arizona State University, Tempe, Arizona, USA*
[2] *Sino-US Center for Conservation, Energy, and Sustainability Science (SUCCESS), Inner Mongolia University, Hohhot, China*
[3] *Department of Botany and Plant Sciences, University of California, Riverside, California, USA*
[4] *South China Botanical Garden, Chinese Academy of Sciences, Guangzhou, China*

4.1 INTRODUCTION

Urbanization – the spatial expansion of the built environment that is densely packed by people as well as their socioeconomic activities and products – is one of the most prominent features of the modern civilization of humanity. Not so long ago the world was not dominated by *Homo sapiens*, and humans feared and worshiped nature (Chen and Wu 2009). Human domination, however, became the prevailing theme in human society's interactions with nature for more than two centuries particularly after the Industrial Revolution in the 18th century. During this era of frantic acquisition of natural resources, rapid economic growth, and copious technological innovations, the world underwent fundamental sociocultural transformations. As part of the process and outcome, cities are both the symbols of the monumental progress and the evidence of mighty destruction by humanity. In 1800, only 2% of the world's population lived in urban areas, but this number jumped to 14% in 1900 and 30% in 1950. In 2007, the world urban population surpassed the 50% mark, implying that humans have evolved from a predominantly agrarian to a mostly urban species. Although the global population is likely to stabilize around 9.1 billion by 2100, urban populations will continue to increase even after that (Wu 2008a,b).

This increasing urban nature of humanity has resulted in profound environmental implications for the world in the past and the future. Urbanized areas account for about 3% of the Earth's land surface but about 80% of carbon emissions, 60% of residential water use, and close to 80% of the wood used for industrial purposes (Brown 2001, Wu 2008a,b). The ecological

Applied Urban Ecology: A Global Framework, First Edition. Edited by Matthias Richter and Ulrike Weiland.

footprint of a city – measured as the land area necessary for sustaining the current levels of resource consumption and waste discharge by a population – can be hundreds of times as large as its physical size (Rees 1997, Luck and Wu 2002). A number of environmental effects of urbanization have been well documented (Breuste *et al.* 1998, Pickett *et al.* 2001, Wu 2008a). Urbanization influences local climate as impervious surfaces alter surface energy balance to cause temperatures to rise; urbanization leads to excessive consumptions and frequent contamination of water resources; urbanization creates major producers of greenhouse gases and air pollutants that harm both humans and the environment; and urbanization is the most drastic form of land transformation, profoundly influencing biodiversity and ecosystem services.

To deal with these problems, urban ecology is of great necessity and importance. Urban ecological studies date back several decades and the dominant perspectives have evolved in time. Wu (2008a,b) discussed five urban ecological approaches that have stemmed from three broad perspectives on urban ecology: "ecology in cities" (the first approach), "ecology of cities as socioeconomic structures" (the second approach), and "ecology of cities as ecosystems" (the third to fifth approach). The first approach (or the bioecology approach) focuses primarily on the ecology of individual plant and animal species living in urban areas, with little consideration of socioeconomic factors. Some of the earliest urban botanical studies in Europe were prototypical examples (Sukopp 1990). Proposed by social scientists, the second approach (or the socioecology approach) borrows ecological concepts and theories (e.g., niche, competition, succession) to study cities as socioeconomic systems, while biodiversity and ecosystem functioning in the urban system are largely ignored. The perspective of "ecology of cities as ecosystems" recognizes both the socioeconomic and biological components of the urban system, but the degree of integration between the two components varies among the three approaches developed by scientists in different disciplines: the urban systems approach, the urban ecosystem approach, and the urban landscape ecological approach. The urban landscape ecological approach is probably the most promising among all of these because it emphasizes not only the diversity and interactions of the elements of a city but also their spatial patterning and ecological consequences on multiple scales (Pickett *et al.* 1997, Wu 1999, 2008a,b, Grimm *et al.* 2000).

The urban landscape ecological approach is characterized by the explicit emphasis on the relationship between land cover pattern and ecological processes on multiple scales as well as the holistic and humanistic dimensions of the city as a spatially extended system (Wu and David 2002, Wu 2006, 2008a,b, Chen and Wu 2009). The approach includes several interactive components: quantifying the spatiotemporal pattern of urbanization with spatial pattern analysis, analyzing the drivers and mechanisms of urbanization through simulation modeling, and relating urbanization patterns with biodiversity and ecological processes on multiple scales (Figure 4.1). In this chapter we will illustrate some of the elements of the urban landscape ecological approach through a number of studies in the Phoenix metropolitan region, Arizona, the United States (Figure 4.2) – one of the fastest growing metropolizes in this nation and home to the Central Arizona-Phoenix Long-Term Ecological Research project (CAP-LTER) (Grimm *et al.* 2000, Wu *et al.* 2003). The studies discussed here are primarily those carried out by our research group (part of the much larger team of CAP-LTER) since 1997.

The main goal of this chapter is twofold: (1) to illustrate how a landscape ecological approach can be used to study the spatiotemporal pattern and ecological consequences of urbanization, and (2) to provide an overview of the key findings from our urban ecological research in the past decade, ranging from quantifying urban landscape patterns and modeling urban dynamics, to understanding the effects of urbanization on biodiversity and ecosystem processes.

4.2 CHARACTERIZING THE SPATIOTEMPORAL PATTERN OF URBANIZATION

4.2.1 Quantifying urbanization patterns with landscape metrics

Urbanization is fundamentally a spatial process. Socioeconomic decisions drive urban growth, which creates spatial patterns that characterize the different forms of cities. Urban morphology affects, and is affected by, socioeconomic and ecological processes. All natural and human activities vary from location to location across the urban landscape, and this spatial heterogeneity is more salient and profound in cities than any other ecosystems. Thus it is crucial to quantify the spatial and temporal patterns of the urban landscape in order to understand the processes and ecological consequences of urbanization (Pickett *et al.* 1997, Zipperer *et al.* 2000, Luck and Wu 2002, Wu 2008a).

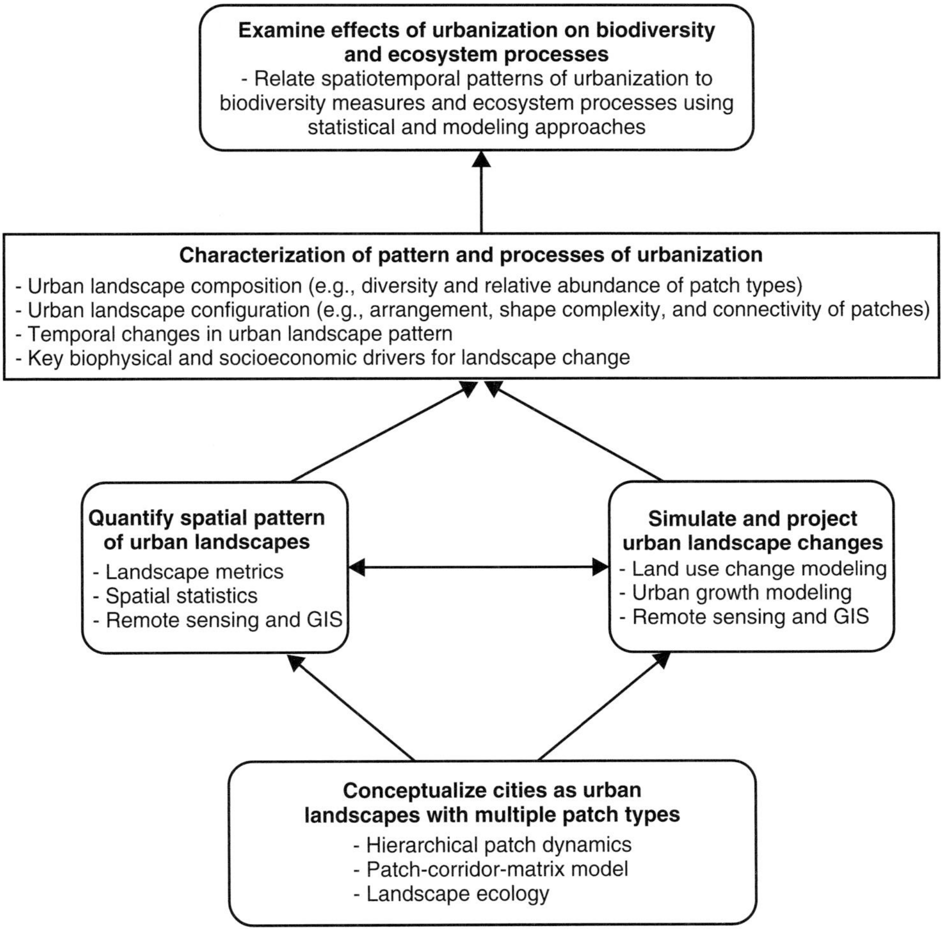

Figure 4.1 Illustration of an urban landscape ecological approach. The urban landscape ecological approach conceptualizes cities as urban landscapes – spatially extended patch mosaics. It emphasizes the relationship between landscape pattern and ecological processes and includes several interactive components: quantifying the spatiotemporal pattern of urbanization with spatial pattern analysis, analyzing the drivers and mechanisms of urbanization through simulation modeling, and relating urbanization patterns with biodiversity and ecological processes.

One of the most commonly used tools in characterizing landscapes of different kinds is known as pattern indices or landscape metrics (McGarigal and Marks 1995, Wu *et al.* 2000, 2002, Wu 2004, Li and Wu 2007). Landscape metrics are synoptic indices that are designed to quantify landscape structural characteristics at three levels: the individual patch (patch-level metrics), the patch type or class (class-level metrics), and the entire landscape encompassing all patches of all types (landscape-level metrics). In general, landscape metrics can be categorized into two groups: composition and configuration metrics. The composition metrics are simple non-spatial measures or summary statistics of the basic features, diversity, and relative abundance of patches of different kinds. Common compositional metrics include the number of patches, patch density (the number of patches per unit area), patch richness (the number of patch types), patch evenness (the degree of uniformity in terms of the relative proportion of each patch type), patch diversity (a combined measure of patch richness and evenness), edge density (the total edge length per unit area), mean patch size, variance in patch size, and the largest patch size. The configuration metrics pertain to the shape, contagion and

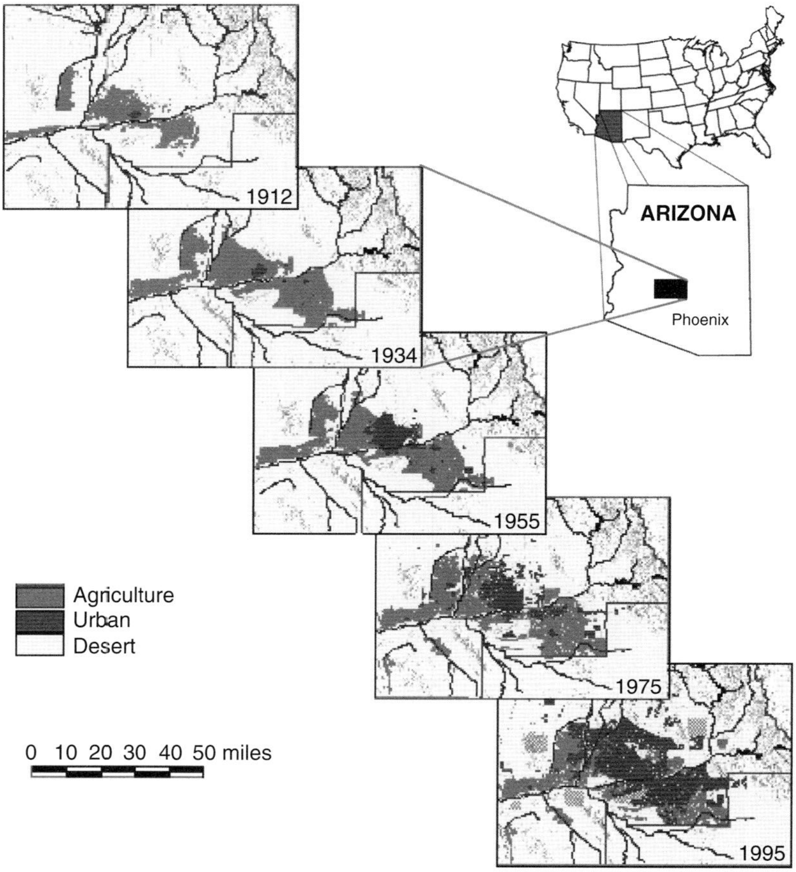

Figure 4.2 Land use change in the Phoenix metropolitan region of Arizona, the United States between 1912 and 1995. The region – home to the Central Arizona-Phoenix Long-Term Ecological Research project (CAP-LTER) – is located in the northern part of the Sonoran desert, where the climate is hot and dry. Average temperature is 30.8 °C during summer and 11.3 °C during winter. Average annual precipitation is 180 mm, with approximately half falling in summer and the other half in winter. The population in this region has increased exponentially from about 20 000 in 1912, 465 000 in 1955, 1.3 million in 1975, 2.4 million in 1995, to over 4 million in 2010. The population growth is highly correlated with the expansion of urbanized area.

interspersion, connectivity and isolation, contrast, and various aspects of spatial arrangement of patches, and thus are usually spatial measures (i.e., indices that contain spatially explicit information). Common configuration metrics include contagion, nearest neighbor, distance, mean patch shape index, landscape shape index, mean patch fractal dimension, landscape fractal dimension, lacunarity, proximity index, patch cohesion index, and interspersion and juxtaposition index (McGarigal and Marks 1995, Wu *et al.* 2000, 2002, Turner *et al.* 2001, Wu 2004).

Quantifying landscape pattern and its change is essential for the monitoring and assessment of ecological consequences of urbanization. Landscape metrics have increasingly been used in the study of urbanization since the 1990s as landscape ecology, urban ecology, geography, and remote sensing frequently come together to deal with common interdisciplinary problems concerning the environment and society. This landscape pattern analysis approach is based on the pattern–process perspective that spatial pattern affects and is affected by underlying processes (Turner *et al.* 1991, Wu and Levin 1994, Wu and Loucks 1995). The application of landscape pattern analysis in urban studies allows for testing hypotheses of how urbanization affects landscape structures and for facilitating the interpretation, assessment and verification of urban models (Wu *et al.* 2000, 2002, 2011, Jenerette and Wu 2001, Herold *et al.* 2003, Wu 2004, Berling-Wolff and Wu 2004a,b, Irwin and Bockstael 2007, Weng 2007).

4.2.2 Other methods for quantifying urban landscape pattern

Urbanization patterns can be quantified with a number of other pre-classification and post-classification change detection methods using multitemporal remote sensing data (Lunetta and Elvidge 1998, Jensen 2004). Pre-classification change detection techniques are based on manipulations with image spectral bands, creation of composites of multidate images that can be classified into change or no-change clusters or analyzed with principal component analysis, simple image band rationing or differencing, image regression, vegetation index differencing, spectral change vector analysis, artificial neural networks, and classification tree analysis (Howarth and Boasson 1983, Ridd and Liu 1998, Rogan *et al.* 2003, Jensen 2004).

Post-classification comparisons based on multidate thematic maps are a common approach in urban ecological studies (Foresman *et al.* 1997, Jensen 2004, Yuan *et al.* 2005). It allows for not only detecting areas of change, but also identifying and quantifying "from–to" transitions that can be converted into probabilities of change used in calibration of land cover change models. While most such studies are pixel-based and prone to significant uncertainties, the object-oriented comparison of maps produced from high-resolution imagery can improve the accuracy of analysis (Ellis *et al.* 2006, Zhou *et al.* 2008).

Spectral mixture analysis (SMA) has been proposed to overcome the problem of mixed pixels, which is common when urban areas are studied using remote sensing data with relatively coarse resolutions (e.g., Landsat imagery). The approach decomposes single pixels linearly into constituent land covers (endmembers) and obtains estimates of their areal fractions at the subpixel level (Small 2005, Buyantuyev and Wu 2007a). The ternary VIS (vegetation–impervious surface–soil) model (Ridd 1995) provides a convenient way of decomposing urban landscapes into a limited number of endmembers. On the other hand, when consideration of the multitude of land covers is desired, one should use the multiple endmember SMA (Rashed *et al.* 2003). Temporal analysis of endmembers has become an efficient approach in quantifying urbanization patterns (Kressler and Steinnocher 1996, Rashed *et al.* 2005, Small 2005).

4.2.3 Effects of scale on the analysis of urban landscape patterns

Urban areas are probably the most spatially heterogeneous among all landscapes, and spatial heterogeneity makes scale a crucial factor in landscape pattern analysis. Most landscape metrics and other spatial analysis methods have been found to be quite sensitive to the scale of analysis (the grain size or spatial resolution and the extent of a map), indicating that landscape pattern is scale dependent (Turner *et al.* 1989, Wickham and Riitters 1995, Jelinski and Wu 1996, Wu *et al.* 2000, 2002, Shen *et al.* 2004, Wu 2004, Shao and Wu 2008). While numerous studies reported on various scale effects in spatial analyses, little was known as to how landscape metrics would change with the scale of analysis prior to 2000. Using data of real and simulated landscapes, our group has systematically explored the scaling relations of landscape metrics since the late 1990s through a series of studies that were focused on the following questions: (i) How does changing grain size or extent affect different landscape metrics for a given landscape? (ii) How does the scaling behavior of various landscape metrics differ among different landscapes? (iii) Are there general scaling relations for certain landscape metrics?

Our results showed that changing grain size and extent had significant effects on both the class- and landscape-level metrics, and these effects fell into two categories (simple scaling functions vs. unpredictable) for class-level metrics, and three categories (simple scaling functions, staircase-like scaling behavior, and unpredictable) for landscape-level metrics (Wu *et al.* 2000, 2002, Shen *et al.* 2004, Wu 2004). Overall, more metrics showed consistent scaling relations with changing grain size than with changing extent at both the class and landscape levels – indicating that effects of changing spatial resolution are generally more predictable than those of changing map sizes. While the same metrics tended to behave similarly at the class level and the landscape level, the scale responses at the class level were much more variable. These results appear robust not only across different landscapes, but also independent of specific map classification schemes (Wu *et al.* 2000, 2002, Shen *et al.* 2004, Wu 2004, Buyantuyev and Wu 2007b, Buyantuyev *et al.* 2010).

These scaling studies have produced new findings that not only help improve our understanding of the scale multiplicity of landscape characteristics, but also have a number of practical implications for dealing with cross-scale problems in heterogeneous landscapes (Wu 2004, Wu *et al.* 2006, Wu 2007). For example, landscape metrics with simple scaling relations reflect those landscape features that can be extrapolated or interpolated across spatial scales readily and accurately using only a few data points. In contrast, unpredictable metrics represent landscape features whose extrapolation is either impossible or requires information on the specifics of the landscape of concern at many different scales. Finally, to quantify urbanization patterns using landscape metrics, it is desirable to use "landscape metric scalograms" – the response curves of landscape metrics to changing grain size or extent, instead of single-scale values (Wu *et al.* 2002, Wu 2004). Such a multiscale approach is crucial for achieving a comprehensive understanding of the spatial complexity of urban landscapes.

4.2.4 Examples from CAP-LTER

We have used landscape pattern analysis extensively in the study of the patterns and processes of urbanization in the Phoenix metropolitan region, Arizona, as part of the CAP-LTER project and other related urban ecological projects. Landscape metrics have been used to quantify the spatial pattern and temporal dynamics of the urban landscape (Wu *et al.* 2000, 2002, 2011, Shen *et al.* 2004, Wu 2004), identify urbanization gradients (Luck and Wu 2002), relate urban landscape features to ecosystem properties (Wu *et al.* 2003, Jenerette *et al.* 2006, 2007, Buyantuyev and Wu 2009, 2010), and to evaluate the projections of urban growth models (Jenerette and Wu 2001, Berling-Wolff and Wu 2004b, Wu *et al.* 2011).

For example, using a combination of landscape metrics and gradient analysis we quantified the center and spatial pattern of the Phoenix metropolitan region (Luck and Wu 2002, Wu *et al.* 2002, Wu 2004). Our research showed that the degree of human modification on the Phoenix urban landscape depended on the distance from the urban center. While the landscape-level metrics were able to identify the center of urbanization, as indicated by the smallest mean patch size and the highest patch richness, density, size coefficient of variation, and landscape shape index, the class-level indices provided more detailed information on the relative contributions of individual land use types (Luck and Wu 2002, Wu 2004).

Different land use types exhibited distinctive but not necessarily unique spatial signatures with different landscape metrics. For instance, for patch type percent coverage, patch density, patch size coefficient of variation, landscape shape index, and area-weighted mean patch shape index, residential and urban land use types displayed similar patterns along the transect from west to the urban center – a largely monotonic gradient with its peak at the urban core (Luck and Wu 2002). Desert showed a similar pattern for patch density, patch size coefficient of variation, landscape shape index, and area-weighted mean patch shape index, but a rather different pattern for patch type percent coverage and mean patch size. For other landscape metrics, agriculture displayed a multiple-peaked pattern. The distinctive "spatial signatures" as distance-based "landscape pattern profiles" may be used to compare urban developmental patterns between cities and dynamics of the same city over time. Such comparisons may help understand different underlying processes that are responsible for various forms of urban morphology (Luck and Wu 2002, Wu *et al.* 2002, 2011, Wu 2004, Seto and Fragkias 2005).

Using historical land use data over a period of about 90 years, Wu *et al.* (2011) used a selected set of landscape metrics to compare urbanization patterns between Phoenix and Las Vegas, the two fastest growing cities in the United States. They found that the two desert cities exhibited markedly similar urbanization patterns: the urban landscape became increasingly more compositionally diverse, structurally fragmented, and geometrically complex as urbanization unfolded. These results can be used to test theories of urban development. For example, Dietzel *et al.* (2005) hypothesized that urbanization exhibits cyclic patterns in time and space driven by two alternating processes: diffusion that spreads urban growth from existing centers to new development areas and coalescence that is characterized by outward expansion and gap infilling of existing urban areas. During a full diffusion–coalescence cycle of urbanization, urban land area increases monotonically; urban patch density, edge density, and mean nearest neighbor distance all increase first, then each peak at different times, and finally decrease, exhibiting a unimodal shape. Contagion is highest at the beginning of the diffusion process and at the end of the coalescence process and reaches its lowest value in between, thus exhibiting a somewhat mirror image of urban patch density (Dietzel *et al.* 2005). Our results, however, showed a monotonic decrease in landscape contagion and a monotonic increase in urban patch density, edge density, and other fragmentation-related metrics over a period of more than 80 years in both Phoenix and Las Vegas. These discrepancies may be attributable to differences in urbanization stages, the scale of analysis, and data accuracy (Wu *et al.* 2011).

Our more detailed studies of urbanization in Phoenix based on Landsat-derived land use/land cover maps between 1985 and 2005 generally supported the general findings of our previous research, showing that urban land covers, especially xeric residential, increased substantially at the expense of undisturbed desert, resulting in a more fragmented and structurally complex landscape (Buyantuyev and Wu 2007a, Buyantuyev *et al.* 2010). In addition, temporal analysis using landscape metrics allowed us to evaluate their sensitivity to changes in desert vegetation and explore various challenges for quantifying urbanization patterns in arid environments (Buyantuyev and Wu 2009, 2010a, Buyantuyev *et al.* 2010).

4.3 SIMULATING SPATIOTEMPORAL DYNAMICS OF URBANIZATION

4.3.1 Importance of simulation models in urban studies

Using landscape metrics or other spatial statistical methods to quantify the spatial and temporal patterns of urbanization is useful in itself, and is often the first step in urban ecological projects (Wu *et al.* 2000, 2003, Luck and Wu 2002). Urbanization, however, is fundamentally driven by socioeconomic processes. Spatial pattern analysis using landscape metrics and other statistical methods does not get to these processes directly although they can be used to help identify the underlying drivers and link patterns and processes in urban landscapes. Simulation modeling of urban spatiotemporal dynamics provides an indispensable tool for exploring the causes and mechanisms of urbanization as well as for managing urban dynamics.

A quantitative model of urbanization provides an efficient method to organize current understanding of urbanization processes into alternate testable hypotheses. Several recent syntheses of urbanization models have documented the progression in both methodology and applications (Jenerette and Wu 2001, Berling-Wolff and Wu 2004a,b, Batty 2005, 2008, Milne *et al.* 2009). This extensive history includes models developed by urban planners, geographers, and landscape ecologists, among others; modeling urban landcover change has a long tradition as an interdisciplinary activity. Early applications of urban growth models were focused on transportation planning (Putnam 1983), followed by applications to better project locations of specific land uses (Harris 1985). Integrations of these two applications were later developed (Berechman and Small 1998), which spawned much of the modern urban spatiotemporal modeling activities (Wu *et al.* 2003, Berling-Wolff and Wu 2004a,b). Current applications of urban growth models include generating improved understanding and projections of urban landcover, potential requirements for water and impacts on water discharges, interactions between urbanization and fire, and effects of urbanization on biodiversity, species invasion, and ecosystem processes (Urban 2000, Jenerette and Wu 2001, Syphard *et al.* 2007, Shen *et al.* 2008, Milne *et al.* 2009).

4.3.2 Approaches to simulating urban dynamics

In developing models of the urbanization process a diverse number of theoretical and computational approaches have been explored. Modeling approaches to urban landscape dynamics range from highly complex descriptions of urban growth with many parameters describing multiple levels of decision making (Landis 1995, Waddell 2002), to cellular automata (CA) models with multiple scales of constraints (Jenerette and Wu 2001, Herold *et al.* 2003, Berling-Wolff and Wu 2004b, Batty 2005), to highly simplified models using a minimal number of parameters (Batty 1991, Makse *et al.* 1995, Fagan *et al.* 2001). Many environmental underlying patterns are important determinants of urbanization trajectories, for example, topography, and socioeconomic patterns, such as transportation corridors and poverty. In addition to these location specific determinants, neighborhood effects of nearby urbanized patches are also important to urban landscape changes. The parameterization schemes vary with theoretical assumptions, and are often based on empirical regressions between predictor variables and landcover patterns or through data-model inversion procedures (Jenerette and Wu 2001, Berling-Wolff and Wu 2004b).

One of the simplest modeling frameworks for urbanization is the spatial Markovian approach that projects land-cover change as a probabilistic outcome (Turner 1987, Wear *et al.* 1996, Jenerette and Wu 2001). Each landscape unit, either a parcel, patch or raster grid cell, is assigned a probability for change based on regression analyses with potential drivers of change. The probabilities can be derived using static variables (e.g., slope) or dynamic variables (e.g., proportion of urbanized area within a defined neighborhood). Another common urban modeling approach is represented by CA models (Batty 1998, Jenerette and Wu 2001, Berling-Wolff and Wu 2004a). CAs are defined by a lattice, a state space, a neighborhood template, and a set of local transition rules. The state space defines the potential states for each cell of the lattice (e.g., urbanized, agricultural, or wildland). The neighborhood template defines the area of influence affecting the transitions in each cell. The local transition rules define the behavior of each cell, which is usually a function of the current state of the cell and the cells in the neighborhood. For urbanization models these rules are often probabilistic.

Hybrid approaches for modeling urban landcover transitions that blend Markov chains with CAs have proliferated in recent decades. In many cases the hybrid approaches allow for explicit inclusion of multiple factors and neighborhoods, which has had much success in generating application-oriented projections and improving understanding of key constraints to urbanization. A number of hybrid models have been developed for specific cities, including UrbanSim for Seattle (Waddell 2002), CUF for San Francisco (Landis 1995), and the SLEUTH model for Santa Barbara with several additional applications globally (Clarke *et al.* 1997, Herold *et al.* 2003). In recent years, agent-based models of urbanization have become increasingly common because of their abilities to simulate directly the decision making processes (Wu and David 2002, Xie *et al.* 2007, Fontaine and Rounsevell 2009, Irwin *et al.* 2009). Each of these hybrid approaches blends multiple theories and techniques for model implementation. They can be tailored to maximize future forecasts, identify consequences of alternative decision scenarios, or improve understanding of the patterns and processes of urbanization.

4.3.3 Examples from CAP-LTER

Several modeling studies of urbanization in the Phoenix metropolitan region have been conducted as part of the CAP-LTER and other related urban projects (Jenerette and Wu 2001, Wu and David 2002, Wu *et al.* 2003, Berling-Wolff and Wu 2004b). The conceptual framework for our modeling work is the hierarchical patch dynamics paradigm (Wu and Loucks 1995, Wu 1999), which is well suited for studying the spatiotemporal dynamics of urbanization and its effects on ecological processes. Our urban development models combined CA-based and spatial Markovan approaches, with extensions to include hierarchical levels of constraints on model dynamics. Defining neighborhood rules for urban models has been challenging, as the rules do not necessarily correspond to easily measured empirical patterns. To ameliorate this problem, Jenerette and Wu (2001) developed a genetic algorithm-based inversion approach that estimates appropriate parameters by iteratively comparing modeled and observed spatial patterns. This technique can be used for parameter estimation and calibration for other land use change models.

Our modeling work has shown that the choice of an appropriate scale (in terms of spatial and temporal resolutions) in urban growth modeling is critically important (Jenerette and Wu 2001, Berling-Wolff and Wu 2004b). If the scale is too fine, data-related uncertainties and overwhelming computational demands can considerably reduce the accuracy and usefulness of the models. At the other extreme, if the scale is too coarse, the model does not have sufficient spatial and mechanistic details that are relevant to socioeconomic drivers and decision making processes. For a given modeling objective, a proper scale (or a range of scales) can be determined by a limited number of simulations on different scales, whose performance is evaluated using landscape metrics and multiple resolution goodness-of-fit (Berling-Wolff and Wu 2004b). Our simulations based on different scenarios suggest that much of the Phoenix area will be urbanized in about 30 years unless dramatic actions are taken soon to slow down the population growth and to build up instead of building out (Berling-Wolff and Wu 2004b). Much of our previous work has emphasized developing useful frameworks for describing the dynamic spatial structure of the Phoenix metropolitan region and developing projections of future patterns. Next generation models of urban change in this region are being developed for extending science applications to better gauge and explore urban sustainability (Wu 2008a,b).

While these examples signify the large strides in urban landcover change modeling, many challenges to describing urban trajectories still exist and may represent barriers for predicting the future of coupled social and biophysical systems (Wu and Marceau 2002, Irwin *et al.* 2009, Milne *et al.* 2009). One major challenge is the inherent multiple-scale nature of land-cover change in urbanizing environments. The decision for converting a given parcel of land to urbanized landcover is directly associated with individuals, developers, municipalities, and banking organizations. For each of these unique agents, the decision making process can vary in response to individual preference to global economic trends. While much urbanization may be contagious at some scales, it may also be dispersive at others. Reconciling the range of scales in the urbanization process and the decision making associated with the process is a fundamental challenge for developing appropriate urban land-cover change models.

Another challenge is the inherent self-organizational, contingent, and adaptive nature of societies and coupled socioecological systems. The rules governing land-cover change may vary systematically and abruptly. The 2008–2009 housing market collapse in the United States epitomizes the occasionally rapid transformation in the process of urban land-cover change. Individual localities may develop rules independent of neighbors or larger systems in which they are embedded. Many of these uncertainties are intractable and can only be poorly predicted at best. Nevertheless, urban land-cover change modeling still has an important role in developing projections of scenarios to improve understanding of how the unpredictable future may unfold. In other words, the inability to predict with confidence does not deplete the importance of developing decision support tools based on reliable science.

4.4 EFFECTS OF URBANIZATION ON BIODIVERSITY AND ECOSYSTEM PROCESSES: EXAMPLES FROM CAP-LTER

4.4.1 Effects of urbanization on biodiversity

Urban habitats are not a random sampling of regions on earth, but tend to be areas of inherently high species richness, making potential losses disproportionately high (Marzluff *et al.* 1998). Urbanization has drastically changed the land use and land cover pattern in the Phoenix metropolitan region, decreasing average patch size and increasing patch density and the juxtaposition of highly contrasting patches (Luck and Wu 2002, Buyantuyev *et al.* 2010, Wu *et al.* 2011). Generally urbanization has the effect of decreasing the amount of habitat for native species and increasing habitat fragmentation for most if not all species. The actual effects of these habitat changes on biodiversity, however, are variable. Humans impose both top-down and bottom-up controls on urban biodiversity. At the CAP-LTER, biodiversity studies have focused mainly on the patterns of three taxonomic groups: perennial plants, arthropods and birds.

Plants

In wildlands, geomorphic and climatological factors generally determine the composition of the vegetative community. In cities, however, humans decide what and where to plant, and seasonal and yearly fluctuations are largely diluted by artificial irrigation practices (Buyantuyev and Wu 2007a, 2009). In studies of perennial plant diversity in the region, alpha diversity was similar among desert and urban sites, but urban sites contained twice as many exotic genera (Hope *et al.* 2006). Beta diversity (turnover) was significantly higher among urban sites leading to an increase in regional gamma diversity (Hope *et al.* 2003, 2006).

Spatial autocorrelation of plant diversity was present among undeveloped Sonoran Desert sites, probably due to factors underlying plant distribution such as slope, aspect, soil type and water availability, but was absent from urban sites (Hope *et al.* 2003, 2006). In urban areas, the best predictors of plant diversity were both ecological and socioecomonic, including family income, median housing age, and land use history (whether it had ever been farmed). Particularly strong was the positive relationship between biodiversity and family income, termed the "luxury effect" (Hope *et al.* 2003).

Arthropods

Arthropod communities have also been found to be affected greatly by changes to their habitats caused by urbanization in the Phoenix metropolitan region. Again, alpha diversity was found to be relatively similar between desert and urban sampling sites, but the species composition was different (McIntyre *et al.* 2001). Ants, springtails, and mites were ubiquitous and abundant across all sites, but the species making up the remainder of each community seemed to be dictated by percent ground cover of a variety of habitat features such as agricultural crops, gravel and lawn, exotic trees and shrubs, native plants and built structures (McIntyre *et al.* 2001).

Related to species composition, the trophic structure of communities in different land uses was also different. Predators, herbivores and detritivores were more common in agricultural sites, whereas omnivores were more abundant in desert, residential and industrial sites (McIntyre *et al.* 2001, Cook and Faeth 2006). The increased productivity in agricultural areas and urban yards clearly had an effect on the temporal patterns observed in arthropod communities in this desert city. Arthropods, being ectotherms, were impacted by seasonal temperature changes, but artificial irrigation in mesic areas seemed to buffer the communities from the effects of seasonal fluctuations in water availability (McIntyre *et al.* 2001).

Birds

Many studies have shown that generally bird diversity is lower but species abundances are higher in urban land uses than in surrounding undeveloped areas (Blair 1996, Cam *et al.* 2000, Johnston 2001, Marzluff and Ewing 2001). In the Phoenix metropolitan region, bird communities were found to be dominated by a few abundant species adept at utilizing urban resources and dealing with urban stressors (Hostetler 1999, Shochat *et al.* 2004, Fokidis *et al.* 2009). Similarly to arthropods, the composition of bird communities was significantly affected by elements of the local habitat structure, particularly volume of woody vegetation (Green and Baker 2003, Hostetler and Knowles-Yanez 2003). While native and non-native woody vegetation affected different guilds differently, native residents and neotropical migrants were negatively affected by urbanization factors such as housing and road density (Green and Baker 2003).

Some of these species have found refuge in desert habitat fragments which consist of more than 20 mountain reserves throughout the Phoenix metropolitan region. However, as in the case of arthropods, these habitat fragments do not completely mimic the outlying desert, and the bird communities seem to be slightly different (Litteral and Wu unpublished data). Even among the fragments themselves, avian community composition differed and appeared to be affected by the size of the fragment, its isolation from other fragments and the type of urban land use surrounding it. The presence of artificial water in the city seemed to shift the balance of competitive relationships towards the proliferation of synanthropic species, such as house sparrows and various species of doves, who are more efficient foragers, but may be less drought tolerant (Litteral and Wu unpublished data).

An integrated approach to studying the changes in the distribution of biodiversity of different taxa has also allowed CAP-LTER researchers to explore trophic interactions that emerge from these changing patterns. For example, the decrease in predation pressure on birds may have been responsible for the increase in the abundance of certain avian species (Faeth *et al.* 2005, Anderies *et al.* 2007). In the same time, the top-down control of arthropod herbivores by avian predators was much stronger in urban than rural areas (Faeth *et al.* 2005, Marussich and Faeth 2009). We also attempted to improve our understanding of how urbanization factors would influence biodiversity patterns by correlating species and community measures with land use and land cover data. We found that the categories used by planners often were not the most relevant categories for biodiversity conservation planning (Hostetler and Knowles-Yanez 2003, Litteral and Wu unpublished data).

4.4.2 Effects of urbanization on soil biogeochemical patterns

Urbanization has a substantial effect on soil biogeochemical patterns and processes (Pouyat *et al.* 2002, Kaye *et al.* 2004, Zhu and Carreiro 2004, Lorenz and Kandeler 2005, Jenerette *et al.* 2006, Hall *et al.* 2009, Pavao-Zuckerman and Byrne 2009). In general, urbanization is associated with increases in soil carbon, organic matter, nitrogen, and other elemental pools. Many human made compounds without native analogs can be found in urban soils ranging from biological control agents, novel pollutants, hormones, to nanoparticles. These materials can be incorporated into soils through irrigation or atmospheric deposition. Gaseous fluxes of carbon and nitrogen compounds from soils are often elevated in urbanized regions. Urban soils are affected by alternate physical substrate, altered climates, altered species, compaction, aeration, novel deposition loads, and other driving factors affecting soil formation (Jenerette *et al.* 2006, Pavao-Zuckerman and Byrne 2009).

While these findings describe a general response pattern of urbanization on soil biogeochemical properties, urbanization also introduces strong heterogeneities in the spatial distribution of urban soils. Strong patterns of urban impacts occur within individual cities and across continental climate gradients of cities. Within an individual city, while mean pools and fluxes increase from background patterns, perhaps more striking are increases in the variability of soil biogeochemical patterns (Jenerette *et al.* 2006). Urban soils within a city range from completely denuded, covered with concrete or buildings, remnant patches of native soils, minimally managed lawnscapes, to highly fertilized, irrigated, and organic matter amended garden soils. This variation can have a distinct spatial structure from the native soil biogeochemical patterns. Within the Phoenix metropolitan region soils were homogenized at small scales and diversified at broader scales when compared to native soils. These contrasting scale-effects may be a signature of urbanization on soil biogeochemical patterns within a city.

Across multiple cities the effects of urbanization on soil biogeochemical patterns may also vary, and describing these broad gradients in urbanization is a growing research direction (Grimm *et al.* 2008a, Pavao-Zuckerman and Byrne 2009). A hypothesis with some support suggests urbanization will cause larger changes to soil biogeochemical patterns in more arid environments than in humid environments (Grimm *et al.* 2008a). However, hypotheses describing urbanization patterns across gradients are only beginning to be developed.

4.4.3 Effects of urbanization on net primary production

Vegetation cover plays a key role in sustaining ecosystem functioning in urban landscapes and providing important ecosystem services (Wu 2008a,b). Net primary production (NPP), the rate at which plant biomass accumulates in an ecosystem, has widely been used as an integrative measure of ecosystem functioning (Lieth and Whittaker 1975, McNaughton *et al.* 1989, Bai *et al.* 2004). While most previous studies reported that urbanization decreased NPP (Milesi *et al.* 2003, Imhoff *et al.* 2004, Xu *et al.* 2007), we found that urbanization actually increased aboveground biomass production in the Phoenix metropolitan region. In general, urbanization enhanced aboveground NPP due to highly productive irrigated plant communities (Buyantuyev and Wu 2009).

The effects of urbanization on vegetation and NPP varied both spatially and temporally. While urban and agricultural land covers exhibited higher aboveground production per unit area in normal and dry years, natural desert and riparian vegetation together contributed more to the regional aboveground NPP in wet years. In particular, during wet years NPP of desert communities dominated by Creosote bush (*Larrea tridentata*) and Bursage (*Ambrosia dumosa*) nearly doubled, whereas urban and agricultural land covers did not respond greatly to rainfall changes. Our correlation analysis confirmed that human supplementation of resources (water and nutrients) in the Phoenix metropolitan region effectively decoupled the usually tight relationship between vegetation growth and precipitation commonly found in arid and semiarid environments (Buyantuyev and Wu 2009). These findings have important implications for predicting long-term environmental impacts in the face of accelerating urbanization and future climate changes.

4.4.4 Effects of urbanization on vegetation phenology

Landscape phenology or vegetation phenology, which studies vegetation development phases and environmental triggers at multiple scales in heterogeneous landscapes, is important for understanding how urbanization affects ecological processes and climate (Liang and Schwartz 2009, Morisette *et al.* 2009). Among common environmental triggers are temperature and moisture, and water availability is particularly important in arid regions (Bowers and Dimmitt 1994, Schwinning *et al.* 2004, Neil and Wu 2006). Excessive heating in urban areas may promote earlier winter-spring growth while simultaneously shortening the overall growing period. Urban land transformations directly affect hydrological flowpaths and consequently vegetation phenology in cities. Also, urban vegetation frequently includes exotic species that may exhibit phenological patterns different from native plants. Studies based on remote-sensing analyses revealed earlier green-up, delayed dormancy, and the extension of growing period by as many as 15 days in some cities of the Northern Hemisphere (Zhang *et al.* 2004a,b).

In the Phoenix metropolitan region our studies have shown that urbanization has affected both leafing and flowering phenology (Buyantuyev and Wu unpublished data, Neil *et al.* 2010, and unpublished data).

Urban vegetation covers (except cultivated grasses) tend to green-up faster than natural desert, and stay photosynthetically active for significantly longer periods. Also, urbanization has added a greater diversity of phenological patterns, some of which are little influenced by climate. Based on herbarium records of plants, we also found that about 19% of plant species examined either advanced or delayed their flowering, and that quite a few species showed significant differences in flowering phenology between urban and non-urban areas (Neil *et al.* 2010). In addition, Neil *et al.* (unpublished data) further showed that landcover types, which were correlated with surface temperatures, had a much stronger effect on plant flowering phenology than water availability in the Phoenix metropolitan region.

4.4.5 Urban heat islands and ecological effects

The urban heat island (UHI) refers to the phenomenon that cities tend to have higher air and surface temperatures than their rural surroundings (Oke 1982, 1997, Arnfield 2003). UHI develops during urbanization as natural vegetation is replaced by impervious surfaces (concrete, asphalt, roof tops, and building walls). This land transformation modifies the near-surface energy budget by reducing evapotranspiration, crowding solar energy absorbing surfaces, and creating heat-trapping canyon-like urban morphology. UHI can be a significant factor for local and regional climatic and environmental changes, leading to a number of social and ecological consequences (Wu 2008a,b, Grimm *et al.* 2008b).

Rapid urbanization in the Phoenix metropolitan region has resulted in increases in the mean daily air temperature by 3.1 °C and in nighttime minimum temperature by 5 °C during the past several decades (Brazel *et al.* 2000, Baker *et al.* 2002). Several studies have shown that surface temperatures are correlated with land use and land cover pattern as well as socioeconomic factors, such as household income (Hsu 1984, Stabler *et al.* 2005, Brazel *et al.* 2007, Jenerette *et al.* 2007). Based on the previous studies, Buyantuyev and Wu (2010) further quantified diurnal and seasonal surface temperature variations at two spatial scales, and explored the biophysical and socioeconomic factors responsible for temperature variations. Our study revealed the existence of the archipelago structure of night-time surface UHI (SUHI). Although it has typically been portrayed at the spatial scale of an urban–rural gradient over which temperature monotonically decreases with distance away from the urban center, UHI is really a multi-scaled phenomenon. Within a city or metropolitan region, many UHIs of different size, shape, intensity, and temporal dynamics may form over sufficiently large impervious surface patches, and similarly many "urban cool islands" (UCIs) may also exist over vegetated patches in the urban landscape. UHIs and UCIs are multiple-scaled, patchy, and dynamic, together forming a hierarchical patch dynamic system (Wu and Loucks 1995, Wu 2008a).

We also were able to identify the daytime heat sinks in the Phoenix region in both the early summer and the late autumn (Buyantuyev and Wu 2010). The formation of the morning heat sinks may be attributed to a variety of factors, including high thermal inertia of built areas and moisture differences between urban and rural areas. In addition, our analysis using geographically weighted regression confirmed the important role of vegetation and pavements in explaining spatiotemporal variations of surface temperatures. The relationship between surface temperature and landscape features were mediated by socioeconomic factors, so that richer neighborhoods tended to be greener and cooler – another example of the "luxury effect." At night, however, the socioeconomic status of neighborhoods was much less important to surface temperatures (Buyantuyev and Wu 2010). In the Phoenix region, urban warming not only has increased energy consumption for cooling and heat stress on biological organisms, but has also resulted in detrimental social consequences such as elevated crime rate (Baker *et al.* 2002). Our ongoing research will address questions of how best to mitigate UHI effects through integrating ecological, economic, and architectural considerations (Wu 2008a,b).

4.4.6 Ecosystem responses to urbanization-induced environmental changes

Urbanization can alter atmospheric chemistry and climate at local and regional scales. Urban areas are characterized by increasing air temperatures, elevated CO_2 concentration, and enhanced nitrogen deposition (Shen *et al.* 2008, Grimm *et al.* 2008b). In the Phoenix metropolitan region, the near surface CO_2 concentration is 470–555 ppm in the city center and 345–370 ppm at the city outskirts (Idso *et al.* 2001, 2002). The nitrogen deposition rate varies from 7 kg N ha^{-1} yr^{-1} in the upwind southwest of Phoenix to 26 kg N ha^{-1} yr^{-1} in the downwind northeast, with the urban core in between (Fenn *et al.* 2003). How do native plants and ecosystems respond to these urbanization-induced environmental changes? Addressing this question is not only crucial to understanding the ecology of urban ecosystems, but also has immediate implications for predicting the ecological consequences of global climate change.

Because of the complexity of multifactor interactions and the lack of field data, we have taken a simulation modeling approach to tackling this question. We adapted a process-based ecosystem model, originally developed by Reynolds and his associates (Reynolds *et al.* 1997, 2006), and parameterized and validated it for the dominant ecosystem of the Sonoran Desert (Shen *et al.* 2005). The model simulates the dynamics of carbon, nitrogen, and water cycling of a desert ecosystem at a daily time step, with explicit consideration of plant functional types of shrub, subshrub, C_3-winter and C_4-summer annual grasses, perennial grasses, and forbs. Through a series of simulation experiments using the model, we have examined how changes in air temperature, CO_2 concentration, and nitrogen deposition may affect ecosystem processes, such as aboveground net primary productivity (ANPP), soil organic matter (SOM), and soil nitrogen content (N_{soil}) along an urban–rural gradient (Shen *et al.* 2008).

Our simulation results showed that these urbanization-induced environmental changes could lead to a 12–120% increase in ANPP and a 69–180% increase in SOM for the native desert remnants in the urban core, with the largest responses occurring in wet years and the smallest in dry years (Shen *et al.* 2008). Conversely, N_{soil} content was higher in the suburban area than in the outside desert and the urban core, a pattern that is consistent with field observations in New York City (Pouyat *et al.* 2002, Gregg *et al.* 2003). We also found that these environmental changes could lead to a shift in species composition of the native ecosystem. Specifically, the urban core environment with higher air temperature, CO_2 concentration, and N deposition generally favored the growth, and thus increased the abundance, of species that were more responsive to CO_2 enrichment and more capable of using winter rains (e.g., C_3 winter annuals). In contrast, a suburban environment with moderate CO_2 concentration and relatively high nitrogen deposition rate promoted the dominance of evergreen and deciduous shrubs that were less responsive to CO_2 enrichment and more drought resistant. Overall, precipitation controlled the magnitude of ecosystem responses to environmental changes in this arid region.

Based on all that we know about the past and present of the world, our planet in future will be warmer in temperature, surrounded by thicker CO_2, and hit by heavier N-containing pollutants. Conceivably, modern cities provide "living laboratories" for studying possible ecological consequences of future climate change. While field manipulative experiments and long-term observations are absolutely necessary, integrating the urban–rural gradient approach with process-based simulation modeling provides a useful way for interfacing urban ecology with global climate change.

4.5 CONCLUDING REMARKS

Although it has a long history, particularly in Europe, urban ecology traditionally has been characterized by approaches that focus on disciplinary inquiries, single-scale investigations, or systems studies without explicit considerations of spatial heterogeneity. In reality, however, cities are the most spatially heterogeneous ecosystems of all, and they are indeed landscapes that are composed of quite conspicuous patches with different sizes, shapes, contents, and dynamics. As we have demonstrated through a series of studies in the Phoenix metropolitan region, therefore, a landscape ecological approach that emphasizes heterogeneity, pattern–process relations, and scale provides an effective way of studying urban systems.

The ultimate goal of urban ecology is to help achieve urban sustainability. This requires that urban ecological studies not only investigate the "ecology" of cities but also the "sustainability" of cities; not only "research" cities in theory but also "shape" them in action. To achieve this goal, we need to integrate natural and social sciences and adopt a transdisciplinary paradigm, and the landscape ecological approach to urban studies is a promising way to move forward (Wu 2006, 2008a,b, 2010). Built on our previous research, our current and future urban ecological studies will further extend this landscape approach, as illustrated in Figure 4.1, to include key elements of urban sustainability in a landscape ecological framework (Figure 4.3). This extended urban landscape

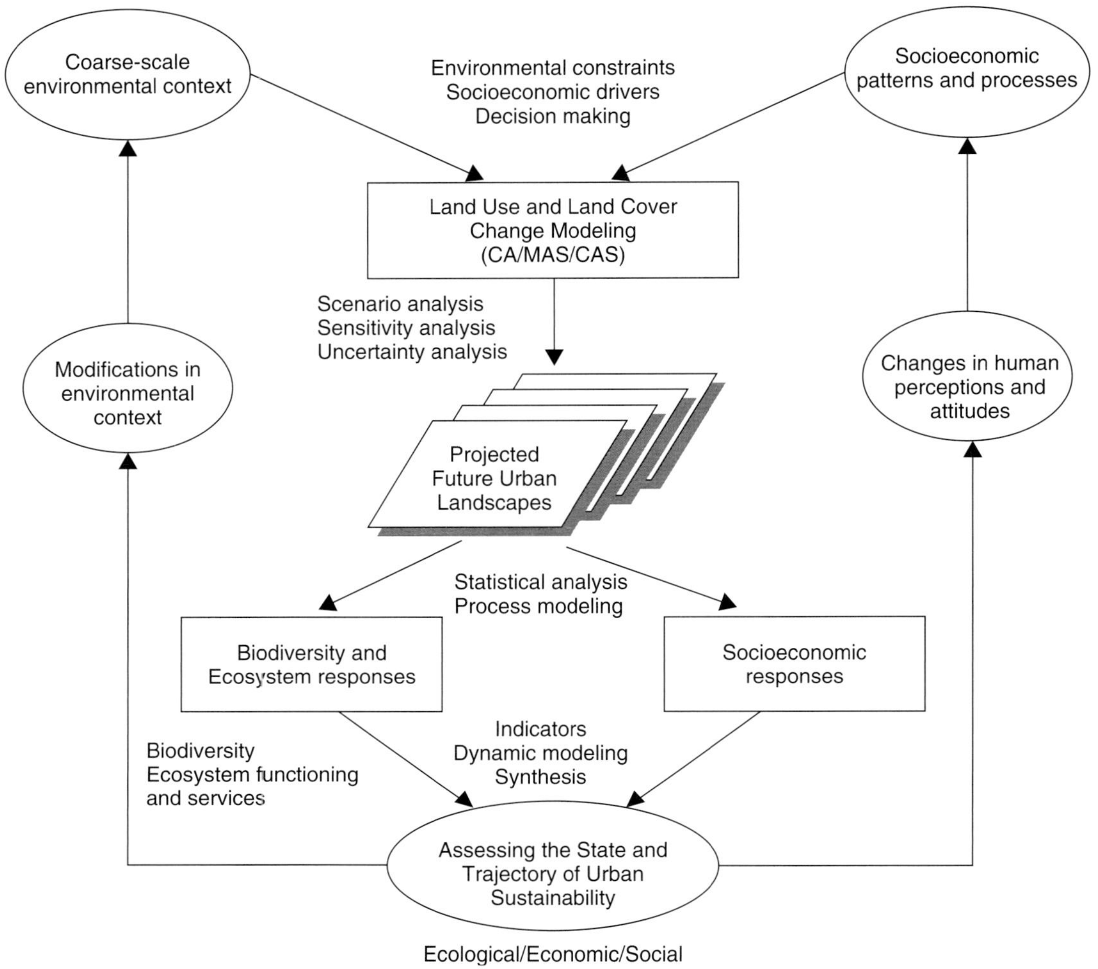

Figure 4.3 An extended urban landscape ecological approach, which integrates ecology with key elements of urban sustainability.

ecological approach will be more effective for coupling natural and social sciences, incorporating feedbacks between urbanization and ecology, and thus providing alternative solutions for decision making.

Cities have been the engines of economic development, cradles of innovation and knowledge production, and centers of sociocultural transformations. Cities also have a lower per capita cost of providing clean water, sanitation, electricity, waste collection, and telecommunications, while offering better access to education, jobs, health care, and social services than rural areas. Cities take up merely 3% of the Earth's land surface, but accommodate more than half the world's population. The potential to increase population density in existing cities without further urban sprawl is great if future urban development focuses on compactness and quality. All of these are important factors for the development of urban sustainability. As the human population continues to rise, the world will be increasingly urban and our well-being and prosperity will increasingly depend on the health of cities. Urban ecology is expected to play an instrumental role in improving existing cities and developing new ones that are more sustainable ecologically, economically, and socially. To realize this goal, urban ecology needs to go beyond the city to consider broader landscapes and go beyond ecology to embrace design sciences (Wu 2008, 2010).

ACKNOWLEDGMENTS

We thank all the current and past members of the Landscape Ecology and Sustainability Laboratory (LESL, formerly LEML) at Arizona State University for their help with and participation in the research reviewed in this chapter. Our research in urban ecology has been supported in part by the National Science Foundation under Grant No. DEB-0423704 (Central Arizona-Phoenix Long-Term Ecological Research) and Grant No. BCS-0508002 (Biocomplexity/CNH), and the US Environmental Protection Agency's Science to Achieve Results (STAR) program (R827676-01-0).

REFERENCES

Anderies, J.M., Katti, M., Shochat, E. (2007) Living in the city: Resource availability, predation, and bird population dynamics in urban areas. *Journal of Theoretical Biology* 247, 36–49.

Annan, K. (2002) United Nations Press Release SG/SM/8261.

Arnfield, A.J. (2003) Two decades of urban climate research: A review of turbulence, exchanges of energy and water, and the urban heat island. *International Journal of Climatology* 23, 1–26.

Bai, Y., Han, X., Wu, J., Chen, Z., Li, L. (2004) Ecosystem stability and compensatory effects in the Inner Mongolia Grassland. *Nature* 431, 181–184.

Baker, L.A., Brazel, A.J., Selover, N., *et al.* (2002) Urbanization and warming of Phoenix (Arizona, USA): Impacts, feedbacks and mitigation. *Urban Ecosystems* 6, 183–203.

Batty, M. (1991) Generating Urban Forms from Diffusive Growth. *Environment and Planning A* 23, 511–544.

Batty, M. (1998) Urban evolution on the desktop: simulation with the use of extended cellular automata. *Environment and Planning A* 30, 1943–1967.

Batty, M. (2005) Agents, cells, and cities: new representational models for simulating multiscale urban dynamics. *Environment and Planning A* 37, 1373–1394.

Batty, M. (2008) The size, scale, and shape of cities. *Science* 319, 769–771.

Berechman, J., Small, K.A. (1998) Research policy and review 25. Modeling land use and transportation: An interpretive review for growth areas. *Environment and Planning A* 20, 1285–1309.

Berling-Wolff, S., Wu, J. (2004a) Modeling urban landscape dynamics: A review. *Ecological Research* 19, 119–129.

Berling-Wolff, S., Wu, J. (2004b) Modeling urban landscape dynamics: A case study in Phoenix, USA. *Urban Ecosystems* 7, 215–240.

Blair, R.B. (1996) Land use and avian species diversity along an urban gradient. *Ecological Applications* 6, 506–519.

Bowers, J.E., Dimmitt, M.A. (1994) Flowering phenology of 6 woody-plants in the northern Sonoran Desert. *Bulletin of the Torrey Botanical Club* 121, 215–229.

Brazel, A.J., Selover, N., Vose, R., Heisler, G. (2000) The tale of two climates – Baltimore and Phoenix urban LTER sites. *Climate Research* 15, 123–135.

Brazel, A., Gober, P., Lee, S.J., *et al.* (2007) Determinants of changes in the regional urban heat island in metropolitan Phoenix (Arizona, USA) between 1990 and 2004. *Climate Research* 33, 171–182.

Breuste, J., Feldmann, H., Uhlmann, O., (eds) (1998) *Urban Ecology*. Springer, Berlin.

Brown, L.R. (2001) *Eco-Economy: Building an Economy for the Earth*. W.W. Norton & Co., New York.

Buyantuyev, A., Wu, J.G. (2007a) Estimating vegetation cover in an urban environment based on Landsat ETM+ imagery: A case study in Phoenix, USA. *International Journal of Remote Sensing* 28, 269–291.

Buyantuyev, A., Wu, J.G. (2007b) Effects of thematic resolution on landscape pattern analysis. *Landscape Ecology* 22, 7–13.

Buyantuyev, A., Wu, J.G. (2009) Urbanization alters spatiotemporal patterns of ecosystem primary productivity: A case study of the Phoenix metropolitan region, USA. *Journal of Arid Environments* 73, 512–520.

Buyantuyev, A., Wu, J.G. (2010) Urban heat islands and landscape heterogeneity: linking spatiotemporal variations in surface temperatures to land-cover and socioeconomic patterns. *Landscape Ecology* 25, 17–33.

Buyantuyev, A., Gries, C., Wu, J. (2010) Multiscale analysis of the urbanization pattern of the Phoenix Metropolitan Landscape of USA: Time, space and thematic resolution. *Landscape and Urban Planning* 94, 206–217.

Cam, E., Nichols, J.D., Sauer, J.R., Hines, J.E., Flather, C.H. (2000) Relative species richness and community completeness: birds and urbanization in the Mid-Atlantic states. *Ecological Applications* 10, 1196–1210.

Chen, X., Wu, J. (2009) Sustainable landscape architecture: implications of the Chinese philosophy of "unity of man with nature" and beyond. *Landscape Ecology* 24, 1015–1026.

Clarke, K.C., Hoppen, S., Gaydos, L. (1997) A self-modifying cellular automaton model of historical urbanization in the San Francisco Bay area. *Environment and Planning B: Planning and Design* 24, 247–261.

Cook, W.M., Faeth, S.H. (2006) Irrigation and land use drive ground arthropod community patterns in an urban desert. *Environmental Entomology* 35, 1532–1540.

Dietzel, C., Herold, M., Hemphill, J.J., Clarke, K.C. (2005) Spatio-temporal dynamics in California's Central Valley: empirical links to urban theory. *International Journal of Geographical Information Science* 19, 175–195.

Ellis, E.C., Wang, H.Q., Xiao, H.S., *et al.* (2006) Measuring long-term ecological changes in densely populated landscapes using current and historical high resolution imagery. *Remote Sensing of Environment* 100, 457–473.

Faeth, S.H., Warren, P.S., Shochat, E., Marussich, W.A. (2005) Trophic dynamics in urban communities. *Bioscience* 55, 399–407.

Fagan, W.F., Meir, E., Carroll, S.S., Wu, J.G. (2001) The ecology of urban landscapes: modeling housing starts as a density-dependent colonization process. *Landscape Ecology* 16, 33–39.

Fenn, M.E., Haeuber, R.A., Tonnesen, G.S., *et al.* (2003) Nitrogen emissions, deposition, and monitoring in the western United States. *Bioscience* 53, 391–403.

Fokidis, H.B., Orchinik, M., Deviche, P. (2009) Corticosterone and corticosteroid binding globulin in birds: Relation to urbanization in a desert city. *General and Comparative Endocrinology* 160, 259–270.

Fontaine, C.M., Rounsevell, M.D.A. (2009) An agent-based approach to model future residential pressure on a regional landscape. *Landscape Ecology* 24, 1237–1254.

Foresman, T.W., Pickett, S.T.A., Zipperer, W.C. (1997) Methods for spatial and temporal land use and land cover assessment for urban ecosystems and application in the greater Baltimore-Chesapeake region. *Urban Ecosystems* 1, 201–216.

Green, D.M., Baker, M.G. (2003) Urbanization impacts on habitat and bird communities in a Sonoran desert ecosystem. *Landscape and Urban Planning* 63, 225–239.

Gregg, J.W., Jones, C.G., Dawson, T.E. (2003) Urbanization effects on tree growth in the vicinity of New York City. *Nature* 424, 183–187.

Grimm, N., Grove, J.M., Pickett, S.T.A., Redman, C.L. (2000) Integrated approaches to long-term studies of urban ecological systems. *Bioscience* 50, 571–584.

Grimm, N.B., Foster, D., Groffman, P., *et al.* (2008a) The changing landscape: ecosystem responses to urbanization and pollution across climatic and societal gradients. *Frontiers in Ecology and the Environment* 6, 264–272.

Grimm, N.B., Faeth, S.H., Golubiewski, N.E., *et al.* (2008b) Global change and the ecology of cities. *Science* 319, 756–760.

Hall, S.J., Ahmed, B., Ortiz, P., Davies, R., Sponseller, R.A., Grimm, N.B. (2009) Urbanization alters soil microbial functioning in the Sonoran Desert. *Ecosystems* 12, 654–671.

Harris, B. (1985) Urban simulation models in regional science. *Journal of Regional Science* 25, 545–567.

Herold, M., Goldstein, N.C., Clarke, K.C. (2003) The spatiotemporal form of urban growth: measurement, analysis and modeling. *Remote Sensing of Environment* 86, 286–302.

Hope, D., Gries, C., Casagrande, D., Redman, C.L., Grimm, N.B., Martin, C. (2006) Drivers of spatial variation in plant diversity across the central Arizona - Phoenix ecosystem. *Society & Natural Resources* 19, 101–116.

Hope, D., Gries, C., Zhu, W., *et al.* (2003) Socioeconomics drive urban plant diversity. *Proceedings of the National Academy of Sciences (USA)* 100, 8788–8792.

Hostetler, M. (1999) Scale, birds, and human decisions: a potential for integrative research in urban ecosystems. *Landscape and Urban Planning* 45, 15–19.

Hostetler, M., Knowles-Yanez, K. (2003) Land use, scale, and bird distributions in the Phoenix metropolitan area. *Landscape and Urban Planning* 62, 55–68.

Howarth, P.J., Boasson, E. (1983) Landsat digital enhancements for change detection in urban environments. *Remote Sensing of Environment* 13, 149–160.

Hsu, S.I. (1984) Variation of an urban heat-island in Phoenix. *Professional Geographer* 36, 196–200.

Idso, C.D., Idso, S.B., Balling, R.C. (2001) An intensive two-week study of an urban CO_2 dome in Phoenix, Arizona, USA. *Atmospheric Environment* 35, 995–1000.

Idso, S.B., Idso, C.D., Balling, R.C. (2002) Seasonal and diurnal variations of near-surface atmospheric CO_2 concentration within a residential sector of the urban CO_2 dome of Phoenix, AZ, USA. *Atmospheric Environment* 36, 1655–1660.

Imhoff, M.L., Bounoua, L., DeFries, R., *et al.* (2004) The consequences of urban land transformation on net primary productivity in the United States. *Remote Sensing of Environment* 89, 434–443.

Irwin, E.G., Bockstael, N.E. (2007) The evolution of urban sprawl: Evidence of spatial heterogeneity and increasing land fragmentation. *Proceeding of the National Academy of Sciences of the United States of America* 104, 20672–20677.

Irwin, E.G., Jayaprakash, C., Munroe, D.K. (2009) Towards a comprehensive framework for modeling urban spatial dynamics. *Landscape Ecology* 24, 1223–1236.

Jelinski, D.E., Wu, J.G. (1996) The modifiable areal unit problem and implications for landscape ecology. *Landscape Ecology* 11, 129–140.

Jenerette, G.D., Wu, J.G. (2001) Analysis and simulation of land-use change in the central Arizona-Phoenix region, USA. *Landscape Ecology* 16, 611–626.

Jenerette, G.D., Wu, J., Grimm, N.B., Hope, D. (2006) Points, patches, and regions: Scaling soil biogeochemical patterns in an urbanized arid ecosystem. *Global Change Biology* 12, 1532–1544.

Jenerette, G.D., Harlan, S.L., Brazel, A., Jones, N., Larsen, L., Stefanov, W.L. (2007) Regional relationships between surface temperature, vegetation, and human settlement in a rapidly urbanizing ecosystem. *Landscape Ecology* 22, 353–365.

Jensen, J.R. (2004) *Introductory Digital Image Processing: A Remote Sensing Perspective. Third Edition.* Prentice Hall, Inc., Old Tappan, NJ, USA.

Johnston, R.F. (2001) Synanthropic birds of North America. In: Marzluff, J.M., Bowman, R., Donnelly, R. (eds) *Avian Ecology and Conservation in an Urbanizing World*. Kluwer Academic Publishers, Boston, pp. 49–67.

Kaye, J.P., Burke, I.C., Mosier, A.R., Guerschman, J.P. (2004) Methane and nitrous oxide fluxes from urban soils to the atmosphere. *Ecological Applications* 14, 975–981.

Kressler, F., Steinnocher, K. (1996) Change detection in urban areas using satellite images and spectral mixture analysis. *International Archives of Photogrammetry and Remote Sensing*. Vol. 31, Part B7, Vienna, Austria, pp. 379–383.

Landis, J.D. (1995) Imagining land use futures: Applying the California urban futures model. *Journal of the American Planning Association* 61, 438–457.

Li, H., Wu, J. (2007) Landscape pattern analysis: Key issues and challenges. In: Wu, J., Hobbs, R.J. (eds) *Key Topics in Landscape Ecology*. Cambridge University Press, Cambridge, UK, pp. 39–61.

Liang, L., Schwartz, M. (2009) Landscape phenology: an integrative approach to seasonal vegetation dynamics. *Landscape Ecology* 24, 465–472.

Lieth, H., Whittaker, R. (1975) *Primary Productivity of the Biosphere*. Springer, New York.

Lorenz, K., Kandeler, E. (2005) Biochemical characterization of urban soil profiles from Stuttgart, Germany. *Soil Biology & Biochemistry* 37, 1373–1385.

Luck, M., Wu, J.G. (2002) A gradient analysis of urban landscape pattern: a case study from the Phoenix metropolitan region, Arizona, USA. *Landscape Ecology* 17, 327–339.

Lunetta, R., Elvidge, C. (1998) *Remote Sensing Change Detection: Environmental monitoring methods and applications*. Ann Arbor Press, Chelsea, MI.

Makse, H.A., Havlin, S., Stanley, H.E. (1995) Modelling urban growth patterns. *Nature* 377, 608–612.

Marussich, W., Faeth, S. (2009) Effects of urbanization on trophic dynamics of arthropod communities on a common desert host plant. *Urban Ecosystems* 12, 265–286.

Marzluff, J.M., Ewing, K. (2001) Restoration of fragmented landscapes for the conservation of birds: A general framework and specific recommendations for urbanizing landscapes. *Restoration Ecology* 9, 280–292.

Marzluff, J.M., Gehlbach, F.R., Manuwal, D.A. (1998) Urban environments: influences on avifauna and challenges for the avian conservationist. In: Marzluff, J.M., Sallabanks, R. (eds) *Avian Conservation: Research and Management*. Island Press, Washington, DC., pp. 283–299.

McGarigal, K., Marks, B.J. (1995) *FRAGSTATS: Spatial Pattern Analysis Program for Quantifying Landscape Structure*. Gen. Tech. Rep. PNW-GTR-351. Pacific Northwest Research Station, USDA-Forest Service, Portland.

McIntyre, N.E., Rango, J., Fagan, W.F., Faeth, S.H. (2001) Ground arthropod community structure in a heterogeneous urban environment. *Landscape and Urban Planning* 52, 257–274.

McNaughton, S.J., Oesterheld, M., Frank, D.A., Williams, K.J. (1989) Ecosystem-level patterns of primary productivity and herbivory in terrestrial habitats. *Nature* 341, 142–144.

Milesi, C., Elvidge, C.D., Nemani, R.R., Running, S.W. (2003) Assessing the impact of urban land development on net primary productivity in the southeastern United States. *Remote Sensing of Environment* 86, 401–410.

Milne, E., Aspinall, R.J., Veldkamp, T.A. (2009) Integrated modeling of natural and social systems in land change science. *Landscape Ecology* 24, 1145–1147.

Morisette, J.T., Richardson, A.D., Knapp, A.K., *et al.* (2009) Tracking the rhythm of the seasons in the face of global change: phenological research in the 21st century. *Frontiers in Ecology and the Environment* 7, 253–260.

Neil, K., Wu, J. (2006) Effects of urbanization on plant flowering phenology. *Urban Ecosystems* 9, 243–257.

Neil, K., Landrum, L., Wu, J. (2010) Effects of urbanization on flowering phenology in the metropolitan Phoenix region of USA: Findings from herbarium records. *Journal of Arid Environments*. 74, 440–444.

Oke, T.R. (1982) The energetic basis of the urban heat island. *Quarterly Journal of Royal Meteorological Society* 108, 1–24.

Oke, T.R. (1997) Urban climates and global change. In: Perry, A., Thompson, R. (eds) *Applied Climatology: Principles and Practices*. Routledge, London, pp. 273–287.

Pavao-Zuckerman, M.A., Byrne, L.B. (2009) Scratching the surface and digging deeper: exploring ecological theories in urban soils. *Urban Ecosystems* 12, 9–20.

Pickett, S.T.A., Burch, W.R., Dalton, S.E., Foresman, T.W., Grove, J.M., Rowntree, R. (1997) A conceptual framework for the study of human ecosystems in urban areas. *Urban Ecosystems* 1, 185–199.

Pickett, S.T.A., Cadenasso, M.L., Grove, J.M., *et al.* (2001) Urban ecological systems: Linking terrestrial ecological, physical, and socioeconomic components of metropolitan areas. *Annual Review of Ecology and Systematics* 32, 127–157.

Pouyat, R., Groffman, P., Yesilonis, I., Hernandez, L. (2002) Soil carbon pools and fluxes in urban ecosystems. *Environmental Pollution* 116, 107–118.

Putnam, S. (1983) *Integrated Urban Models*. Pion Limited, London, UK.

Rashed, T., Weeks, J.R., Stow, D., Fugate, D. (2005) Measuring temporal compositions of urban morphology through spectral mixture analysis: toward a soft approach to change analysis in crowded cities. *International Journal of Remote Sensing* 26, 699–718.

Rashed, T., Weeks, J.R., Roberts, D., Rogan, J., Powell, R. (2003) Measuring the physical composition of urban morphology using multiple endmember spectral mixture models. *Photogrammetric Engineering and Remote Sensing* 69, 1011–1020.

Rees, W.E. (1997) Urban ecosystems: The human dimension. *Urban Ecosystems* 1, 63–75.

Reynolds, J.F., Virginia, R.A., Schlesinger, W.H. (1997) Defining functional types for models of desertification. In: Smith, T.M., Shugart, H.H., Woodward, F.I. (eds) *Plant Functional Types: Their Relevance to Ecosystem Properties and Glogal Change*. Cambridge University Press, Cambridge, pp. 195–216.

Reynolds, J.F., Kemp, P.R., Ogle, K., Fernández, R.J., Gao, Q., Wu, J. (2006) Modeling the unique attributes of dryland ecosystems. In: Huenneke, L.F., Havstad, K.M., Schlesinger, W.H. (eds) *Structure and Function of a Chihuahuan Desert Ecosystem*. Oxford University Press, Oxford, UK, pp. 321–353.

Ridd, M.K. (1995) Exploring a V-I-S (Vegetation-Impervious Surface-Soil) model for urban ecosystem analysis through remote-sensing – comparative anatomy for cities. *International Journal of Remote Sensing* 16, 2165–2185.

Ridd, M.K., Liu, J.J. (1998) A comparison of four algorithms for change detection in an urban environment. *Remote Sensing of Environment* 63, 95–100.

Rogan, J., Miller, J., Stow, D., Franklin, J., Levien, L., Fischer, C. (2003) Land-cover change monitoring with classification trees using Landsat TM and ancillary data. *Photogrammetric Engineering and Remote Sensing* 69, 793–804.

Schwinning, S., Sala, O.E., Loik, M.E., Ehleringer, J.R. (2004) Thresholds, memory, and seasonality: understanding pulse dynamics in arid/semi-arid ecosystems. *Oecologia* 141, 191–193.

Seto, K.C., Fragkias, M. (2005) Quantifying spatiotemporal patterns of urban land-use change in four cities of China with time series landscape metrics. *Landscape Ecology* 20, 871–888.

Shao, G.F., Wu, J.G. (2008) On the accuracy of landscape pattern analysis using remote sensing data. *Landscape Ecology* 23, 505–511.

Shen, W., Jenerette, G.D., Wu, J., Gardner, R.H. (2004) Evaluating empirical scaling relations of pattern metrics with simulated landscapes. *Ecography* 27, 459–469.

Shen, W., Wu, J., Grimm, N.B., Hope, D. (2008) Effects of urbanization-induced environmental changes on ecosystem functioning in the Phoenix metropolitan region, USA. *Ecosystems* 11, 138–155.

Shen, W., Wu, J., Kemp, P.R., Reynolds, J.D., Grimm, N.B. (2005) Simulating the dynamics of primary productivity of a Sonoran ecosystem: Model parameterization and validation. *Ecological Modelling* 189, 1–24.

Shochat, E., Lerman, S.B., Katti, M., Lewis, D.B. (2004) Linking optimal foraging behavior to bird community structure in an urban-desert landscape: Field experiments with artificial food patches. *American Naturalist* 164, 232–243.

Small, C. (2005) A global analysis of urban reflectance. *International Journal of Remote Sensing* 26, 661–681.

Stabler, L.B., Martin, C.A., Brazel, A.J. (2005) Microclimates in a desert city were related to land use and vegetation index. *Urban Forestry & Urban Greening* 3, 137–147.

Sukopp, H. (1990) Urban ecology and its application in Europe. In: Sukopp, H., Hejny, S., Kowarik, I. (eds) *Urban Ecology: Plants and Plant Communities in Urban Environments*. SPB Academic Publishing bv, The Hague, The Netherlands, pp. 2–22.

Syphard, A.D., Clarke, K.C., Franklin, J. (2007) Simulating fire frequency and urban growth in southern California coastal shrublands, USA. *Landscape Ecology* 22, 431–445.

Turner, M.G. (1987) Spatial simulation of landscape changes in Georgia: A comparison of 3 transition models. *Landscape Ecology* 1, 29–36.

Turner, M.G., O'Neill, R.V., Gardner, R.H., Milne, B.T. (1989) Effects of changing spatial scale on the analysis of landscape pattern. *Landscape Ecology* 3, 153–162.

Turner, S.J., O'Neill, R.V., Conley, W., Conley, M.R., Humphries, H.C. (1991) Pattern and scale: Statistics for landscape ecology. In: Turner, M.G., Gardner, R.H. (eds) *Quantitative Methods in Landscape Ecology*. Springer, New York, pp. 17–49.

Turner, M.G., Gardner, R.H., O'Neill, R.V. (2001) *Landscape Ecology in Theory and Practice: Pattern and Process*. Springer, New York.

Urban, D.L. (2000) Using model analysis to design monitoring programs for landscape management and impact assessment. *Ecological Applications* 10, 1820–1832.

Waddell, P. (2002) Modeling urban development for land use, transportation, and environmental planning. *Journal of the American Planning Association* 68, 297–314.

Wear, D.N., Turner, M.G., Flamm, R.O. (1996) Ecosystem management with multiple owners: Landscape dynamics in a southern Appalachian watershed. *Ecological Applications* 6, 1173–1188.

Weng, Y.-C. (2007) Spatiotemporal changes of landscape pattern in response to urbanization. *Landscape and Urban Planning* 81, 341–353.

Wickham, J.D., Riitters, K.H. (1995) Sensitivity of landscape metrics to pixel size. *International Journal of Remote Sensing* 16, 3585–3595.

Wu, J. (1999) Hierarchy and scaling: Extrapolating information along a scaling ladder. *Canadian Journal of Remote Sensing* 25, 367–380.

Wu, J. (2004) Effects of changing scale on landscape pattern analysis: Scaling relations. *Landscape Ecology* 19, 125–138.

Wu, J. (2006) Landscape ecology, cross-disciplinarity, and sustainability science. *Landscape Ecology* 21, 1–4.

Wu, J. (2007) Scale and scaling: A cross-disciplinary perspective. In: Wu, J., Hobbs, R. (eds) *Key Topics in Landscape Ecology*. Cambridge University Press, Cambridge, UK, pp. 115–142.

Wu, J. (2008a) Toward a landscape ecology of cities: beyond buildings, trees, and urban forests. In: Carreiro, M.M., Song, Y.C., Wu, J.G. (eds) *Ecology, Planning and Management of Urban Forests: International Perspectives*. Springer, New York, pp. 10–28.

Wu, J. (2008b) Making the case for landscape ecology: an effective approach to urban sustainability. *Landscape Journal* 27, 41–50.
Wu, J. (2010) Urban sustainability: an inevitable goal of landscape research. *Landcape Ecology* 25, 1–4.
Wu, J., David, J.L. (2002) A spatially explicit hierarchical approach to modeling complex ecological systems: Theory and applications. *Ecological Modelling* 153, 7–26.
Wu, J., Levin, S.A. (1994) A spatial patch dynamic modeling approach to pattern and process in an annual grassland. *Ecological Monographs* 64(4), 447–464.
Wu, J., Loucks, O.L. (1995) From balance-of-nature to hierarchical patch dynamics: A paradigm shift in ecology. *Quarterly Review of Biology* 70, 439–466.
Wu, J., Marceau, D.J. (2002) Modeling complex ecological systems: An introduction. *Ecological Modelling* 153, 1–6.
Wu, J., Jenerette, G.D., David, J.L. (2003) Linking land use change with ecosystem processes: A hierarchical patch dynamics model. In: Guhathakurta, S. (ed.) *Integrated Land Use and Environmental Models*. Springer, Berlin, pp. 99–119.
Wu, J., Jelinski, D.E., Luck, M., Tueller, P.T. (2000) Multiscale analysis of landscape heterogeneity: Scale variance and pattern metrics. *Geographic Information Sciences* 6, 6–19.
Wu, J., Jenerette, G., Buyantuyev, A., Redman, C. (2011) Quantifying spatiotemporal patterns of urbanization: The case of the two fastest growing metropolitan regions in the United States. *Ecological Complexity* 8, 1–8.
Wu, J., Jones, K.B., Li, H., Loucks, O.L. (eds) (2006) *Scaling and Uncertainty Analysis in Ecology: Methods and Applications*. Springer, Dordrecht, The Netherlands.
Wu, J., Shen, W.J., Sun, W.Z., Tueller, P.T. (2002) Empirical patterns of the effects of changing scale on landscape metrics. *Landscape Ecology* 17, 761–782.
Xie, Y., Batty, M., Zhao, K. (2007) Simulating emergent urban form using agent-based modeling: Desakota in the Suzhou-Wuxian region in China. *Annals of the Association of American Geographers* 97, 477–495.
Xu, C., Liu, M., An, S., Chen, J.M., Yan, P. (2007) Assessing the impact of urbanization on regional net primary productivity in Jiangyin County, China. *Journal of Environmental Management* 85, 597–606.
Yuan, F., Sawaya, K.E., Loeffelholz, B.C., Bauer, M.E. (2005) Land cover classification and change analysis of the Twin Cities (Minnesota) Metropolitan Area by multitemporal Landsat remote sensing. *Remote Sensing of Environment* 98, 317–328.
Zhang, X.Y., Friedl, M.A., Schaaf, C.B., Strahler, A.H. (2004a) Climate controls on vegetation phenological patterns in northern mid- and high latitudes inferred from MODIS data. *Global Change Biology* 10, 1133–1145.
Zhang, X.Y., Friedl, M.A., Schaaf, C.B., Strahler, A.H., Schneider, A. (2004b) The footprint of urban climates on vegetation phenology. *Geophysical Research Letters* 31, L12209, doi:10.1029/2004GL020137.
Zhou, W., Troy, A., Grove, M. (2008) Object-based land cover classification and change analysis in the Baltimore metropolitan area using multitemporal high resolution remote sensing data. *Sensors* 8, 1613–1636.
Zhu, W.X., Carreiro, M.M. (2004) Variations of soluble organic nitrogen and microbial nitrogen in deciduous forest soils along an urban - rural gradient. *Soil Biology & Biochemistry* 36, 279–288.
Zipperer, W.C., Wu, J., Pouyat, R.V., Pickett, S.T.A. (2000) The application of ecological principles to urban and urbanizing landscapes. *Ecological Applications* 10, 685–688.

CHAPTER 5

Designing urban systems: ecological strategies with stocks and flows of energy and material

Peter Baccini

*ETH Zürich, Zurich, Switzerland**

5.1 THE CHALLENGE OF A NEW URBANITY

In the cultural evolution of humans urban life is an emerging phenomenon. Urban settlements are, according to anthropological and archaeological findings, inventions of the agrarian societies (Table 5.1). The transfer from nomadic groups to settled tribes led to a new social system. It was a change from small egalitarian groups to larger social entities with hierarchical socioeconomic relations. In analogy to the structure of natural ecosystems: about 80–90% of the people within a regional agrarian society were in the role of "primary producers." They were controlled and protected by 10–20% of the political, economic, and military leaders, surrounded by their religious, administrative, and technical staff. The primary producers had to elaborate a surplus of food and energy carriers to support the secondary and tertiary level; from an ecological point of view it was a new cultural form of symbiosis. The *polis*, as a social concept and as a dense morphological manifestation within the landscape, was born. It started, according to the hypothesis of archaeologists, in different continents and different cultures on this globe, most likely independent from each other, 6000–8000 years ago.

From a physiological point of view the rate-determining step of the growth of cities (meaning primarily the growth in population) was their acquisition of the necessary agrarian "hinterland." Rome for example, in its climax as a world power of western antiquity, had to develop a logistical master plan to feed approximately 1 million inhabitants daily, based mainly on the use of relatively small vehicles and an average transport speed of less than 5 km h^{-1}. In the first half of the 19th century cities were still perceived in the old paradigm of a feudal hierarchy. An impressive illustration for this is von Thünen's model of the "City State" (von Thünen 1826). It is mainly an economic model of the city that is "driven by solar energy." The consequent spatial arrangement is the radial concentric pattern of the primary and secondary producers of goods with the city in its center (Thünen's circles). The urban people lived within the walls, the rural people outside of them. In contemporary terms the agrarian culture was a "solar system" and, in principle, a sustainable one.

In the 20th century a fundamental and dramatic change occurred in urban development. The large-scale exploitation of fossil energy and the technical inventions and innovations in the transport infrastructure (railways, individual road vehicles with combustion engines, large vehicles in air transport, electronic telecommunications) allowed, on the one hand, a rapid liberation from the limits of renewable biomass and on the other hand a high exchange rate of people and goods over large distances. At the beginning of the 21st century it seems that this form of human life is to be copied on a global scale. According to demographic forecasts it is most probable that, by the middle of this century, 80% of the world's population (or approximately 8 billion inhabitants) will live in an urban

*Postal address: Kriesbühl, CH-6027 Römerswil, Switzerland.

Applied Urban Ecology: A Global Framework, First Edition. Edited by Matthias Richter and Ulrike Weiland.

Table 5.1 Culture periods, defined anthropologically (after Groh 1992).

	Hunter–gatherer	Agrarian society	Transition period	Urban society
Type of settlement	Nomadic	Settled 80% rural 20% urban	Rural → urban	urban >80%
Time period in years Before present b.p. After present a.p.	>12 000–7000 b.p.	7000–200 b.p.	200 b.p.–100 a.p.	100 a.p. – ?
Global population size (order of magnitude, in 10^9 capita)	0.001–0.005	0.005–0.5	0.5–8	8–?
Societal organization	Egalitarian small groups	Feudal invention and expansion of the city	Human rights "dissolution of the city" formation of urban systems	Democratic regional urban systems
Type of economy	Subsistence economy	Planned and market economy	Mainly market economy	"Sustainable economy"
Energy supply	Solar	Solar, partly cultivated biomass	Growth-oriented fossil fuels	New solar technology

environment (Table 5.1). This is a dramatic change in the evolution of human settlement within three to four generations. Common to all cultural periods is the strategy to minimize the risks, especially with regard to fatal shortages of the basic resources.

In the 20th century urbanization led to a "dilution" of urban settlement from a dense center into a network with a large variety of nodes and connections (Baccini and Oswald 1998). The distinct separation of rural and urban segments within a cultural landscape disappeared. The once concentric and regional "hinterland" diffused into a global set of "hinterlands." This new form of urbanity started within the culture of Europe and North America. Today the great majority of the population lives an urban life in settlements outside of the classical centers of the 19th century. The "New Urbanity" is defined by access to all the relevant goods and services a city can offer within half an hour's travel time (be it on foot or by private or public transport); therefore the term "urban system" is introduced. The name of the model to qualify and quantify an urban system is *Netzstadt* (Oswald and Baccini 2003) (Box 5.1).

With regard to the demand on resources to run urban systems, the focus is placed on the main physical resources. It is taken for granted that the general term "resources" also includes non-physical resources. The following section concentrates on engineering concepts for long-term oriented reconstruction processes

Box 5.1 Definition of the "urban system" as *Netzstadt*

An urban system is composed of open geogenic and anthropogenic networks that are connected with each other. The nodes of these networks are places of high densities of people, physical goods (geogenic included) and information. These nodes are connected by fluxes of people, goods and information. The system's boundary is given by political conventions in the case of anthropogenic subsystems, by climatic properties for geogenic subsystems.

in urban systems. With regard to physical resources two questions are to be discussed:

- What are the main characteristics of urban systems?
- In what way should they be reconstructed?

5.2 URBAN SYSTEMS AND THEIR RESOURCE MANAGEMENT

5.2.1 Methodology applied investigating resource management of complex systems

A system's physical resource management is characterized by a material management system. Material management systems are investigated and evaluated using the method called material flow (or flux) analysis (MFA) (Baccini and Brunner 1991, Baccini and Bader 1996, Brunner and Rechberger 2003). These systems consist of discrete processes connected to flows of "goods" as carriers of matter and energy. The mathematics to describe such a system is based on the balance equations for the "processes", treated as discrete volumes for which the principle of matter and energy conservation holds (Baccini and Bader 1996). The subsequent modeling uses specific parameters to define certain process properties (e.g., transfer coefficients). Such systems can be handled in various states regarding their time dependence. The simplest one is the steady-state situation. The most complicated are those systems in which stocks, flows and transfer coefficients are time dependent. Furthermore one has to consider feedback mechanisms in such systems, a property that is well known since system theory and cybernetics have been in use. In contemporary literature the term "metabolism" is most often applied to comprehend the physiological processes of anthropogenic ecosystems.

On a global scale the material management system can be characterized by three main processes and 12 fluxes (Figure 5.1). In a first approximation planet Earth is considered as a physically closed system. The external energy source, the sun, emits radiation energy to the planet. Transformed energy is re-emitted to the Universe. Within the system there are three distinct processes interconnected with material fluxes: water from the geosphere to the biosphere, organic carbon from the biosphere to the anthroposphere, metals from the anthroposphere to the biosphere. Each system has its own material or resource stock. "Sustainable development" expresses a normative position with regard to man's use of these resources (WCED 1987). On a long-term scale, the anthroposphere should function

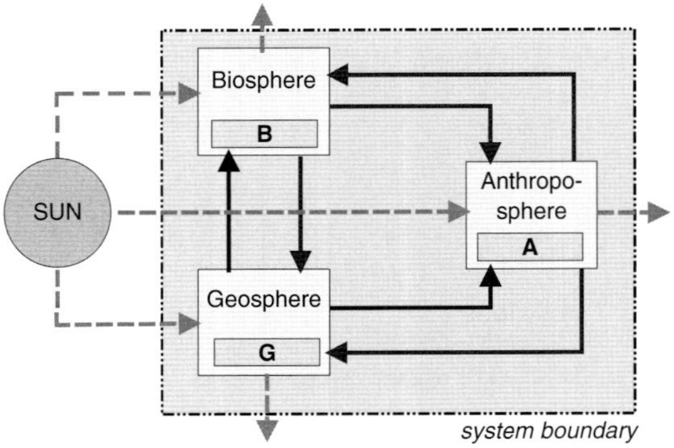

Figure 5.1 Scheme of main physiological processes of planet Earth. The dashed lines indicate energy flows, the letters A, B, G indicate the stocks.

on the basis of renewable resources in such a way that the stocks of the other spheres are not diminished. In this respect "stocks" include not only material resources such as fresh water or timber in forests, but also the wealth of biodiversity in the ecosystems of the biosphere and the economic and social quality of the anthroposphere. Future generations should have the same opportunities to develop their anthroposphere with respect to the wealth of resources of the other two spheres. An illustration of the problems to be solved, with regard to physical resources, is given in Section 5.3.

5.2.2 Relevant differences between agrarian and urban systems on a regional scale

Two cases in very different cultural situations, namely in a peasant society of Nicaragua (Pfister and Baccini 2005) and in an urban society in Western Europe, namely an urban system in the Swiss lowlands (Faist 2000), serve to illustrate the distinct differences of the activity "to nourish" (definition see Baccini and Brunner 1991). In both cases economically extended MFA was used (Kytzia *et al.* 2004). The method is already well equipped to elucidate the relevant physiological characteristics of the systems studied. The peasant society (A) consists mainly of farmers, grouped into families, practicing a subsistence economy. The region they are living in belongs to one of the poorest ones on earth. In contrast the urban society is rooted in one of the richest countries in Europe. The synoptic presentation of some results is given in Table 5.2.

Table 5.2 Comparison of a peasant society in Nicaragua (Pfister and Baccini 2005) and an urban society in Western Europe (Faist 2000) with regard to their energy demand in the activity "to nourish" and their economic effort. (Energy flows in gigajoules per capita and per year).

	Agrarian society A		Urban society U	
	GJ/cap&yr	Self-sufficiency %	GJ/cap&yr	Self-sufficiency %
Energy total	19	80	180	10
Energy to nourish	17		30	
Regional supply		>90		60
Ratio of total income	>80%		10%	

The total energy consumption of the urban society (U) is approximately 10-fold that of the agrarian society (A). For the activity "to nourish" (including agricultural production, upgrading and distribution, consumption) A needs roughly 90% of its total energy demand, whereas U can manage with only 20% of its total. A's energy source is from local forests (80% self-sufficiency). U's energy carriers are mainly fossil fuels and are imported. U needs 80% of its energy demand for the activities "to reside and work" and "to transport and communicate." A's food is mainly produced within the region (90% self-sufficiency). U needs, on the bottom line (taking import/export flows into account), a "global hinterland" for approximately 40% of its food demand. In A the peasant household has to spend roughly 90% of its income on food, whereas the urban household can manage this activity with only 10% (average values) of its total income. In the comparison of the two cases the following additional insights must be stated:

1 The peasants in A have a reasonable strategy in their agricultural production. Due to their increasing population their system is neither economically nor ecologically sustainable. They need a "hinterland" where they can sell their labor force. However their national "hinterland" cannot offer enough labor. A second source of income is the cash crop coffee, a product to be sold on the global market, which is mostly out of reach for poor farmers. Their main problems are thus population growth and the lack of a strong complementary region offering labor and/or goods prices for agricultural products.

2 The people in U are economically successful in a global market, mostly due to their products in the tertiary sector. However, due to their strong dependence on non-renewable energy sources (see Figure 5.2) their system is not sustainable in the long-term (see also Table 5.3). Without reconstruction of their physical infrastructure from a fossil to a solar system within the next two to three generations, U will collapse.

The starting positions of an A and a U society to enter an SD process are completely different, only seen from

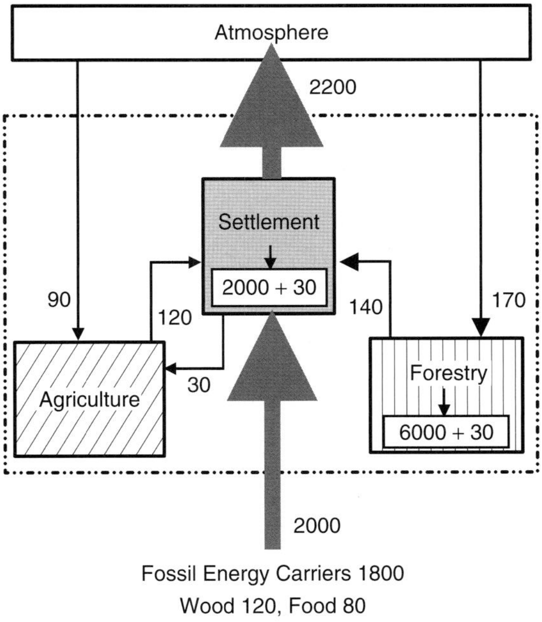

Figure 5.2 Carbon flow in an urban system (Swiss lowland region). Flow units in kg carbon per capita and year; stock units in kg per capita. "Agriculture" covers 55%, "Forestry" 33% and "Settlement" 12% of the total area (after Müller *et al.* 1995, Baccini and Bader 1996).

Table 5.3 A selection of estimated resource reservoirs per capita for a world population of 8 billion, compared with corresponding stocks and actual consumption rates in developed urban systems.

	Global reservoirs for 8 billion people	Stocks in developed urban systems	Consumption of developed urban systems
Territories	ha/capita	ha/capita	ha/capita and year
Agriculture	**0.5**		
Forestry	**0.3**		
Settlement		**0.03**	**0.0001**
Timber	m^3/capita **50**	m^3/capita **10**	m^3/capita and year **0.4**
Oil	GJ/capita **800**	GJ/capita **40**	GJ/capita and year **100**
Copper	kg/capita **300**	kg/capita **300**	kg/capita and year **10**

The data, giving only order of magnitudes, is based on published resource estimations (such as Global 2000, Buitenkamp *et al.* 1992, Zeltner *et al.* 1999).

an ecological and an economic point of view. Here the various differences in political and social culture are not yet considered. Since the developed countries with their new urbanity consume about 80% of the demand of the total anthroposphere, the focus in resource management for a sustainable development is on these urban systems.

5.2.3 The resource management perspectives on a global scale

Within the "transition period" (Table 5.1) urbanization covers the whole globe, meaning that the A societies move towards U societies (Table 5.2). Each urban system has its idiosyncrasies with regard to the essential mass goods and the corresponding "hinterlands", illustrated with a European example given below. To reach a "physiological status of sustainability" the system has to measure its resource demand on the "global scale of scarcity." There are several studies available to illustrate the ecological footprint of large urban systems (see for example "sustainable Netherlands" by Buitenkamp *et al.* 1992). An illustration is given in Table 5.3.

As with the territories a globalization of the urban system would not reduce the available productive area for agriculture and forestry by more than a few percent. If the forestry management is limited to the present timber demand of urban systems, the global reservoir could satisfy the annual need by its annual increment of 1–2% of the stock. Even if there are still uncertainties about the size of fossil energy reservoirs, the oil example illustrates that a globalization of the urban system type would not be possible for 8 billion people. The copper example illustrates the following phenomenon: the theoretical copper ore reservoir in the earth crust per capita for 8 billion people is equal to the per capita stock copper metal already accumulated in the developed urban system. It is obvious that urban systems are about to become secondary mining sites for copper. In conclusion the following four hypotheses can be stated:

1 There is *enough territory* "on reserve" to continue the urban growth (1 m^2 per capita and year for the next 50 years). It is not a question of size, but of design.
2 The actual *design of the "cultural landscape" decreases biodiversity* (Rio 92 declaration to save biodiversity, one important indicator for sustainable development).
3 The *dependence on non-renewable energy* (fossil fuels) is the *most critical physiological aspect concerning the limited reservoirs and the effects on climate change*, followed by the fresh water shortage for crop production in some regions.
4 The *rate-determining step is the reconstruction of the "built anthroposphere"* (not a specific technological innovation) with a drastic improvement of its ecological quality.

5.2.4 The essential mass resources in the development of urban regions

The anthroposphere of the 21st century (Figure 5.1) can be seen as a heterogeneous patchwork of urban systems. In order to meet the ecological criteria of sustainable development the metabolism of urban systems has to be evaluated with regard to the regional and global stocks. In the following the metabolism of a typical urban system is exemplified with the Swiss lowland region (Baccini 1997). In this region a communication network was built within 40 years (1950–1990) allowing practically every inhabitant, independent of his place of residence, to reach any urban activity within half an hour. The region has become a "compact" urban system (5 million inhabitants, population density 500 per km^2). The forest area stays constant due to a forest conservation law. The agricultural land is constantly reduced, but steadily increases its productivity. Neither territorial planning nor environmental protection measurements (since 1970) have significantly influenced the growth of the urban settlement area and its metabolic rate (consumption of joules and kilograms per capita or per km^2 and per year). There is enough territory "on reserve" to continue this regional growth (3 m^2 per capita and year for the next 20–30 years).

From a physiological point of view, the survival of urban regions increasingly depends on a continental or even global "hinterland" (back country). Whenever an urban region expands a corresponding resource area must be found "on the global market."

Three working hypotheses are chosen for evaluating the sustainability of urban systems (Baccini 1997). The metabolism of an urban system is "sustainable", if the following conditions hold:

1 The demand for essential "mass goods" such as *water, biomass, construction materials (e.g., stones) and energy carriers* can be satisfied *autochthonously by more than 80% in the long term*. The degree of self-sufficiency, arbitrarily chosen with respect to a set of essential goods, determines the ecologically defined border of the urban system.

2 The remaining needs can be covered from the "external market" in such a way that *the global resource stock is not reduced significantly*. The mean fluxes of available goods per capita from the external market are determined by the annual growth rate of the "global hinterland" divided by the world population.

3 The *outputs* (emissions) are *no burden for future generations*. No environmental system (hydrosphere, atmosphere, lithosphere) may be used as a sink for anthropogenic fluxes in such a way that the receiving system or parts of it becomes a "hazardous site" for man and the biosphere.

Four essential mass goods of the urban metabolism (i.e., water, biomass, construction materials nd energy carriers), were investigated in a region of the Swiss lowlands (Baccini and Oswald 1998). The results reflect the actual situation in a quasi steady state situation. The following conclusions can be drawn:

• There are quantitative and qualitative indications that the *use of groundwater* does not meet the criteria of "sustainability."

• There is a large *disparity between the ecological potential and the actual economic value of the two subsystems "agriculture" and "forestry."* Within the urban system, agriculture and forestry have become a new type of common land (due to the high degree of subsidy) whose role has to be newly defined under the aspects of sustainability.

• Presently the management system of *construction materials* exhibits a low efficiency of the material stock representing the infrastructure and a high dependence on non-renewable resources. However, enough alternatives are available on a long-term scale.

• With regard to *energy* the region strongly depends on imported fossil fuels (Figure 5.2). With respect to the criteria of regional sustainability a new set-up of energy transformation represents the greatest challenge (see next paragraph). It is related to the transformation of the settlement area.

A carbon flux analysis for a typical urban region is given in Figure 5.2. The process "Settlement", using only about 12% of the total area of the region, is responsible for the predominant carbon flux (2000 kg carbon per capita and per year). The input stems mainly (90%) from imported fossil fuels. Approximately 60% of the demand for food is covered by the regional agriculture. The rest has to be imported. The carbon turnover of the process "Agriculture" (covering 55% of the regional area) is three to four times higher than that in "Forestry" (33% of the total area). Their combined net contribution to the target process "Settlement" is only about 10% of the total. This is mainly due to the relatively high levels of meat production and consumption. The largest carbon stocks are found in the soils of the latter two processes. However the concentration of the organic carbon in soils is relatively low (a few percent). The most concentrated forms are in the forest trees and in the wooden goods of the buildings. Both are in the same order of magnitude (10^3 kg per capita).

From a physiological point of view an urban system of the type "20th century", exemplified by the Swiss lowlands, has the following main characteristics: the settlement stock and the type of energy transformation are the key factors to gain a "status of sustainability". The other three mass goods (water, biomass and construction materials) are less critical for the region studied (i.e., less effort would have to be applied to reach their desired status in material management). From an engineering point of view it follows that a transformation to a "sustainable status" is only possible by a reconstruction of the urban system (i.e., buildings and transportation network). In other words: it is not sufficient to construct only "new ecological cities" as additions to the existing stocks. However the process of reconstruction is a cultural one and has to include all relevant properties of human society, namely the political, socioeconomic and ecological ones. In history it is a new challenge never encountered before. Former generations were mainly oriented towards growth processes in urban development.

5.3 STRATEGIES OF RECONSTRUCTION

5.3.1 The 2000 watt society

Macroeconomic models predict that the global primary energy demand will increase by a factor of 2 to 4 by the year 2050. In contrast, climate analyses made by the Intergovernmental Panel of Climate Change (IPCC) claim that CO_2 emissions in 2050 should not exceed the values of 1990 or should even be 20% lower. By 2100 emissions should be reduced to one third of the present value. The common wisdom to deal with these opposing trends is the concept of decarbonization, that is the continuous decrease of the carbon emission per unit energy utilization. The decarbonization rates needed to compensate for the growing demand while keeping the CO_2 emissions constant should be at least 2% per year compared to actual values of 0.3% (Grübler and Nakicenovic 1997). It is argued that the target can only be met if per capita energy consumption in the industrialized countries is lowered from their typical level of 5000 to 10 000 W to about 2000 W, the present global average. This would leave expansion capacity for the developing countries, which presently have a per capita demand of between 300 and 1000 W. Based on the example of Switzerland it can be shown (Baccini and Imboden 2001) that the two key issues to attain this goal are the quality of buildings and the demand for mobility. The conversion of the present energy system to a 2000 W system is neither limited by technology nor by finances but by the acceptance of a new lifestyle in which energy is used more efficiently and more intelligently than today. In Table 5.4 this statement is exemplified for the case of Switzerland.

These numbers are typical for most countries of Western Europe as well as for other industrialized countries. In 1995, 45% of the Swiss end energy consumption was related to the operation of buildings, another 33% to transportation of people and goods.

Table 5.4 Consumption of end energy per capita in Switzerland by sectors. Values in percent of the average per capita consumption of end energy in 1997 (3600 W per capita).

	1980	1995	2050	Potential savings in a 2000 watt society
Buildings Private housing, industry, public sector (incl. lighting and warm water)	47	45	15	Reduction of energy for heating/cooling by a factor 3 relative to the present average standard
Transportation People and goods	25	33	16	Change to energy-efficient modes of transportation. Saturation of demand for mobility
Production Agriculture, industry, services	22	22	11	Based on a recent study by the Swiss Academy of Engineering Sciences
TOTAL	94	100	42	

Both sectors bear a tremendous potential for energy savings. Whereas the transportation sector reflects primarily what people consider to belong to a modern lifestyle (private instead of public transportation, big and heavy cars instead of small ones, etc.), the reduction of the energy demand of buildings meets less resistance from the consumers, although it needs much time.

Over the past two centuries Switzerland has been transformed into a gigantic system of constructed facilities. Currently, some 2 million buildings are interconnected by a sophisticated network of infrastructure comprising transportation systems (railways, roads, waterways, and airports), systems for water distribution, for waste water and solid waste management, for the production and distribution of energy, and for information exchange and telecommunication. The totality of this infrastructure, including all public and private buildings ("Constructed Switzerland" – CS), is valued at about 2000 billion USD or 300 000 USD per capita (present rebuilding costs). This corresponds to the total output of the Swiss economy of about one decade.

In the coming century CS is not likely to grow as fast as it has in the past. Rather, the main challenge in the next century will be to efficiently and sustainably manage (i.e., operate, maintain and renew) the vast assets of constructed facilities which have been created in this century. This genuinely new challenge must be met within given spatial, economic and natural resource constraints in a world which faces rapidly changing technological opportunities, social structures and values. Presently CS's maintenance and renovation costs are about 30 billion USD annually; this is insufficient to meet sustainability criteria in many cases.

Thus the task of the coming generations will be very different from the challenges faced by the past generations who built CS. On the one hand, the next generations will inherit a solid infrastructure and a powerful economy. On the other hand, this legacy sets significant constraints on future economic and social development that can only be changed over a period of decades or generations. For example today the average energy demand per square meter and year is more than three times greater than what can be achieved presently, virtually without extra costs. One recipe for the 2000 watt society must thus be to lower this value by applying the best available standards whenever new buildings are built and existing ones renovated. Other potential savings are listed in Table 5.4. The energetic basis of Switzerland as a 2000 watt society would be simple and flexible. According to the IPCC-scenarios, between 900 and 1100 W per person could still be produced from fossil fuels in the year 2050. Later this number would decrease to about 300 W. Another 700 W per person would come from already existing hydroelectric power plants, while the rest (200 W increasing to about 1000 W) can easily be provided by new renewable resources (solar, wind, geothermal). Nuclear power would no longer be needed in this picture.

5.3.2 Transformation of urban regions in a "time of safe practice"

The example of energy demonstrates that the concept of sustainable development, in its strictest interpretation, will remain an unattainable goal for many years to come. The world's present dependence on non-renewable energy (and other) resources is so strong that quick changes would lead to considerable disturbances in the economic and social sectors. Therefore it is important to have an alternative, less absolute criterion to measure the relative progress towards sustainability. The concept of *constant time of safe practice* (Imboden 1993) is based on the time during which a given policy (or practice) can be sustained without severe danger. Constant safe practice means that the development of alternative strategies (such as the one described above for the case of energy) should at least keep pace with the exploitation of the non-renewable resources. Thus the concept of constant (or increasing) time of safe practice serves as a guideline through the period of transition from today to the time of real sustainability.

5.3.3 The exploration of urban stocks

An urban system in growth (Figure 5.3a) shows a mean stock increment of 1–3% per year (Baccini and Bader 1996) mainly from primary sources (e.g., gravel, clay, timber, metal, glass, etc.). During this period, the secondary resources could not nearly satisfy the overall demand. In this case most of the engineering activity is found in the so-called production branches. If the population becomes stabilized and the per capita consumption reaches an optimal level (a quasi steady-state situation), the urban stock becomes the main mining site and the secondary resources are the important source for the further transformation of the urban infrastructure (Figure 5.3b). In this way the process waste management becomes much more important in

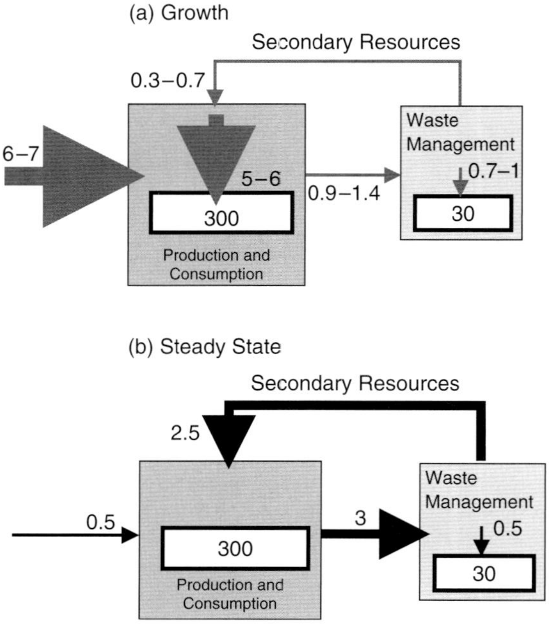

Figure 5.3 Material management scheme of an urban system in the (a) growth phase and (b) steady state phase. The flow unit is in tons per capita and year; the stocks in tons per capita.

the primary and secondary sectors of the urban economy. The built stock of the anthroposphere serves as the main source (urban mining). There are already some regional examples of this type of materials management (e.g., paper, glass, iron), for which the consumption flux shows relatively low growth rates and a small stock increment. In these cases the secondary sources have become the dominant ones, due to economic reasons. However, with regard to the overall fluxes, their quantitative contribution is still very small.

For long-term oriented urban mining it is indispensable to dispose of a method of exploration of urban stocks (Lichtensteiger and Baccini 2008). Investigations on the level of construction elements allow an understanding of the relationship between materials and their function and provide a basis for tailored steps of mining or environmental management decisions. The ark-house method, combined with the MFA method, gives an insight into regional dynamics. In addition this procedure enables a comprehensive comparison with further stocks (infrastructure, landfills, etc.) and finally a validation by production data and/or foreign trade statistics. Two examples are shown from these investigations (Lichtensteiger and Baccini 2008). Gravel/sand is taken as an example for bulk materials (Figure 5.4), copper as an example for trace materials (Figure 5.5). In Figure 5.5 alloys are included by their copper content.

The whole genesis path of gravel/sand includes cement, concrete, and broken concrete as further genesis stages. The import rate is low. Cement production in Switzerland started in the year 1871. Statistical data on the annual amount of cement production has been available since 1902. The materials for cement and concrete are well known. Therefore material fluxes triggered by cement production since 1871 may be easily calculated and also the stock diminution of primary resources (lime/marl and gravel/sand). Since Switzerland has not been affected by war or earthquakes during the 20th century, this production data is also suitable for the validation of the urban stock evaluation described above. Compared to the high input the mortality (output) of buildings and infrastructure in the 20th century was low. The mortality for buildings was ~0.1% per year. Based on waste management experiences of the 1990s, the output of "buildings and infrastructure" is estimated to be about 10–15% of the input. Thus in the year 2000 the stock in "buildings and infrastructure" amounted to 375 t per capita: 40 t cement and chemically combined water, 175 t fixed gravel/sand, 160 t loose gravel/sand. For loose gravel/sand this stock is equal to the input; the output of 25 t per capita is recycled. Regarding an uncertainty range of ±10 to ±30%, the validation is already adequate for many requirements and research questions of urban exploration. The 95 t in the building stock, which were found by the ark-house method, are close to the 110 t found by a top-down approach (statistical areal data) in a former study of the Swiss lowlands. The advantage of the bottom-up approach of the ark-house method is the good knowledge about the inner structure of the stocks, which is of vital importance for exploration.

Due to the genesis path of copper movables (motor vehicles, machines, etc.) have to be considered. Dissipative losses occur in all three stocks in use. From a quantitative view of urban exploration the copper flowing into soils and aquatic systems is negligibly small. However from an environmental point of view this flow is ecologically relevant. Within the system boundary of Switzerland there are neither copper mills (apart from semi-manufacturers) nor primary copper resources. However the copper trade is a key business

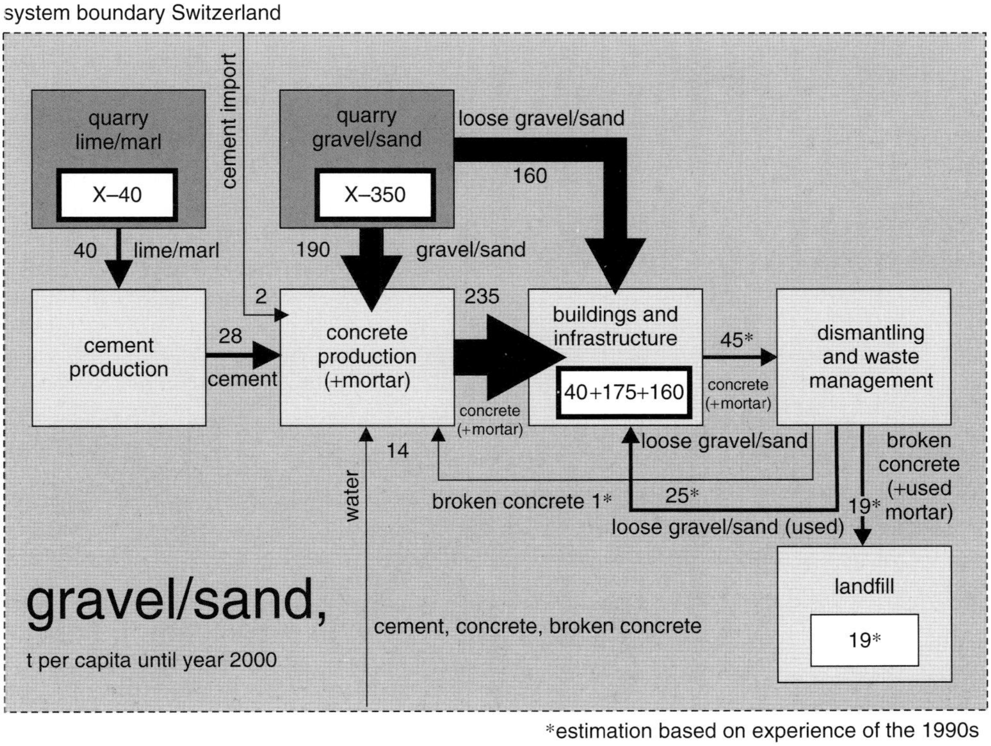

Figure 5.4 System gravel/sand, including cement, concrete (and mortar), (Lichtensteiger and Baccini 2008). All data are calculated per capita for the year 2000 (7.2 million inhabitants) since the start of cement use.

and the urban stocks per capita are among the highest worldwide. Import and export are the predominant fluxes. In the case of copper exploration in Switzerland validation is carried out using foreign trade statistics. Buildings and infrastructure are the dominant urban stocks in use. Movables are not negligible. The deposit in landfills is already in the same order of magnitude as each of the stocks in use. The annual stock increase is indicated by small arrows. The uncertainty range is ±5 (buildings and landfills) respectively to ±30% (infrastructure and movables). The building stock amounts to 80 kg per capita.

On a global perspective, it is reasonable to assume that urban exploration is first an option for rich countries, which tend to transform their constructed environment rather than expand it, due to economic and demographic reasons. Transformation is also a chance to improve ecological, esthetic, and social qualities of urban landscapes, which arise from the 20th century. Regarding environmental qualities, the knowledge of the growth of urban stocks and their future dynamics will facilitate the control and regulation of dissipative losses from urban surfaces. With respect to long-term resource management and urban development, modeling and simulation based on urban exploration and "if–then scenarios" will support the debate on sustainable development options including urban mining.

5.4 DEVELOPING STRATEGIES FOR THE DESIGN OF URBAN SYSTEMS

As already stated in Section 5.1, urban systems have reached a high degree of complexity. Any developing strategy for and within the system touches and influences a range of elements. Therefore it has to show in a transparent way which models and methods comprise the basis of a strategy selection process. The physical aspects emphasized in Sections 5.2 and 5.3 are just one facet. The relatively young research field of

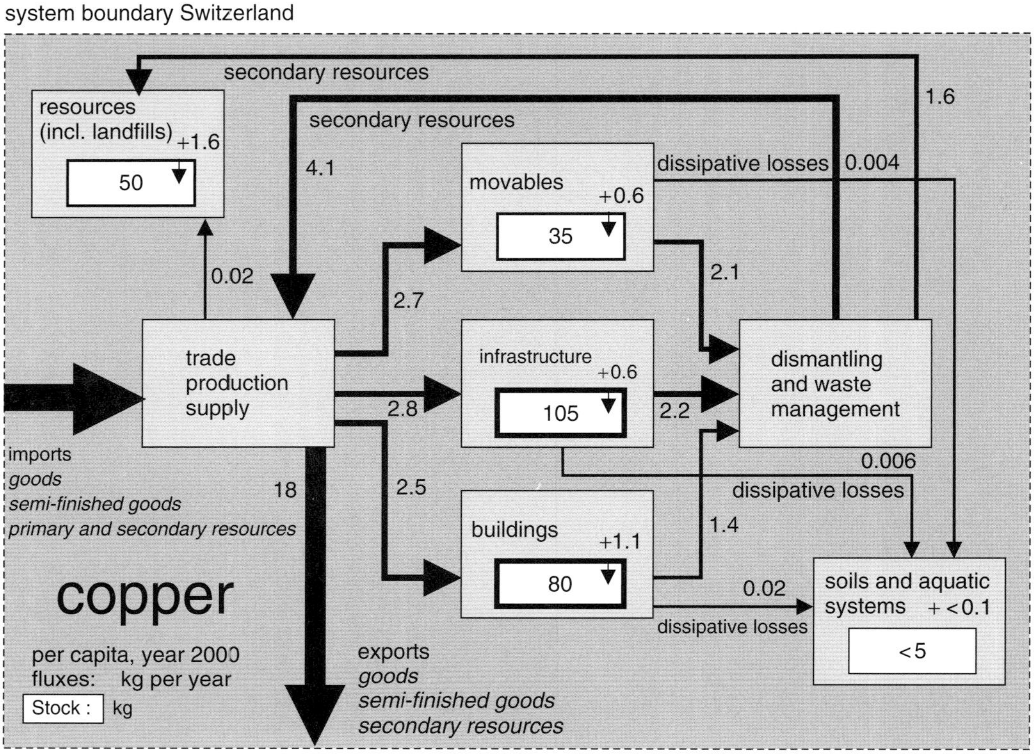

Figure 5.5 Refined exploration of trace materials, exemplified by metallic copper at the end of the 20th century (Lichtensteiger and Baccini 2008).

transdisciplinary research was chosen to "grasp the relevant complexity" (Pohl and Hirsch Hadorn 2007) of urban life (Baccini and Oswald 1998). The methodological history of this project is described elsewhere (Baccini and Oswald 2008).

The urban system is modeled as a *Netzstadt* (Oswald and Baccini 2003). It consists of a limited set of elements that allow at first morphological and physiological analysis and synthesis to be carried out. It can be extended to socioeconomic studies (Box 5.2).

Box 5.2 Elements of the *Netzstadt Model*

Six territories:	waters, forest, settlement, agriculture, infrastructure, fallow land
Four key resources:	water, biomass, construction materials, energy
Four network properties:	nodes, connections, borders, scales

The urban system is being generated and transformed by *four basic activities*:

- to nourish and recover
- to clean
- to reside and work
- to transport and communicate.

This allows comparisons of human settlements independent of their cultural existence in time and space. *Five essential urban quality criteria* are chosen:

- Identification
- Diversity
- Flexibility
- Degree of self-sufficiency
- Resource efficiency.

The quality goals for a concrete urban project have to be elaborated by a participative political process and should not be given a priori by the professional urban designers. Therefore urban design in the *Netzstadt* method consists of three steps:

1 the analysis, elucidating the essential properties of the urban system in question, helping the inhabitants to formulate the strengths, weaknesses, opportunities, and threads (SWOT) of their urbanity (it is the challenge to develop tools),

2 sketching of the long-term orientated urban goals by the inhabitants, and

3 the design for reconstructing urban systems meeting the target qualities of step two, starting from the state defined in step one.

It follows from this that the proposals for the reconstruction of urban systems in Section 5.3 have to be integrated in such a model that allows the design work to be undertaken for a concrete project in a tailor-made process.

REFERENCES

Baccini, P. (1997) A city's metabolism: Towards the sustainable development of urban systems. *Journal of Urban Technology* 4(2), 27–39.

Baccini, P., Bader, H.P. (1996) *Regionaler Stoffhaushalt*. Spektrum Akademischer Verlag, Heidelberg.

Baccini, P., Brunner, P.H. (1991) *Metabolism of the Anthroposphere*. Springer, New York.

Baccini, P., Imboden, D. (2001) Technological strategies for reaching sustainable resource management in urban regions. In: *Our Fragile World: Challenges and Opportunities for Sustainable Development, Forerunner to Encyclopaedia of Life Support Systems*. EOLSS Publ., Oxford, pp. 2153–2173.

Baccini, P., Oswald, F. (1998) *Netzstadt: Transdisziplinäre Methoden zum Umbau urbaner Systeme*. vdf Hochschulverlag, Zurich.

Baccini, P., Oswald, F. (2008) Designing the urban: Linking physiology and morphology. In: Hirsch Hadorn, G., Hoffmann-Riem, H., Biber-Klemm, S., Grossenbacher-Mansuy, W., Joye, D., Pohl, C., Wiesmann, U., Zemp, E. (eds) *Handbook of Transdisciplinary Research*. Springer Science + Business Media B.V., pp. 79–88.

Brunner, P.H., Rechberger, H. (2003) *Practical Handbook of Material Flow Analysis*. Lewis Publishers.

Buitenkamp, M., Venner, H., Wams, T. (eds) (1992) *Action Plan Sustainable Netherlands*. Friends of the Earth Netherlands, Amsterdam.

Faist, M. (2000) *Ressourceneffizienz in der Aktivität Ernähren*. Diss. ETH Zürich Nr. 13884, Zurich.

Global (2000) *The Global 2000 Report to the President*, 1981, Washington DC.

Groh, D. (1992) *Anthropologische Dimensionen der Geschichte*. Suhrkamp, Frankfurt/Main.

Grübler, A., Nakicenovic, N. (1997) Decarbonizing the global energy system. *IIASA Report* RR-97–6, Laxenburg, Austria (reprinted from *Technological Forecasting and Social Change* 53, 97–110 (1996)).

Imboden, D.M. (1993) The energy needs of today are the prejudices of tomorrow. *GAIA* 2, 330–337.

Kytzia, S., Faist, M., Baccini, P. (2004) Economically extended-MFA: a material flow approach for a better understanding of food production chain. *Journal of Cleaner Production* 12, 877–889.

Lichtensteiger, T., Baccini, P. (2008) Exploration of urban stocks. *Journal of Environmental Engineering and Management* 18(1), 41–48.

Müller, D., Oehler, D., Baccini, P. (1995) *Regionale Bewirtschaftung von Biomasse (Regional Management of Biomass)*. vdf Hochschulverlag, Zurich.

Oswald, F., Baccini, P. (2003) *Designing the Urban*. Birkhäuser, Basel.

Pfister, F., Baccini, P. (2005) Resource potentials and limitations of a Nicaraguan agricultural region. *Environment, Development and Sustainability* 7, 337–361.

Pohl, C., Hirsch Hadorn, G. (2007) *Principles for Designing Transdisciplinary Research*, proposed by the Swiss Academies of Arts and Sciences. Oekom, Munich.

von Thünen, J.H. (1826) *Der isolierte Staat in Beziehung auf Landwirtschaft und Nationalökonomie*. G. Fischer, Stuttgart (1966, reprint).

WCED (World Commission on Environment and Development) (1987) *Our Common Future*. Oxford University Press, Oxford.

Zeltner, C., Bader, H.-P., Scheidegger, R., Baccini, P. (1999) Sustainable metal management exemplified by copper in the USA. *Regional Environmental Change* 1(1), 31–46.

B. SOCIOENVIRONMENTAL THREATS

CHAPTER 6

Environmental and ecological threats in Indian mega-cities

Surinder Aggarwal[1] and Carsten Butsch[2]

[1] *Department of Geography, University of Delhi, Delhi, India*
[2] *Institute for Geography, University of Cologne, Cologne, Germany*

6.1 URBANIZATION DYNAMICS AND EMERGENCE OF MEGA-CITIES

Urban growth in India happened largely after national independence in 1947 (see Figure 6.1) and the growth remained substantially confined to large cities of more than 100 000 inhabitants (Aggarwal 2008). Such cities have shared close to two-thirds of the total urban population. In recent years, the distribution and growth is becoming even more skewed in favor of larger cities exceeding one million inhabitants (see Figure 6.1). As evident from Figure 6.1, even within million cities the growth and share of 5 million and above sized mega- and emerging mega-cities is much higher than that of lower sized cities. As per United Nations (UN) (2007) estimates for 2005, India has around 40 urban agglomerations with a population of more than 1 million, including the three mega-cities of Mumbai (18.2 million), Delhi (15.9 million) and Kolkata (14.2 million). These three mega-cities, with a total population of 48 million, have a share of about 15% of the total urban population in India. The five emerging mega-cities, including Chennai (6.9 million), Bangalore (6.5 million), Hyderabad (6.1 million), Ahmedabad (5.7 million) and Pune (4.4 million) are catching up fast to attain the title of a "mega-city" within the next 5–10 years, as their present growth rates, ranging between 3 and 4%, are higher than those of the present three mega-cities. According to the UN (2007) estimates Mumbai crossed the 20 million threshold by 2010 and Delhi 5 years later. With such a phenomenal urban growth in favor of large cities, it is fundamental to understand the accompanying environmental and social threats they pose towards a sustainable and harmonious urbanization. It is obvious that Indian (emerging) mega-cities will continue to play a decisive role in the economic and social development of India.

Indian urbanization is characterized by a slow growth rate, but its present large volume of close to 330 million people makes it a major urban player in a globalizing environment. Urbanization for that matter is a powerful and spatial–temporal restructuring process. Such a huge urban population, almost equivalent to the combined populations of the United States and the United Kingdom, presents many opportunities for the generation of livelihood and integration into global markets. It is noticed that the contribution of urban areas, particularly the large cities, to the country's national income has been steadily increasing from about one-third in the early 1950s to about 60% in

Applied Urban Ecology: A Global Framework, First Edition. Edited by Matthias Richter and Ulrike Weiland.

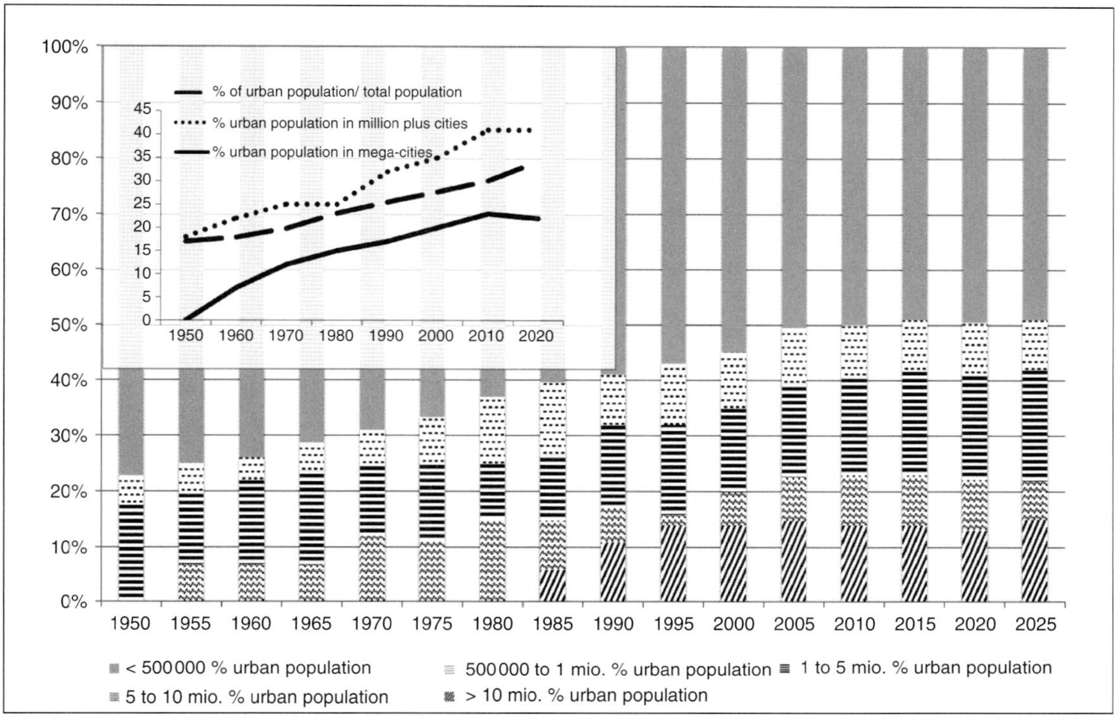

Figure 6.1 Trends of urban and large cities' growth in India.
Source: UN (United Nations Secretariat, Population Division of the Department of Economic and Social Affairs) (2007) *World Population Prospects: The 2006 Revision and World Urbanization Prospects: The 2007 Revision.* http://esa.un.org/unup (accessed 22 March 2011).

2001, while they contain only 28% of the population (Kumar 2005). At the same time economic capital limitations are evident for their faster development. This is really a paradoxical and an ambivalent situation for a developing country to tackle both opposing forces. Today, and more than ever before, cities are home to humanity's great expectations (UNCHS 2008). Likewise hope appears from the large cities of India, which in recent years have shown unprecedented growth, not only in numbers but also as the foci of investment, production and consumption. Many of the emerging mega-cities such as Bangalore, Hyderabad and Pune have also become global hubs of information and communication technology (ICT) or automotive production. Likewise, Mumbai and Delhi are growing centers of financial and quaternary services. Chennai is emerging as an advantageous location for automobile and engineering industries. Calcutta (now Kolkata), the oldest mega-city of India, is regaining its lost shine with the generation of a new industrial and service sector investment environment. Apparently there is economic buoyancy in most big cities of India.

Paradoxically, the same cities also generate a feeling of despair when we examine the degraded environment, including the impacts of climate change and the damage to the ecosystems, huge ecological footprints, mounting gaps and disparities in environmental and economic infrastructure and services, social and economic insecurities, spatial fragmentation, poverty and crime. It seems that cities are losing their harmony with nature and social systems. With growing vulnerabilities and insecurities, environmental degradation, and careless energy/resource use, sustainability of such cities is at risk. Considering the last mentioned environmental and social concerns, the present chapter is an attempt to examine the environmental and social threats that compromise the sustainability and the harmonious development of Indian mega-cities.

6.2 ENVIRONMENTAL THREATS

Mega-urbanization, as a major driving force with colossal fossil energy use, urban sprawl, land use/cover changes exhibits local, regional and national/more widespread environmental impacts. Local and regional impacts are visible through inefficient water services, deteriorating water quality, groundwater depletion, and air and land pollution. On another scale global changes, also mediated through mega-urbanization and globalization processes, produce impacts like climate change, natural and man-made disasters, loss to ecosystem services and biodiversity. Both dimensions of mega-urbanization threaten human security and urban ecosystems health (Figure 6.2).

The impacts, of course are not equal on all social and economic groups as these are mediated by human sensitivities, vulnerabilities and thresholds. Factors such as governance, policies, institutional roles and civil society involvement also produce differential impacts on the communities. Nevertheless, the vulnerable and marginalized groups suffer most from such influences as their sensitivities are high and preparedness is low. In the following sections we examine largely the localized threats produced on the city environment.

6.2.1 Environmental threats from waste water and sewerage disposal

Waste water and sewage in mega-urban areas is generated from point and non-point sources. Point sources include domestic, municipal and industrial sources, and raw sewage, while the typical non-point sources cover bank side, defecation fields, septic tanks, sewer leakages, stagnant raw sewage surrounding unauthorized

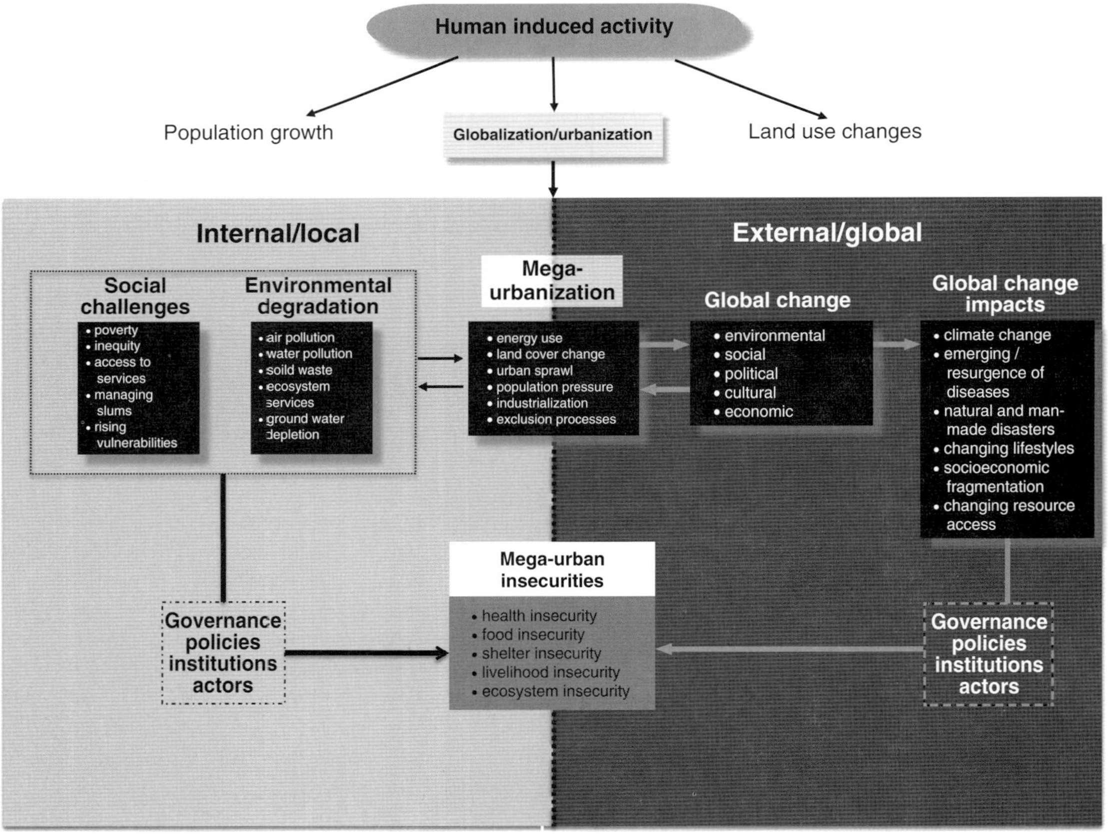

Figure 6.2 Mega-urbanization and threat to human security.

settlements (slums), raw sewage in open drains, and polluted water flowing from landfill sites. Many small industrial units located in the non-conforming areas do not have individual or common effluent treatment facilities and hence the industrial effluents are largely disposed of in open drains or other water bodies.

Although availability and coverage of municipal drinking water supply in mega-cities has significantly improved (90–95%) during the last few years, waste water disposal has not shown commensurate improvement. Mumbai, Kolkata, and Chennai generate 2228 million liters per day (MLD), 1383 MLD, and 276 MLD of domestic wastewater respectively, but only about two-thirds of it gets collected, and only close to half is effectively treated (Maiti and Agrawal 2005). For example, Delhi, the capital city of India, is able to provide effective treatment before disposal only to half of its waste water (including waste water from industries) (Central Pollution Control Board; CPCB 2004). This gap in generation, collection, and treatment of raw sewerage is largely attributed to the insufficient number of sewage treatment plants (STPs), the poor performance of the existing STPs and the variety of non-point sources of waste water (CPCB 2007). This leaves a huge volume of wastewater uncollected and untreated by all the mega-cities in India, in varying proportions. Untreated waste water is largely disposed of on agricultural land, or in water bodies and degraded valleys.

The waste water issue challenges all the mega-cities with respect to its contamination of drinking water through leaking pipes, scattered cesspools, exposure of vulnerable groups (bathing, washing and irrigation practices), pollution of the surface and ground water, and ecological and health threats to the downstream communities. Mega-cities are the major polluters of not only its freshwater resources, but of those in the downstream areas, too. Since a large volume of partially treated and untreated wastewater finally drains into rivers, which are the major source of municipal water supply for the towns downstream, it is believed that every consumer has, over the years, been exposed to unknown quantities of pollutants in the water. Additionally, incapacity of the existing city drainage has resulted in increased incidences of urban flooding followed by epidemics of waterborne diseases (Gupta 2006). The cholera epidemic of Delhi in 1989 resulted in more than 10 000 deaths after the heavy monsoons, largely within the poorly drained and sewered low-income settlements. In her recent study, Singh (2008) has associated a predominant incidence of waterborne diseases, such as diarrhea and skin diseases, in slum settlements of Delhi where people are frequently exposed to waste water. Eutrophication, associated with algal bloom and drying out of water bodies within the cities of Delhi, Bangalore, Kolkata, and Ahmedabad, is also due to wastewater disposal into them. As all mega-cities of India exhibit a huge urban sprawl, they are degrading peri-urban water fields either directly through *in situ* sanitation, industrial discharges and leaking sewers or because of infiltration of polluted surface watercourses. If we examine the population distribution of all the major cities of India based on catchment areas, we find close to 80% of their population is concentrated in major river basins. Around 80% of the population of these major cities is possibly at greater risk, directly or indirectly, due to their location in polluted river basins.

In brief, inadequate sewerage facilities, poor management of wastewater, and prolonged exposure to non-point sources remain the cores explanations for increasing environmental and public health risks in Indian mega-cities.

6.2.2 Deteriorating air quality

Air pollution in Indian big cities is largely contributed to by motorized vehicles. Their number has swelled since economic liberalization and globalization. Currently there are more than 50 million motor vehicles and about 20% of them are concentrated in the four mega-cities – 55.6% are registered in Delhi, 19.5% in Chennai, 15.3% in Mumbai and 9.5% in Kolkata (Nagdeve 2007). Increased motorization and personalized modes of transport, due to poor public transport, have produced automobile crowding and accompanied air pollution without a commensurate growth and coverage in road capacity. Poor fuel quality, ill-maintained vehicles and poor transport management further exacerbate the pollution levels. Of the total air pollution load from various sources, vehicular pollution contributes to an alarming 64% in Delhi, 52% in Mumbai, 50% in Ahmedabad (Government of India 2005–2008) and 30% in Kolkata. Among different type vehicles two-stroke vehicles, including auto-rickshaws, share the maximum pollution load, in particular CO emissions and suspended particular matter (SPM). About 20% of poorly maintained vehicles contribute to 60% of vehicular pollution in India (Pundir 2001). Non-motorized transport, which needs a major boost in Indian cities, is

Table 6.1 **(a)** Classification of pollution levels

	Annual mean concentration range ($\mu g\ m^{-3}$)				
	Industrial (I)			Residential (R)	
Pollution level	**SO_2 & NO_2**	**RSPM**	**SPM**	**SO_2, NO_2 & RSPM**	**SPM**
Low (L)	0–40	0–60	0–180	0–30	0–70
Moderate (M)	40–80	60–120	180–360	30–60	70–140
High (H)	80–120	120–180	360–540	60–90	140–210
Critical (C)	>120	>180	>540	>90	>210

(b) Pollution levels of major pollutants in mega-cities of India in 2005

STATE, UT/CITY	SO_2		NO_2		RSPM		SPM	
AREA CLASS	**I**	**R**	**I**	**R**	**I**	**R**	**I**	**R**
Hyderabad	L	L	L	M	M	C	M	C
Ahmedabad	L	L	L	L	H	C	M	C
Bangalore	L	L	M	M	M	H	L	M
Mumbai	L	L	L	L	M	H	M	C
Pune	L	L	L	M	M	C	M	C
Chennai	L	L	L	L	M	M	L	M
Kolkata	L	L	M	M	H	C	M	C
Delhi	L	L	M	H	C	C	H	C

Source: Central Pollution Control Board, Annual Report 2007.

only used by the poor and for short distances. Thus the poor are lower air polluters when compared with other income groups. Unfortunately they pay the maximum health penalty, as they are exposed the most in their working and living environments (Krafft *et al.* 2003).

Industries contribute to a lesser degree as these are largely located on the fringe of big cities or have been displaced or closed. Industrial pollution is confined mainly to the urban peripheral regions within planned/organized industrial sectors of Ahmedabad, Pune, Delhi, New Mumbai, and Kolkata. Indoor air pollution is also a minor contributor to pollution load and does not have a major impact on the environment, but rather affects in a major way the health of the poor as they use unsafe biofuels for cooking and other domestic needs. India has among the largest disease burdens due to the use of dirty household fuels; roughly substantial percentage of the national burden of disease is due to indoor air pollution (Smith 2000).

The level and trend of major pollutants across all the mega-cities illustrates a serious situation overall (Table 6.1a,b). Comparison of CPCB data for at least four major mega-cities (Delhi, Mumbai, Kolkata, and Chennai) for the 1996–2004 period indicates that SPM, and respirable suspended particular matter (RSPM) have shown levels consistently exceeding those set as the safe standards for human health (CPCB 2000, 2003–2007, Government of India 1999). Both pollutants are of critical importance for human health as they still exceed three to four times the prescribed safe levels in all mega-cities except Bangalore and Chennai. Delhi at one time was infamously known as the asthma capital of the world until a few years back, due to alarming levels of both SPM and RSPM (Aggarwal 2009). SO_2 levels, even though they are below the safe levels, have even shown consistent decline in values over the last 10-year period. NO_2 levels are at an unsafe level in many cities (Table 6.1a,b) and have also shown rising trends, largely due to growth in the automobile population. Delhi confirms this trend as it has the largest share in per capita vehicular population in India.

The decreasing trend in most pollutants may be due to various interventions in recent years such as the reduction of sulfur content in diesel, use of cleaner and unleaded fuel (compressed natural gas – CNG) in Delhi and Mumbai, high pollution norms in automobile certification, and displacement or closure of polluting industries in the central cities. As discussed later, the rising pollution load through greenhouse gases (GHGs) production is seriously contributing to the changes in local climate. Heat island effects and heat stress related diseases are on the rise in most cities.

6.2.3 Urban waste mismanagement and environmental degradation

Unlike wastewater and air pollution, littered municipal waste is the most visual blight in Indian cities. Despite the very low amount of waste produced per capita (Table 6.2), in comparison to industrial countries (almost one-tenth/capita), its mismanagement has produced very serious environmental hazards, ecological damage and health threats, largely to disadvantaged groups. Economic development, fast food habits, consumer lifestyles, and more recently the information technology and globalization forces, have further intensified the waste management problem in the booming mega-cities. Dumping of toxic wastes on the soils of developing countries, including India, by the industrialized nations is another serious concern for the environmental and human safety in such countries. Major ports like Mumbai handle and distribute such toxic wastes within the country. Production and exposure to e-waste and medical waste are new threats with their unsafe and unsegregated collection and deposit practices. On the social side, waste handling and collection by the very poor provides livelihood in a variety of ways (Köberlein 2003) despite the serious health risks they encounter with hazardous exposures (Aggarwal and Sandhya 2004).

The increase in the generation of non-segregated solid wastes and the inadequacy of waste collection, transportation and disposal infrastructure have created distress in solid waste management across all the big cities. The clearing efficiency of all mega-cities ranges from as low as 62% in Delhi to the highest in Kolkata (90%) (Table 6.2). Municipal agencies spend about 5–25% of their budget on solid waste management, but still the present level of service in many urban areas is inadequate to secure against the environmental and health threats (Kumar and Gaikwad 2004).

Pressure of waste generation has far outpaced the capacity of most municipalities and a substantial part of the solid waste remains unattended, particularly in poor neighborhoods, unplanned and unauthorized residential areas and squatter settlements. Health threats are even more serious and critical for those communities living close to landfill sites. In Hyderabad and Delhi, for example, many unauthorized settlements have developed on the fringe of landfill sites due to the low threat of eviction by the municipal government. The dwellers have maximum exposure to the nuisances of waste and are thus extremely vulnerable to health risks. A study conducted on the Bhalswa landfill site in Delhi found that the groundwater within 2 km of the landfill site is contaminated with magnesium, calcium, chloride, and sulfate to highly critical levels due to leaching effects (Sandhya 2004). The study further noticed that thousands of illegal residents living around the landfill site are exposed to the dangers of the site, either through direct contact while collecting waste material (injuries, skin diseases) or indirectly by consuming leachate-contaminated ground water (diarrhea, cholera, hepatitis). Rodents, flies, and other disease vectors emanating from the site remained a constant health threat for the residents. Women and

Table 6.2 Solid waste management issues in mega-cities of India.

Waste issue/City	Mumbai	Delhi	Kolkata	Chennai	Bangalore	Ahmedabad	Pune
Solid waste generated (in metric tonnes/day)	6000	6000	2500	2500	2500	1273	1200
Garbage generated (kg per capita and per day)	0.50	0.61	0.55	0.58	0.58	0.36	0.47
Clearing efficiency (%)	86	62	90	80	85	80	70

children were observed to be the most affected residents as they stay near the site for a maximum period. The catastrophic outbreak of plague in Surat during 1994 was also associated with uncollected waste from the poor neighborhoods over a long period.

Likewise, mounting medical waste with the phenomenal growth of private hospitals and other health care units in the big cities has produced serious environmental and health hazards with unsafe management practices. Most hospitals and other health establishments are weak in segregating medical waste and do not have enough capacity and resources to install incinerators either individually or collectively. Under such conditions biomedical waste is also dumped along with other domestic waste, which further threatens the health of the poor and those handling the generated waste.

On the other hand uncollected waste, plastic waste in particular, blocks the storm water drains during the rainy times and most flood situations in the mega-cities are now associated with such conditions. The land burden that the increase in waste generation would impose is evident from the fact that the cumulative requirement of land for disposal of untreated solid waste (USW) in India would reach around 169.6 km^2 by 2047 compared to 20.2 km^2 in 1997 (CPCB 2000). All the mega-cities in India are finding it difficult to find suitable and sufficient land for safe disposal of their growing waste. With the strong denial of the fringe rural communities to allow disposal of urban waste on their land a serious future challenge has emerged for the mega-cities' administrations.

6.2.4 Ecosystem damages and ecological footprints

The massive transformation of landscapes in the process of mega-urbanization has severe effects on the state of ecosystems (forest, land, water) and indirect effects on human well-being. Mega-cities influence the ecosystems as these consume an enormous amount of energy and other natural resources – air, water, land, forest, etc. and release ordinary and toxic wastes into their natural sinks that comprise the urban ecosystems (United Nations 2005). India is already running into ecological deficits, more so in the urban areas. Mega-cities are the biggest sufferers of ecological deficits, that is they consume more than can be provided by the ecosystems within their borders. These effects can be visualized and experienced in all Indian (emerging) mega-cities through surface and ground water pollution and depletion, ecological footprints on their surrounding regions, loss of forests and encroachment on water bodies for land acquisition.

Water quality monitoring results obtained between 1995 to 2006 (CPCB 2007) indicate that organic and bacterial contamination continue to be critical in rivers that pass through mega-cities like Delhi (Yamuna River), Hyderabad (Musi River), Ahmedabad (Sabarmati River). For example, the level of fecal coliforms, another important indicator of pollution, was found to be highest in the Yamuna River in Agra, Nizamuddin, and Okhla in Delhi (most probable number, MPN 5.2×10^6–3.7×10^6) as well as in Sabarmati at Ahmedabad (MPN 1.1×10^6–4.6×10^5). About 15 MLD of industrial effluents (mainly nitrates, phenols and cyanides) from about 100 industrial units under hazardous category are released into Hussainsagar Lake in the center of Hyderabad City (EPTRI 1996). This damage is mainly due to discharge of untreated wastewater, largely from non-point sources, and also untreated sewerage from the treatment plants. The receiving water bodies also do not have adequate water for dilution and this accentuates the pollution process. Likewise, the ground water pollution situation in most mega-cities is becoming critical owing to leachate and untreated wastewater seepages (CPCB 2003). Most of the irreversible groundwater damages occur with the activation of geohydrochemical processes triggered by over pumping. Arsenic levels in the 49 ground water samples investigated from different parts of Delhi were found to be in the range 0.0170–0.1000 ppm, which is more than the WHO recommended value of 0.010 ppm in drinking water (Lalwani *et al.* 2004). The ecological health of most water bodies is therefore seriously threatened and these are unable to provide free ecological services of bathing, performing rituals, livelihood generation, etc., especially for the poor.

Mega-urbanization has led to a phenomenal decrease of natural (groundwater) recharge due to surface sealing and soil compaction, thus promoting an imbalance in the overall groundwater budget. Also consistent withdrawal of groundwater by various sectors is seriously depleting the groundwater resource, which probably cannot be replenished for future generations. Ahmedabad and Delhi represent classic examples. In Ahmedabad, high dependence on groundwater, in the absence of perennial water sources, has resulted in depletion of the water table at an annual rate of 2 to 3 m for the last 40 years (Government of India (GOI)

2005–2008, Ahmedabad City Development Plan (CDP) 2008). The existing water table is at a depth of 120 m. The south zone of Delhi is experiencing groundwater depletion in the quartzite zone and water table ranges of 40–60 meters in depth within post- and pre-monsoon periods (GOI 2005–2008, Delhi CDP). In some blocks of South Delhi no more water can be extracted as it is enriched in nitrates in the lower aquifers.

As not all the resource requirement comes from the city itself, nor do all the wastes produced stay there, mega-cities have large ecological footprints that threaten not only their own well-being but that of many other adjoining areas, sometimes even far reaching areas (Ainsworth 2008). Faced with the problems of river water pollution and rising upstream and city demands, the existing local/regional water resources of all mega-cities are becoming inadequate. For easy solutions, the city administration prefers augmentation solutions from distant river basins, without considering their ecological footprints. The same is happening in view of forest and land resource demand for the affluent lifestyles in the fast growing economies of such cities. Delhi and Bangalore, for example, meet their excess water demand from distant river basins located in the Himalayas and Deccan plateau respectively, disregarding *in situ* ecosystem damages and producing conflicts with riparian states. For Ahmedabad, a new reservoir Dharoi, located 150 km upstream of Ahmedabad on the Sabarmati River, was commissioned in 1978 and now fails to serve the summer months' demand. The city is looking for another river basin such as Mahi for its new rising demand.

Also surface water systems are heavily disturbed by human activities affecting both the drinking water supply as well as the retention capacity in storm events. The case of Bangalore exemplifies these issues. From the 262 lakes in Bangalore 30 years ago, only 81 are left now (CPCB 2000, p.76). Hussainsagar Lake, used as a source of drinking water, has declined by 40% largely due to urban encroachment. The proposed international airport near Shamshabad is going to affect 11% of the catchment area of Himayatsagar reservoir (Ramachandraiah and Prasad 2004). Shrinking wetlands within the Indian mega-cities, like Kolkata, provided an important ecosystem service for filtering the pollutants and habitat for fish species, plants, insects and birds. Now the wetland is being encroached on for building a suburban Salt Lake City and is infested with mosquitoes as well as being home to other environmental nuisances.

The destruction of eco-capital, understood as those ecosystems that provide humans with ecosystem services, is currently a great concern in the course of mega-urbanization (Hu 2002). Due to a lack of acknowledgment of the important functions ecosystems provide, they are not valued. Furthermore, the lack of governance and planning makes an effective protection of these common goods practically impossible. Sound action plans are needed fast to prevent the further loss of ecosystem services and the negative effects on the well-being of millions of mega-urban dwellers. Especially with respect to the regulative services ecosystems also play a dominant role for disaster prevention, which will be discussed in the next section.

6.2.5 Threats from natural hazards, disasters, and climate change

Mega-cities are not only large agglomerations of people but also a huge accumulation of goods and values, both mobile and immobile. Due to this enormous concentration of human life, global lifelines of communication, and values the potential effect of disasters on mega-cities is enormous. Kraas (2003) therefore addresses mega-cities as global risk areas, which are exposed on the one hand to environmental hazards and on the other hand to man-made hazards. A third dimension is the connectedness of the different types of hazards to global change. In the following sections some potential threats to Indian mega-cities will be discussed, starting with earthquakes as a natural event, and followed by a section on the disaster effects of global environmental change.

Although not situated in a hot spot region for earthquakes, Indian cities have been affected by tectonic events in the past. Especially Kolkata, with a thick sediment deposit, is an area of great concern as it lies over the Bengal Basin and at the boundary of the seismic zones III and IV of the zonation map of India. Kolkata has been affected by the earthquakes of 1897, 1906, and by the 1964 Calcutta earthquake (Mohanty and Yanger 2008). The consequences of a tectonic event hitting high-rise buildings and other properties built with compromised building regulations is an impending disastrous situation for human and property losses. Both Ahmedabad and Delhi are earthquake-risk cities as they are located within moderate to high-risk seismic zones.

Natural disasters related to climatic phenomena are very likely to increase in frequency and magnitude in the course of climate change. Mega-cities as major pollutants, however, are both culprits as well as victims of climate change. The Intergovernmental Panel on Climate Change's (IPCC) fourth assessment report, summarizing the research on climate change on a global scale (Meehl *et al.* 2007), predicts two major trends for the South Asian subcontinent. On the one hand the monsoon regime will change leading to an increase in the overall precipitation (statistically) as well as an increase in the interannual variability of the monsoon (Meehl *et al.* 2007, p.778). On the other hand the surface temperature will generally rise according to different scenarios by between 2.5 and 5 °C by 2100. These two major trends have clear implications and impacts of climate change in the Indian mega-cities. The increasing variability of the monsoon will not only increase the water crisis in cities like Chennai, it will also endanger the energy security of cities like Mumbai and Pune, which are nearly completely dependent on hydroelectric power. Also, Delhi's water supply will be endangered, as it relies on water stored in the Himalayan glaciers – a reservoir that is decreasing fast. Furthermore, the increase in the variability of rainfall also leads to an increase in flooding events. Especially in combination with a global sea level rise, major coastal cities of India, Mumbai, Chennai, and Kolkata in particular, are at a higher risk. In a 1996 study by The Energy and Resources Institute (TERI), Mumbai was found to be the most vulnerable city to a 1 m sea level rise. In fact, it is both the physical as well as the built environment of Mumbai that makes it vulnerable to sea level rise.

The massive land use and land cover changes in favor of built-up areas, and rising GHG concentration are already producing urban heat island effects in most mega-cities of India. Already heat island effects of 6 K for Delhi, 9.5 K in Mumbai, 4 K in Kolkata and Chennai and 10 K in Pune have been observed (Santamouris 2001). Aggarwal and Sen (2007), in their study on climate change and its health impacts for Delhi, demonstrate that the climate of Delhi has become warmer by 1 K and the monthly ranges in temperature extremes are fast disappearing, indicating more of a warming trend. The intensification of the heat island effect is causing heat stress and the rising precipitation regime is largely responsible for the re-emergence of vector-borne diseases like dengue and malaria. A scientific report prepared by the Earth System Science Partnership also supports the fact that rising temperatures along with stronger rainfall events and surface sealing (land use change) will lead to improved conditions for vector breeding (Confalionieri and McMichael 2007).

Flooding is another phenomenon associated with climate change that is going to impact big cities in India with more vengeance, as cities are already affected by poor drainage systems. Schiller *et al.* (2001) has long ago raised concern about the possible climate hazards and disasters that can hit global cities like Mumbai, especially from the rise in sea level and flooding. In fact, floods and landslides can affect about 60% of the Mumbai population, which dwells in slums located at marginal locations and is hence most vulnerable to such disasters. More recently Rao (2003) also expressed the threat from storm surges on most parts of the Indian coast, including the mega-cities of Kolkata, Chennai, and Mumbai.

6.3 MEGA-SOCIAL CHALLENGES

Mega-cities are basically man-made environments, which means that a society transforms natural landscapes according to its cultural and social values and norms. Simultaneously to this process social transformations occur within mega-urbanizing societies. A socioecological perspective on these two developments attempts to overcome the dualistic view on nature and society as separated entities. In that capacity, mega-cities as a societal phenomenon also enjoy free ecosystem services of cleansing their polluted environments. Due to this interconnectedness, societal phenomena like poverty have an influence on the environment while, vice versa, degraded environments have an influence on poverty. In the process of mega-urban development in India we notice that not only the urban environment is damaged, including a severe loss of ecosystem services, as discussed in the previous section, but also it has produced linked effects on the social environment. Considering the poor financial base and bad urban governance, the metropolitan cities get burdened with acute housing shortages and emergence of slums, poverty, high crime rate, scarcity of safe drinking water, and prevalence of environment related diseases (Aggarwal 2008, xxiii). In the following sections we discuss some of the above social issues from urban fragmentation, rising vulnerabilities, insecurities, and inequities perspectives.

6.3.1 Poverty and fragmentation

Poverty is a condition in which certain societal groups cannot maintain their livelihoods. Poverty in India is defined and measured by specifying a poverty line based on a particular level of per capita consumption (2200 kcal per capita for urban areas) or the equivalent income and assessing what proportion of the population falls below this line (Bapat 2009, p.2). Urban poverty rate – measured by the official definition – is especially high in the (emerging) mega-cities. The extent of the population below the poverty line (BPL) varies from 9% in Chennai, to 11.42% in Delhi, 13.48% in Hyderabad, and 26.1% in Ahmedabad (according to the respective City Development Plans). A comprehensive report on urban poverty in India (UNDP 2009) shows that the share of BPL population has been declining in the last 40 years in all Indian cities – though progress is far from satisfactory. Since most mega-cities in India are presently strongly influenced by globalization (external influences) and restructuring of the national economy (internal forces), we discuss below how poverty is linked with internal (access to resources) and external (globalization and fragmentation) forces within the large cities. In the present course of globalization, the global societal fabric is subject to change, leading on the one hand to closer economic relations across the globe and on the other hand this development is not an inclusive one. This excluding development manifests itself in the form of massive societal disconnections of urban societies – a process that is often addressed as fragmentation (Coy 2006, Dittrich 2003, Scholz 2000). India's (emerging) mega-cities in particular, serve as an example here, especially in the fast growing urban agglomerations of the south. Globalization results in steep economic growth for many mega-cities like Pune and Bangalore, while at the same time urban poverty remains a visible marginalization issue. Castells (2000), in his famous work *End of Millennium* maintains that globalization proceeds selectively wherein it includes and excludes segments of societies from the works of information, wealth and power. What is most striking is that the poor are actually disconnected from the benefits of globalization to which India has been exposed since the beginning of the New Economic Policy of the 1990s. This societal disconnection is addressed by the concept of fragmentation, which suggests that the urban societies actually are becoming separated into three groups. The first group, the emerging upper class – "winners of globalization" – is connected to global economic processes and profits from the new economic opportunities. A second group represents the new middle class – "pseudo-winners of globalization" – who benefit marginally and with risk through the exploitation of their knowledge and skill acumen. For example, the whole sector of business process outsourcing (BPO) offers temporarily well-paid job opportunities in ITC hub cities like Delhi and Bangalore. A third group in the discourse are the "losers of globalization", which can be divided into two subgroups. The first subgroup consists of those who have lost their previously government secured and subsidized jobs in the traditional industries, such as the workers in the textile mills in Mumbai or elsewhere. On the other hand there are those largely marginalized, vulnerable and unskilled groups (i.e., the poor) of the big cities who have no chances to improve their living as they are not at all integrated into the global knowledge economy. The influence of several multilateral international development agencies and some bilateral development agencies is significant to promote privatization in urban areas. The conditionality of lending for urban development often includes clauses such as allowing transnational or multinational utilities to enter the city. Such policies lead to exclusion within the city through the process of privatization and also increase debt-liability, reducing the city's capability to address the issue of sustainable poverty reduction.

Fragmentation essentially stresses the parallel existence of different social groups (winners, pseudo-winners and losers of globalization) and networks in a direct spatial neighborhood with no or few interactions. Furthermore, these few interactions are limited to the exchange of goods and services rather than being true social interactions. Although there are contact points between the included and the excluded (the latter, for example, working as housemaids or drivers for the former), their societal relations and social action spaces differ substantially. Coy (2006) developed a scheme to address the urban fragmentation in Latin America. Keeping some limitations in mind this can also be applied in the Indian context. Gated communities, for example, are a relatively new phenomenon in the Indian context, whereas they are established concepts of middle and upper class housing in the Americas. Also the issue of inner-city slums has not been regarded as being particularly relevant in the Indian context so far. However within cities as a whole different living realities exist, which are addressed by Coy as "islands of wealth" and "oceans of poverty". Global forces like neoliberalism and deregulation lead to marginalization and disconnections of large segments of the urban societies. The poor are forced into informal activities and are also spatially marginalized in their living conditions. These fragments of the poor are surrounding the fragments of the rich who live in gated communities a westernized lifestyle, largely disconnected from the everyday life of the poor, even if mostly in a close spatial neighborhood. With its' huge ITC enclaves of the knowledge economy and professionals working at national and global ITC companies – located on the periphery of the metropolitan area – Bangalore is a classic example of a self-contained gated landscape to illustrate excluding growth happening within the mega-cities of India. Likewise during recent fieldwork in Pune, for example, one of the authors observed, with respect to health care, that the sources of health care vary completely between a middle-income housing colony and an adjacent slum. The realities of life in both fragments are completely different, even though there are certain functional connections as some of the female slum dwellers are employed in middle income households. This fragmentation leads to several inequalities, which induce severe vulnerabilities, insecurities and inequities for the excluded.

6.3.2 Rising vulnerabilities and insecurities

As Watts and Bohle (1993) point out poverty and vulnerability must not to be generally equated with each other. Even if the poor are very often vulnerable it is not only their poverty, which makes them vulnerable as vulnerability actually is an assessment from a systematic perspective. Insecurities of entitlements, like a house or education, are one component of vulnerability, which can be a permanent stress factor or can lead to acute crisis if surrounding conditions change and the ability to earn a livelihood becomes lost. In mega-urbanizing India the share of the vulnerable population is growing as the cities are heavily congested and not all have the ability and resources to satisfy their basic needs (cf. the example of water supply in Chennai).

The most visible phenomenon of vulnerability in the mega-urban context are slums (unauthorized squatter settlements), which often appear as small interspersed urban fragments in the different city structures. Nevertheless these fragments are home to 40 million Indians who live in areas officially classified as slums – a number roughly equal to the total population of Spain. Especially in the mega-cities and the emerging mega-cities slum dwellers form a large part of the total urban population. The highest figure (Sivaramakrishnan *et al.* 2007) is reported for Mumbai with a share of 48.9% of the total population living in slums – not taking into account the large number of completely shelterless pavement dwellers. Also for Kolkata (32.6%), Chennai (25.6%), Pune (20.9%), and Delhi (18.9%), the official estimates are far above the national average, whereas unofficial estimates are usually even higher. In the case of Pune, for example, even the Municipal Corporation estimates that the share of the slum population is at least twice as high as the official census figures (Pune Municipal Corporation 2006). Mumbai, the largest mega-city of India, is in an abysmal state. More than half of the population is forced to live in slums and shanties. If dilapidated chawls (one-room tenements) are included, about 75% of the population is living in dilapidated housing. According to conservative estimates by the Census of India (2001), the eight (emerging) mega-cities alone contained nearly 20 million slum dwellers. Hyper-urban growth, denial to access to environmental services, insecurity of tenure, and inaccessibility to affordable land and credit are among the major reasons for the large construction and continuity of slums across all the mega-cities of India. Demand for extra land by the globalization processes and market driven economy in the large cities of India is further distancing the poor from an affordable shelter. City level interventions have not produced acceptable results to house the poor, except upgrading and resettlement in a limited way. NGOs have played a significant role towards the upgrading of notified slums in partnership with various stakeholders, including the slum families (GOI, Ahamadabad City Development Plan (CDP), not dated).

Besides the stresses from the poor living environments, slum dwellers are also exposed to shelter insecurity if their dwellings are situated in an undeclared or objectionable slum. Bohle and Sakdapolrak (2008), for example, describe the case of an objectionable slum in Chennai, which was partially destroyed as it was situated on a plot designated for a road-widening project. This constant insecurity is, of course, an enormous stress factor for the inhabitants of these settlements. Also well established and economically and socially integrated large-scale slum settlements, such as Mumbai's infamous slum Dharavi, situated next to the city's airport and at a premium location has been subjected to threats of clearance and eviction of residents. There are mixed views about vertical development of slums in Indian mega-cities like Mumbai. It is believed by urban planners that vertical development of slums would allow new space for the crowded metropolises. On the contrary, vertical development erodes social fabrics and physical structures established by the poor families over the years and should not be sacrificed for the ideas of a globalized mega-city.

In the connection of fragmentation and the vulnerability framework, the phenomenon of the working poor also needs to be addressed. It is one of the most obvious appearances of socioeconomic exclusion. The UNDP defines the working poor in its report on poverty in urban India (UNDP 2009) as those employees (in the formal as well as in the informal sector) that belong to a household below the poverty line. Their income is therefore not sufficient to secure their family's livelihoods. This is due to the fact that their workforce is being exploited in a globalized economy with a loss of regulation as a result of neoliberalism. This problem is also addressed in the Delhi CDP. While an average Delhi BPL household has an income of between 1500 to 2500 Indian rupees per month (Euro 22–38) their expenditure is usually higher than that. This results in a permanent dependency on local moneylenders, which highly increases their vulnerability. The case of Delhi is, however, not an isolated one. As the (emerging) mega-cities are also becoming globalized and fragmented, the phenomenon can be suspected to be especially present here.

Directly linked to these livelihood and poverty insecurities is food insecurity in urban settings. Urban food insecurity essentially is linked to poverty, discrimination, lack of livelihood, unhygienic living, and lack of literacy, basic amenities, and health care. However, food insecurities are the most severe manifestation of absolute poverty. With the concentration of marginal and vulnerable groups (casual and contractual labor, street children, slum dwellers, homeless families) and other forms of vulnerabilities (degraded environments, indebtedness, child and women labor, lack of social security) in the big cities of Delhi, Mumbai, and Kolkata, food security is becoming a big equity issue with the fluctuations of the market within the globalizing economy framework. Furthermore, in the sprawling mega-cities, loss of agricultural lands in fringe areas and the shift to floriculture in urban fringes also affect food security for those living on the urban fringe. Delhi has a large agricultural fringe and with its rapid sprawl and demand for floriculture and other commercial land, the marginal farmers are becoming more vulnerable to such land use changes. Food chains by private sector investors in most mega-cities are affecting the conventional poor street vendors. In urban settings, lack of income translates more directly into lack of food than in a rural setting, as cash is needed to buy food. Global environmental change and the subsequent degradation of ecosystems, climate change and desertification are also affecting food access for the urban poor.

Quoting the 2005 round of the National Sample Survey Organisation (NSSO), the UNDP (2009) states that four in a thousand urban households in India do not have sufficient food for several months of the year, one in a thousand even for the whole year. According to the *Food Insecurity Atlas of Urban India* by the M.S. Swaminathan Research Foundation and the World Food Programme (2002), more than 38% of children under the age of 3 in India's cities and towns are underweight and more than 35% of children in urban areas are stunted. According to the report, the poor in India's expanding urban areas do not get the requisite amount of calories or nutrients specified by accepted Indian Council of Medical Research (ICMR) norms. It also suggests that absorption and assimilation of food by the urban poor is further impaired by non-food factors such as inadequate sanitation facilities, insufficient housing and woeful access to clean drinking water.

6.3.3 Inequities and inequalities in urban services

"Equity" is a normative term to address a condition in which inequalities are perceived as being unjust. Under the conditions of urban fragmentation, increasing vulnerabilities and insecurities, the distribution of resources and the lack of entitlements of the vulnerable have to be addressed as inequities. The severe societal differences manifest themselves especially in inequities in access to basic services. As Table 6.3 shows, the availability of basic amenities and services varies substantially between the seven (including emerging) mega-cities. Most obvious are the gaps in the availability of essential sanitary infrastructure such as drinking water (with water-scarce Chennai leading the doubtful negative list) or toilets show that the population's coverage with basic amenities is far from being satisfactory. These statistical figures however, do not illustrate the inequity in the distribution. The lack of drinking water in Chennai, for example, is mostly experienced by the poor. In the summer months, when

Table 6.3 Access to urban amenities and social services (%).

	Mumbai	Delhi	Kolkata	Chennai	Bangalore	Ahmedabad	Pune
Waste water treated 1998	37.8	–	–	70.00	82.9	–	–
Households with electricity 1991	89.77	81.37	77.17	81.91	81.82	80.26	87.46
Households with drinking water 1991	95.68	96.27	92.24	59.90	80.89	92.01	94.56
Households with toilets 1991	74.45	66.81	89.19	77.25	81.54	72.66	73.10
Households with electricity, drinking water and toilet 1994	71.69	62.17	68.80	44.95	62.69	65.35	68.62
Net enrolment ratio primary schools 2005	75.50	65.12	37.70	76.90	68.50	63.40	–
% scheduled caste/scheduled tribe enrolment primary schools 2005	9.8	13.2	7.75	25.9	28.4	15.4	–

Sources: Census 1991 and UNDP 2009.

the water reservoirs of Chennai's water supply are drying out, the rich can still secure their water supply as they can afford to purchase water from private vendors, who deliver water from peri-urban areas to their homes (Ruet *et al.* 2007). At the same time the poor do not have other means to cope with this water crisis other than to reduce their consumption far below prescribed norms.

Also with respect to access to basic services such as education huge gaps exist. Whereas the general enrolment for primary education varies for Mumbai, Delhi, Chennai, Bangalore, and Ahmedabad between 63 and 77% (see Table 6.3) the corresponding figure for Kolkata is 38%. However in each of the mentioned cities the primary education enrolment rates for members of vulnerable scheduled casts or scheduled tribes are significantly lower. They vary from a poor 7.8% in Kolkata to a mediocre 28% in Bangalore, revealing huge inequities between different societal groups. This is relevant in so far as inequities in education are not only a manifestation of fragmentation; indeed they are a continuation of fragmentation, as the future generations will also be excluded. A third inequity to be mentioned here is the inequality in the access to health care services. Although it is difficult to illustrate health inequities by secondary data, the case of Ahmedabad shall serve as an example here. Figure 6.3 illustrates the population growth of Ahmedabad since 1970, the growth in the number of public sector hospital beds and the development of the number of indoor and outdoor treatments. The number of hospital beds – shown in the bar diagram – nearly doubled while the population grew by the factor 2.3. The number of treatment institutions (not shown in the diagram) first increased from 31 in 1971 to 37 in 1992 to decline again to 26 in 2004. These developments have to result in an overburdening of the existing treatment institutions with a decline in the quality of care delivered, which is reflected in the decrease of absolute treatment numbers. This is in fact an illustration of inequities as it is especially the poor who rely on the subsidized health care services. They do not have the financial resources to afford private medical care. Therefore the lack of investment in the public health care sector leads to an increase in health inequities and health vulnerabilities.

6.4 CONCLUDING REMARKS

It is obvious from the above account that Indian mega-cities are under acute stress because of demographic, environmental/ecological and social challenges. In the Indian situation such large environmental and social challenges are obvious in the mega-cities due to their strong demographic dynamics, including those of massive poor migrants. Understandably, their growing economies, under restructuring and globalizing influences, do not exhibit commensurate growth in environmental, economic and social infrastructure resulting in huge social–environmental costs and

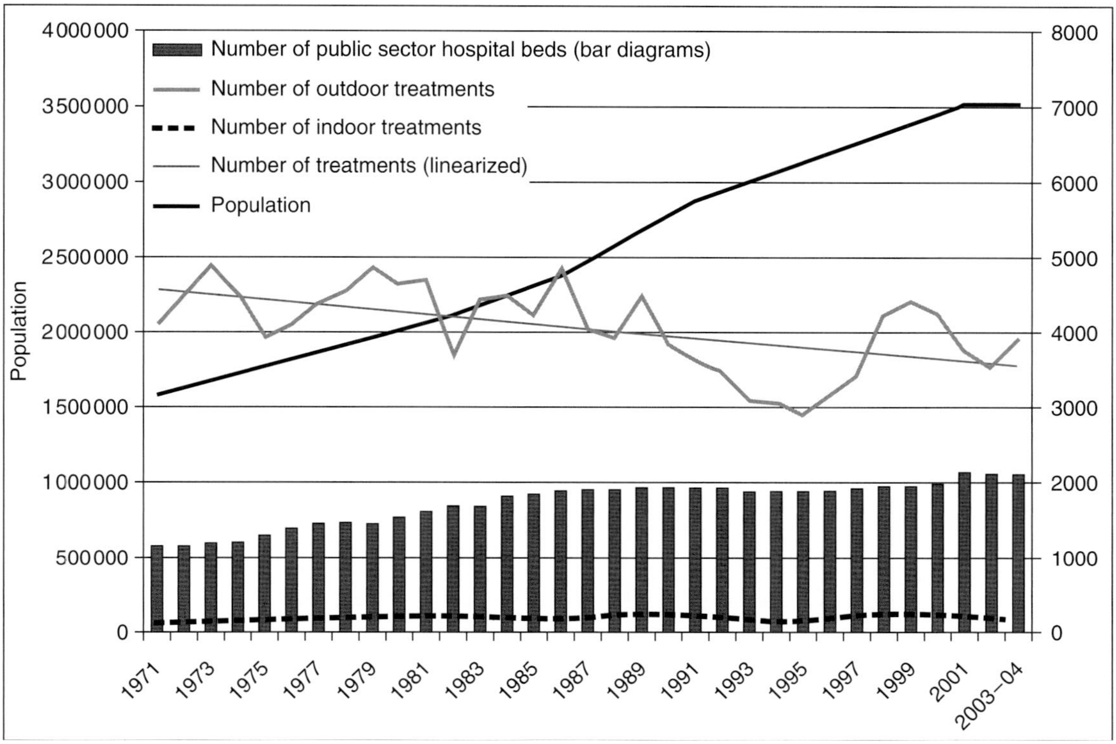

Figure 6.3 Healthcare gaps in Ahmedabad.
Source of data: Ahmedabad City Development Plan.

deficits. Poor urban governance and an inflexible conservative urban planning approach are largely responsible for such an outcome, rather than financial failure, as is normally blamed. Cities are competing to attract investments, whereas they do not compete to produce good conditions for quality of life. Unless such a healthy investment climate is produced, the situation may not change much even though investment from outside is assured. Cities need to build their own capacities in cooperation with other stakeholders and actors for a sustainable and harmonious growth.

For most mega-cities future challenges include degraded environments mainly caused by air, water, and land pollution and access to urban services and housing. Biological, chemical, and physical pollutants have become new threats to human life in big cities due to poor environmental management. Excessive concentration of heavy metals, such as lead, and automobile and industrial pollutants, such as hydrocarbons, SPM, oxides of nitrogen and carbon monoxide, will increasingly pollute the ambient air. It is rather a daunting task to address all the environmental challenges that mega-cities in India are facing, due to its complexity, scale and urban dynamism. The brief overview on natural hazards and disasters associated with environmental change illustrates that mega-cities, due to their size and concentration of people are relatively more vulnerable to such natural and hybrid events. Also, the insecurities arising from the disaster proneness of mega-cities are not equal and the marginalized groups are more vulnerable to such impacts of environmental change. However, especially due to the double role of mega-cities as causers and sufferers of climate change, they need to take a lead in the development of sound mitigation and adaptation strategies. Unfortunately there are hardly any actions in this direction on the local level yet, although Delhi (as India's international metropolis on the political level) is showing at least some initiative.

The social situation in India's (emerging) mega-cities therefore has to be addressed as being unjust from a social equity perspective. Discussions on the underlying causes of the fragmented development of mega-urban societies find strong linkages with poverty, vulnerability and equity, which are interconnected phenomena. As the causes of this aggregate problem are manifold, complex and appear on different scales, though manifested in the local setting of mega-cities, there are no simplistic solutions. What is clear is that many environmental and social problems of today's mega-urban India are largely related to technological deficits, bad governance and lack of administrative capacities. In order to address and tackle disparities, equity issues, vulnerabilities, insecurities and poverty in such mega-cities, sound policies, administrative reforms and financial empowerment are desired steps. In order to overcome the societal fragmentation, development has to become inclusive, which could be achieved by installing security systems that empower all strata of society to fulfil their basic needs. The need for reform has also been recognized by the Government of India, which launched the Jawaharlal Nehru National Urban Renewal Mission across 262 selected cities of India, including all the mega-cities. This Union program tackles the issues of urban infrastructure, governance and accessibility of services for the poor, acknowledging the fact that these problems are beyond the capacities of individual urban local bodies. However, at the present extent it can only be a starting point to solve India's mega-urban social problems. From a socioecological perspective societal reform is also urgently needed to overcome the vicious circle of environmental degradation and poverty. In brief, to produce environmentally sound, sustainable, inclusive and harmonious mega-cities, there is a need to reinvent urban planning, to bring about reforms in urban administration, and to have good governance, strong political will and a persistent civil society engagement.

ACKNOWLEDGMENTS

The authors sincerely acknowledge the technical and critical inputs provided by Reena Singh, University of Cologne, during the progress of this chapter.

REFERENCES

Aggarwal, S.K. (2008) Urban development in India: An introduction. In: Aggarwal, S.K. (ed.) *Perspectives on Urban Development in India: Essays by V. Nath*. Concept Publishers, New Delhi.

Aggarwal, S.K. (2009) Delhi: Towards a healthy city to live. In: Voguet, A. (ed.) *Globalization and Health*. Manohar Publications, New Delhi, pp. 101–130.

Aggarwal, S.K., Sandhya, T. (2004) Exposure to health risks of a solid waste site in North Delhi squatter settlements. *Region, Health and Health Care* 9(1/2), 18–27.

Aggarwal, S.K., Sen, S. (2007) *Climate Change and Risks to Human Health in Delhi*. Paper presented at Beijing ESSP Conference, November 2007, p. 82.

Ainsworth, D. (2008) Mega-cities: A new dynamic organism. *USC News*. University of Southern California. http://uscnews.usc.edu/science_technology/megacities_a_new_dynamic_organism.html (accessed 21 December 2009).

Bapat, M. (2009) *Poverty Lines and Lives of the Poor. Underestimation of Urban Poverty – the Case of India*. IIED Working Paper No. 20. London.

Bohle, H.G., Sakdapolrak, P. (2008) Leben mit der Krise. Vertreibung von Slumbewohnern in der Megastadt Chennai. *Geographische Rundschau* 4, 12–21.

Castells, M. (2000) *End of Millennium: The Information Age*. Economy, Society and Culture, vol. 3, second edition. Blackwell, Oxford.

Confalionieri, U., McMichael, A.J. (2007) *Global Environmental Change and Human Health: Science Plan and Implementation Strategy*. Earth System Science Partnership (DIVERSITAS, IGBP, IHDP, and WCRP) Report No. 4. http://www.essp.org (accessed 22 March 2011).

Coy, M. (2006) Inner-city development and strategies for urban renewal: the case of the mega-city Sao Paolo. In: Kraas, F. Gaes, H., Kyi, M.M. (eds) *Mega-city Yangon*. Lit Verlag, Münster, pp. 63–78.

CPCB (Central Pollution Control Board) (2000) *Management of MSW*. New Delhi, India.

CPCB (2003) Groundwater. *Parivesh*. A newsletter from ENVIS Centre, Central pollution Control Board, Ministry of Environment and Forest, New Delhi, India.

CPCB (2004) *Status of Sewerage and Sewage Treatment Plants in Delhi*. CUPS/57/2004–05, Central Pollution Control Board, Ministry of Environment and Forest, New Delhi, India.

CPCB (2007) *Annual Report 2006–2007*. Central Pollution Control Board, Ministry of Environment and Forest, New Delhi, India.

Dittrich, C. (2003) Bangalore: Polarisierung und Fragmentierung in Indiens Hightech-Kapitale. *Geographische Rundschau* 55(10), 40–45.

EPTRI (Environment Protection, Training and Research Institute, Hyderabad) (1996) *State of Environment for Hyderabad Urban Agglomeration*. EPTRI, Hyderabad.

Government of India (1999) *Compendium of Environment Statistics*. Central Statistics Organisation, Ministry of Statistics and Programme Implementation, New Delhi, India.

Government of India (2005–2008) *Jawaharlal Nehru Urban Renewal Mission (JNURM). City Development Plans for Ahmedabad, Chennai, Delhi, Hyderabad, Kolkata, Mumbai, Pune*.

Gupta, K. (2006) Wastewater disposal in the major cities of India. *International Journal of Environment and Pollution* 28(1/2), 57–66.

Hu, D. (2002) Eco-capital and its services in urbanisation. *IHDP Newsletter* 3, 16.

Kraas, F. (2003) Mega-cities as global risk areas. *Petermanns Geographische Mitteilungen* 147(4), 6–15.

Krafft T., Wolf, T., Aggarwal, S. (2003) A new urban penalty? Environmental and health risks in Delhi. *Petermanns Geographische Mitteilungen* 147(4), 20–27.

Köberlein, M. (2003) *Living from waste: livelihood of the actors involved in Delhi's informal waste economy*. Verlag für Entwicklungspolitik, Saarbrücken.

Köberlein, M. (2005) Waste and the city: Public response to the problem of municipal solid waste management in Indian metropolitan cities. In: Hust, E., Mann, M. (eds) *Urbanization and Governance in India*. Manohar, New Delhi, pp. 177–199.

Kumar, A. (2005) *Sustainable Transport Environment in Indian Mega-cities: Problems and Remedies*. http://iussp2005.princeton.edu/download.aspx?submissionId=51017 (accessed 22 March 2011).

Kumar, S., Gaikwad, S.A. (2004) Municipal solid waste management in Indian urban centres: An approach for betterment. In: Gupta, K.R. (ed.) *Urban Development Debates in the New Millennium*. Atlantic Publishers & Distributors, New Delhi, pp. 100–111.

Lalwani, S., Dogra, T.D., Bhardwaj, D.N., Sharma, R.K., Murty, O.P., Vij, A. (2004) Study on arsenic level in groundwater in Delhi using hydride generator accessory coupled with atomic absorption spectrophotometer. *Indian Journal of Clinical Biochemistry* 19(2), 135–140.

Maiti, S., Agrawal, P.K. (2005) Environmental degradation in the context of growing urbanization: A focus on the metropolitan cities of India. *Journal of Human Ecology* 17(4), 277–287.

Meehl, G.A., Stocker, T.F., Collins, W.D., *et al.* (eds) (2007) *Climate Change* 2007: *The Physical Science Basis. Contribution of Working Group I to the Fourth Assessment Report of the Intergovernmental Panel on Climate Change*. Cambridge University Press, Cambridge.

Mohanty, W.K., Yanger, W.M. (2008) Seismic hazard in mega-city Kolkata, India. *Natural Hazards* 47(1), 39–54.

M.S. Swaminathan Research Foundation and the World Food Programme (2002) *Food Insecurity Atlas of Urban India*. Chennai.

Nagdeve, D. (2007) *Air Pollution in Mega-cities of India*. International Institute for Population Sciences. mailto: http://www.cosis.net/abstracts/EGU2007/00121/EGU2007-J-00121.pdf (accessed 22 March 2011).

Pundir, B.P. (2001) *Vehicular Air Pollution in India: Recent Control Measures and Related Issues*. Oxford University Press, New Delhi.

Pune Municipal Corporation (2006) *Environmental Status Report 2006*. Pune.

Ramachandraiah, C., Prasad, S. (2004) *Impact of Urban Growth on Water Bodies: The Case of Hyderabad*. Working Paper No. 60. Center for Economic and Social Studies, Hyderabad.

Rao, A.D. (2003) Sea level rise and coastal hazards along the Indian coasts. In: Dash, S.K., Rao, P. (eds) *Climate Change in India and Mitigation Policies*. WWF, New Delhi.

Ruet, J., Gambiez, M., Lacour, E. (2007) Private appropriation of resource: impact of peri-urban farmers selling water to Chennai Metropolitan Water Board. *Cities* 24(2), 110–121.

Sandhya, T. (2004) Impact of Solid Waste Disposal on Water Quality and Health – A Study of Bhalaswa Landfill Site. Unpublished MPhil Dissertation, submitted to the Department of Geography, University of Delhi, Delhi

Santamouris, M. (2001) Heat-island effect. In: Santamouris, M., Asimakopoulos, D.N. (eds): *Energy and Climate Change in the Urban Built Environment*. Earthscan, London, pp. 48–68.

Schiller, A., de Sherbinin, A., Hsieh, W.-H., Pulsipher, A. (2001) *The Vulnerability of Global Cities to Climate Hazards*. Mega-cities, National Science Foundation.

Scholz, F. (2000) Globalisierung versus Fragmentierung. *Nord-Süd Aktuell* 2. Quartal 2000, 255–271.

Singh, R. (2008) Social vulnerability and wastewater related risks: a case study of Delhi. In: Bohle, H.-G., Warner, K. (eds) *Mega-Cities: Resilience and Social Vulnerability*. Source 10/2008, publication series of UNU-EHS, Bonn, pp. 121–131.

Sivaramakrishnan, K.C., Kundu, A., Singh, B.N. (2007) *Handbook of Urbanization in India*. Oxford University Press, New Delhi.

Smith, K.R. (2000) National burden of disease in India from indoor air pollution. *Proceedings of the National Academy of Sciences of the USA* 97 (24), 13286–13293

UN (United Nations Secretariat) (2007) *World Population Prospects: The 2006 Revision and World Urbanization Prospects: The 2007 Revision*. http://esa.un.org/unup (accessed 22 March 2011).

UNCHS (United Nations Centre of Human Settlements) (2008) *Annual Report 2008*. http://www.unhabitat.org/pmss/listItemDetails.aspx?publicationID=2670 (accessed 16 April 2011).

UNDP (2009) *India Urban Poverty Report 2009*. Oxford University Press, New Delhi.

United Nations (2005) *Ecosystems and Human Well-Being – a Synthesis. A Report of the Millennium Ecosystem Assessment*. http://www.maweb.org/en/Synthesis.aspx (accessed 21 December 2009).

Watts, M.J., Bohle, H.-G. (1993) The space of vulnerability: the causal structure of hunger and famine. *Progress in Human Geography* 17(1), 43–67.

CHAPTER 7

From wasteland to wilderness – aspects of a new form of urban nature

Dieter Rink[1] and Harriet Herbst[2]

[1] *Department of Urban and Environmental Sociology, Helmholtz Centre for Environmental Research – UFZ, Leipzig, Germany*
[2] *Freelance translator, Ablass, Germany*

7.1 INTRODUCTION

In the last two decades wilderness has become a popular concept, one could even say a fashionable ecological term and has been promoted by nature conservation, through NGO campaigns and tourism (Diemer *et al.* 2003, Bauer 2005a). The wilderness idea has also now reached the political level of the European Union; thus the EU Committee on the environment, public health and food safety passed a draft resolution about "wilderness in Europe" which was adopted by the European Parliament in February 2009 (see European Parliament 2009). By the end of the 1990s the wilderness idea had also entered into the urbanistic discussion. In discussions about urban ecology, political nature conservation and planning, for instance in the German speaking regions, a new form of urban nature is talked about using terms and keywords such as "new wilderness" (Ganser and Dettmar 1999), "alternative" or "second wilderness" (Dettmar 2002), "secondary wilderness" (Burkhardt *et al.* 2008), "post-industrial wilderness" (Giseke 2007) or "jungle in the city" (Kowarik and Körner 2005). In Great Britain "urban wilderness" is not such a common term, instead the terms "urban wildspace" or "urban wildlife areas" are more frequently used (Herbst 2003, p.4). The number of adjectives used shows that we are dealing here with an new topic that is difficult to describe. The phenomenon itself is, however, not so new, since urban ecological research has been dealing with spontaneous vegetation in urban areas for some time, in particular on wastelands (Wittig and Zucchi 1993, Rebele and Dettmar 1996, Keil 2002). A series of ambitions are connected with the use of the term "urban wilderness": the implementation of new management concepts and nature conservation approaches, the hope for more public awareness and passing off urban wilderness as a clever and inexpensive idea for urban wastelands. In further proposals urban wilderness is even read as an "expression of the changing relationship between people and nature" which is formed through a "new correspondence between the city and nature" (Becker and Giseke 2004, p.22). This interpretation is, however, controversial; the spreading of wilderness is also seen as a warning sign of the perforated city.

Here we look at the meaning of wilderness and its interpretation and use in an urban context. We investigate wastelands as a source of urban wilderness and discuss the processes of wasteland creation with respect to current urban developments. Different facets of urban wilderness are then brought into the discussion – in relation to planning, ecology and lastly the social and educational aspects of wastelands and wilderness. The chapter is summed up with an overview of the potential and functions of urban wilderness for current and future urban development.

Applied Urban Ecology: A Global Framework, First Edition. Edited by Matthias Richter and Ulrike Weiland.

7.2 URBAN WILDERNESS – SOME ATTEMPTS AT DEFINING THE TERM

"Wilderness" is a term much used in various different ways by numerous organizations for different means – nature conservation, education, recreation, etc. When one thinks of "wilderness" unmanaged nature comes to mind – as one might find in untouched forests or national parks. In the academic and (nature conservation) political discussion the definition of the International Union for the Conservation of Nature and Natural Resources (IUCN) has become used as the main wilderness definition (see IUCN 1994, p.17); in principle all of the current definitions are based on this. Also in Europe in the last 10 or 15 years wilderness has been followed as a new model for nature conservation with all the difficulties that are connected with this in such a densely populated and intensively used region of the world. The problems of locations strongly influenced by man, in particular cities, and the problems related to urban wilderness (isolation, contamination, remains of buildings, neophytes, wild animals or "pests") are, however, rarely considered here. Taking into account the IUCN's definition Burkhardt *et al.* suggest differentiating between areas that are formally defined as wilderness by the IUCN and which portray "primary wilderness" and "areas, which due to lack of human influence appear wild or unkempt due to processes of succession" (Burkhardt *et al.* 2008, p.29). In the urban surroundings stages of succession on wasteland areas are often described as wilderness. The term "secondary wilderness" is applied in order to describe vegetation that has arisen spontaneously on anthropogenic locations and which can develop unhindered (Burkhardt *et al.* 2008, p.29).

Diemer *et al.* (2003) define urban wilderness as unused sites in urban and suburban areas, such as on abandoned industrial areas, rail yards, former borderlines, unused lots and recreational parks that are left to nature. They extend from small to medium size sites, but never achieve the extent of "normal" wilderness areas. Only a few of these sites are formally protected and thus they experience numerous uses and in exceptional cases fulfil the criteria of "non-influenced" development (see Diemer *et al.* 2003, p.9). They differentiate into

1 "Urban wilderness" – designated (nature) conservation areas within or near to urban centers, <1000 ha with "low human impact"
2 "Urban rewilding areas" – abandoned urban, industrial and commercial sites (including rail yards, former coal mining areas) with succession but no conservation status and (different) forms of human uses, <500 ha
3 "Rewilding microcosms" – small areas (<several ha) such as public and private gardens, edges of parks, streams, ponds, harbors and so on, they have no conservation status and (different) forms of human uses (Diemer *et al.* 2003).

The distinction between "urban wilderness" and "rewilding areas" seems vague, because the process of rewilding – that is succession – occurs in both categories. However the long-term objectives are different: "Wilderness areas are tracts of land specifically set aside to evolve without human interference, whereas rewilding areas, or fractions thereof, may never attain this state, due to the multiple management objectives" (Diemer *et al.* 2003, p.10).

The term "urban wilderness" is used to counter the negative image of "wasteland nature", which was assigned only a low ecological value in the past and was merely granted an existence on the fringe. In contrast the concept of urban wilderness deals with a certain type of nature, one which is worth protection.

7.3 WASTELANDS AS A SOURCE OF URBAN WILDERNESS

Urban wilderness may have different origins such as encapsulated countryside or natural areas alongside streams or rivers, but certainly one of the major sources of urban wilderness is wasteland. "Wasteland" is a term that encompasses a range of sites from derelict industrial areas, abandoned gardens or even bomb sites to cleared land awaiting redevelopment. There are many different terms used for this sort of land: derelict land, brownfields, vacant land, neglected land, abandoned land (see Herbst 2003). The word "wasteland" used here is based on that of Zucchi and Flisse and describes "sites of different sizes and in different locations that were formerly used in various ways and are now (in the short or long term) no longer, or only extensively, used and are colonised by natural succession" (see Zucchi and Flisse 1993, p.45, Herbst 2003). The question is whether all wasteland sites can be classed as urban wilderness – in theory the answer must be yes (if they are allowed to become wild) since they are (by definition) sites on which nature is left to take its course and no management and little

Figure 7.1 Stave Hill Nature Park – an urban wildlife area created on a wasteland in the docklands of London. Photo: H. Herbst.

intervention occurs. As Nicholson Lord says "in terms of what one might call naturalness – defined as nature being left alone to get on with it – there's not much in the UK today to beat brownfield land" (Nicholson Lord 2005).

Many wastelands were created in European cities as a result of the destruction from the Second World War (Figure 7.1). These sites were often larger than those small industrial or residential wastelands that previously existed and were first re-used after several years or sometimes only after decades. Thus bigger awareness of "wild" vegetation in cities or the occurrence of wildlife in cities stemmed from these bomb sites after the Second World War. These "wastelands" provided the generation of post war children with readily available (if perhaps sometimes dangerous) urban wilderness on their doorstep. Since about the 1960s and 1970s the creation of urban wastelands in Europe has been due mainly to the results of economic structural change. This was either connected to the closure of certain companies or to the relocation of companies or factories from the inner-city to the city outskirts or sometimes abroad. The movement of companies abroad has continued in the last few years due to the increasing international division of work.

The extent of wastelands is very different in the individual countries and regions of Europe. The range of the known area in each EU country extends from 11 000 ha of wasteland in the Netherlands to 900 000 ha in Romania.[1] The proportion of the total country area of each country varies from 0.01% in Sweden to 3.8% in Romania (Oliver *et al.* 2005, p.279). The existing data also show that, just in a quantitative sense, wastelands pose a considerable challenge. Oliver *et al.* (2005, p.277) estimate for instance that about 100 000 wastelands exist in the United Kingdom, 200 000 in France and 362 000 in Germany. The

[1] However it should be noted that there are very different definitions and classification criteria in individual countries, in addition the statistical recording is insufficient. Thus it is not possible to guarantee an international comparison.

main proportion of wastelands in Europe was formerly used industrially, although this varies regionally. Wastelands have also arisen due to the process of shrinkage – caused by the population decrease through demographic change and emigration. Shrinkage changes the mechanism of spatial use, the pressure of use reduces and wastelands occur successively. Projections within the European–American comparative project "Shrinking Cities" have demonstrated that the "shrinking city" will be a patchwork city of heterogeneous fragments of shrunken and island like city districts, wastelands as well as connections with no landscape structure (Oswalt and Rieniets 2006). According to the population projections this process will affect many European regions in the coming decades; in the regions already affected it will probably increase its dynamic. Wastelands will thus lose their character as a temporary occurrence, they will spread out and wastelands with spontaneous vegetation will accompany the process of shrinkage. This will lead to an increase in wilderness areas and thus a new challenge for urban planners.

7.4 URBAN WILDERNESS IN PLANNING

The planning-related discussions about urban wilderness have their origin and focal points in urban wasteland: the increase in wasteland in the city stimulates the discussion about wilderness (Becker and Giseke 2004, p.22, see Rink 2009). A look at the use of the term "wasteland" shows that for decades wastelands were seen as "eyesores" by the authorities as well as politicians, where the tipping of rubbish or rubble was often tolerated or sometimes even encouraged (compare Rebele and Dettmar 1996, p.125). It is thus no wonder that urban wastelands were perceived negatively as they symbolize decline and loss of value and stand for neglect and "uncontrolled growth" in an otherwise well-ordered city. The history of the term in urban and landscape planning makes it clear that urban wasteland remained a risk category for a long time or was seen as a problem area, a scathe on the urban landscape, as a wasted urban building resource or as a dangerous contaminated area (Keil 2002, p.28).

As already mentioned, a large number of terms exist for wasteland: whilst in the European definitions usually a former use of the area is central to the definition, in the United States the "real or perceived environmental contamination" is the most important criterion impeding the re-use or redevelopment of the site (De Sousa 2003, p.182, Herbst 2003).[2] Here decontamination for the goal of revaluation or for the re-vegetation of the wasteland is central (De Sousa 2008, p.1). It is notable that the planning field completely ignored wildlife on wastelands for a long time. Although the ecological and social value of wildlife on wastelands was proven in the 1970s, even in the 1980s in Germany for instance, every sort of legal status was refused for even well established vegetation on wastelands. Wildlife on wastelands could be sacrificed for new development or commercial uses at any time, the ecological potential or recreational requirements of the population were also completely ignored. Since the 1990s a change of direction has become apparent in the planning field, thus the ecological potential of wasteland (Figure 7.2) is now taken into account and the population is included in planning via stakeholder involvements. The revitalization of wastelands has experienced an enhancement under the model of sustainable urban development (see De Sousa 2008).

In Germany the International Building Exhibition (IBA) Emscher Park in the Ruhr area promoted wasteland to a catch word. The IBA has become a prime example and symbol for the restructuring and greening of former industrial and commercial areas; there are also large wilderness areas found here (see Ganser 1999). In the meantime urban wilderness has been selected as a new approach, in particular in urban restructuring in East Germany where various strategies

Figure 7.2 Industrial wasteland in Leipzig, Plagwitz 10 years after closure. Photo: D. Rink.

[2] An overview of European and American definitions are also offered by Oliver *et al.* 2005, who emphasized the urban character of wastelands.

exist. There is a radical perspective on the process of shrinkage and urban restructuring in East Germany, which could be described as a "wilderness venture." It is based on the assumption that the shrinkage process cannot be steered and controlled completely, but rather that it is characterized by uncontrolled perforations. The ideal pathway for the post-industrial landscape is seen through the "moving in of wilderness", "since it opens up space for particular types of experiments and experiences, which do not burden the public budgets and serve as a contribution to the regeneration of the ecological balance of the area as compensation areas" (Hartz 2007). It is accepted that the relationship between city and nature is basically changed in the process of shrinkage, whereby urban wilderness provokes a new way of viewing urban nature. Taking a more moderate point of view urban wilderness stands for a new design approach that includes spontaneously occurring vegetation in planning or design as it is or as it develops (Giseke 2007). The key word for this is the "use of vegetation development under its own dynamic", which is also directly related to the "minimisation of costs for the site and for management" (Schwarz 2004, p.741). This position is characterized by casting the topic of "wilderness" in a positive light. The term "wilderness" is thus supplemented by adjectives such as "controlled", "steered" or "cultivated", which are supposed to remove its provocative or threatening connotations (see Rink 2009, p.280).

There is evidence that the phenomenon of wilderness is becoming more accepted in the urban planning scene, although perhaps through the use of a different name. In the United Kingdom the urban wild space movement incorporates these sort of wilderness areas, in Germany there are initiatives that try to establish a new class of green space (*Naturerfahrungsräume*) in urban areas. These are small initiatives but perhaps the beginning of a bigger movement to incorporate more wilderness into urban planning.

For the moment, however, it should be maintained that wilderness is conceived differently in the planning discussion and seen ambivalently. Only part of the planning field is undertaking the change in perspective connected to the positive conception of urban wilderness; this ranges from tolerating the existence of spontaneous nature on wastelands to the inclusion of wasteland as an open space category, as in some British cities (e.g., London, Birmingham). The mainstream of the planning field, at least their institutional parts, still retain the model of designed and managed urban green space. At the moment it is not yet foreseeable where and to which extent wilderness will play a role in urban restructuring and which functions this new form of urban nature will take.

7.5 ON THE ECOLOGY OF URBAN WILDERNESS

For a long time wastelands were seen as being useless or a dumping site from an ecological point of view, the vegetation on them was visualized as "improper nature" or "second class nature", everything that grew spontaneously in the city was – where possible – exchanged for garden greenery or sprayed with weed killer (Hard 2001, p.259). The vegetation on wastelands did, however, provoke some interest from urban ecologists as "ruderal or spontaneous vegetation" as early as the 1920s (Lachmund 1999). However it was first through the environmental debates of the 1970s and 1980s and the establishment of urban ecology as a research field that work on this type of urban nature was carried out independently. Urban nature has been systematically recorded through habitat mapping since the 1980s in the United Kingdom or (West) Germany (see Lachmund 1999), although industrial areas and wastelands were often not included in the surveys since they were seen as "negative habitats" (Keil 2002, p.20). The habitat mapping showed that cities – in comparison with other types of landscapes – usually reveal a high biological diversity (Figure 7.3; Sukopp and Wittig 1993, Del Tredici 2010).

Figure 7.3 Industrial ruin and wild urban nature as part of the designed landscape park of Duisburg Nord. Photo: D. Rink.

Looking specifically at wasteland sites their high species diversity contributes to the high ecological value of many of these sites in comparison with similar sized green spaces in urban areas. This high diversity is thought to be due to the types of substrates and soil conditions, different levels of disturbance, varying micro-climate on sites, varied intensities of use, variety of structures on sites (such as walls, old buildings, paved areas rubble and tracks), site histories and sometimes the extreme conditions found on the sites (Vedder and Drecker 1994, Reidl 1998, Chipchase 1999). It also depends on site size and the type of site: the sites with the most diverse flora have been found to be industrial wastelands, followed by commercial wastelands (or light industry), railway wastelands and then derelict green spaces (all of these being over 5 ha) with wasteland sites under 5 ha all having a lower floral diversity (Reidl 1998). The diversity of sites also alters with the age and the different successional stages of sites. The diversity of the site peaks at about six to nine years and then tends to decrease with increasing maturity of the vegetation (Rebele 1988).

Two interesting features of the vegetation of wasteland sites are the high proportion of neophytes and the relatively high number of Red Data Book species. In industrial areas neophytes often occur through the transport of materials from abroad – for instance the transport of seeds with raw materials such as cotton or the distribution of maritime plants along railway lines (see Gilbert 1992, Rebele and Dettmar 1996, Reidl 1998). Many of the Red Data Book species found are those whose natural habitat has been destroyed or disturbed and thus the species colonize wastelands as a secondary habitat. Although the presence of rare species on wastelands is exciting, they actually play a minor role in the whole vegetation picture of wasteland sites; more common (and possibly, at least culturally, more important) are the ruderal flora, grasslands and generalist species – such as cow parsley, herb robert, buttercup, etc. (see Reidl 1998, Chipchase 1999, Evans 2001). Their value is reflected in this quotation: "The rude but beautiful weeds that colonize the forgotten or unofficial places provide a commonwealth of their own" (Evans 2001, p.20). A further important factor relating to the flora of urban wastelands is the adaptation and suitability of the flora to urban conditions. Wasteland (or spontaneous) flora is also often adaptable to the use of the site, for instance for children's play, as work by Hard and Pirner (1988) has shown. For instance in areas of high use low, spreading plants are found, whereas in the less highly used areas herbaceous flora and bushes are able to develop. This provides valuable information for the possible use of such species in landscaping urban green spaces and reducing the maintenance required for the vegetation.

Wasteland sites have also been found to provide habitats for a wide range of animal species. There are various reasons for this including the provision of rich nutritional sources of food such as seeds, nectar and biomass (Klausnitzer and Klausnitzer 1993). This is especially important in winter when wasteland plants, such as dock, thistle, mullein, etc., provide a vital food source for seed-eating birds. Some animal species spend much of their time in other urban habitats but still depend on wasteland areas to provide them with sufficient food (Zucchi and Flisse 1993). Another factor favorable to many animal species is the low intraspecific competition and low number of predators on wasteland sites due to the difficulty of migration to such sites from the surrounding countryside (see Abs 1992, Vedder and Drecker 1994). The lack of disturbance on many such sites may also favor particular mammal and bird species (Vedder and Drecker 1994). However, other species may benefit from local disturbances – for instance, mountain biking ensures that a loose, friable substrate is maintained for ground nesting aculeate Hymenoptera and thermophilic invertebrates (Harvey 2000).

Despite this evidence the question of biodiversity of wastelands is still discussed controversially, thus Heinz Ellenberg talks of the "pseudo-diversity of urban nature" and refers here to the high proportion of neophytes (cited by Körner 2005, p.59). In contrast many authors see overgrown or "greened" wastelands as a trial medium with which urban biodiversity can be increased (De Sousa 2003, p.194, see Haase 2008, p.6). It appears from this that urban wastelands provide a diverse range of habitats within the urban area which in turn offer refuges for plants and animals as well as providing interesting sites for children's play. It is their temporal, spatial and biological diversity that makes them such an interesting and vital part of the network of urban green spaces.

7.6 URBAN WILDERNESS IN A SOCIAL CONTEXT

In contrast to the numerous floral and faunal surveys about urban wilderness (see above) social scientific studies, which also deal with the use and the acceptance of wild – and successional processes in urban areas have been rare up to now (Bauer 2005b, p.50).

These investigations use various methods but usually involve relatively small studies with not many interviewees so that the studies are only partially representative. Three types of studies can be differentiated which touch on the topic of urban wilderness: these are first ones that deal with designated wilderness areas, their acceptance and the natural experiences possible in them. This is mainly the US-American research direction, which deals in particular with wild animals (see Cordell *et al.* 1998). In Europe Nicole Bauer has written a first basic study of the perception of wilderness (Bauer 2005a). Urban nature, in particular nature on wastelands (urban wilderness), is only briefly touched on in this context. Essentially wilderness is associated with a far-off, expansive and exotic nature; in the German-speaking countries widespread images of wilderness are for instance of jungles, the desert or of high mountains (see Bauer 2005a,b).

Second, there are empirical studies, which deal with urban nature – usually with urban green space (Breuste and Breuste 2001, Rink 2005). Urban wilderness is also only a peripheral topic here, the research deals mainly with the assessment and attractiveness as well as the use of parks, green spaces and urban woodland. These investigations have shown how highly designed and managed nature is valued, urban open spaces left to nature are, in comparison, used less and are not so well accepted (Breuste and Breuste 2001, Rink 2005). To distinguish from "real" wilderness urban nature is evaluated as daily nature. It is a type of nature that one perceives and uses every day – in daily contexts – for example on a route, on a sports field, whilst going for a walk, taking the dog out, etc. It is a type of nature on which one does not have high or excessive demands; it is thus entitled "surrogate nature" (Rink 2005). Spontaneous vegetation, wild greenery or nature on wastelands is not necessarily seen as urban nature – because it is not designed and managed, does not serve any (specific) goal and because it often cannot be used. It is perceived to be unattractive and esthetically displeasing (compare Breuste and Breuste 2001, Rink 2005). It is often associated with dirt and rubbish as well as uncertainty and danger, fear is related to it and the danger of injury, particularly for children, is referred to. This type of nature is only granted a value for children – as a location for adventurous play. For the majority of people this type of urban nature is not seen as wilderness, but there are some groups, especially in the middle class, which have incorporated this type of wildlife into their romantic view on nature. However, in comparison with other types of urban nature (parks, urban woodland, etc.) nature on wastelands is evaluated as being "natural" although this does not necessarily follow on to a particular worthiness for protection, the ecological value of urban habitats often not being known.

Third there are studies that explicitly look at urban or industrial wastelands, whereby the main issues are their use and suitability for recreation as well as their function for experiencing nature in cities (Konopka and Wüstendörfer 1995, Harrison 2001, Keil 2002). These studies show a different result: the importance of the recreational and experience value of wastelands and urban wilderness, especially for children and teenagers, but also for adults. Konopka and Wüstendörfer's survey adduced a negative esthetic assessment but a general acceptance of urban wastelands (Konopka and Wüstendörfer 1995, p.767). Wasteland vegetation has obviously long been categorized as belonging to the urban weeds, however it can be determined here that the "weed tolerance" of urban residents has increased (Keil 2002, p.16). In a small survey in the United Kingdom Harrison found out that nearly all respondents are aware of wasteland areas in cities and many of the criteria which define wilderness areas can be applied to urban areas: "Despite some obvious management in urban areas and the fact that they are by no means "untouched by humans", feelings of wilderness and isolation could be identified within these areas" (Harrison 2001, p.8). The most popular definition for wilderness was "an unspoilt area, untouched by man", whereas two third of the respondents agreed that parks could be viewed as wild (Harrsion 2001).

However the studies about the use, perception and acceptance of nature on wastelands or urban wilderness have brought about no clear findings up to now. The question as to whether and how succession will be accepted on urban wastelands cannot be clearly answered through the evaluation of the existing studies. It is, however, clear that it deals with a type of urban nature, which requires further differentiation. The current stage of growth obviously has a decisive influence on the perception of the site: young wastelands can apparently only be reconciled with the usual ideas of nature with difficulty, whilst older wastelands are more similar to these ideas – although perhaps not similar to what one understands by the term "wilderness".

Furthermore the location and the urban context, the size, the usability and last but not least the esthetic picture are all criteria that should be used to assess or evaluate spontaneous vegetation in cities.

Finally the studies make it clear that there are different views of urban wilderness depending on the social group. The idea of wilderness seems here to be part of the romantic view of nature of the middle class, whilst the lower class possibly has a more down to earth view of wild urban nature. This is especially since this group typically lives in a residential area that is characterized by shrinkage and wastelands: thus spontaneous urban nature occurs in a context of decline, dilapidation and neglect. This requires more differentiated studies then have been the case up to now.

7.7 EDUCATIONAL VALUE OF URBAN WILDERNESS

The educational value of wilderness cannot be overstated. In today's cities where free time is organized to such an extent that different age groups and activities are separated from each other (for instance provision of different playgrounds for each age group) wastelands or wilderness areas provide an opportunity for many sorts of people to socialize and play together. These areas offer possibilities for activities by different age groups and interests – for instance mountain biking, building huts, hide and seek, climbing, digging, etc. There are no rules or signs stating what one is or isn't allowed to do: here it is possible to meet and play with people without having to be a member of a formal institution or having to follow distinct rules (Tschäppeler *et al.* 2007).

With respect to the educational value of wasteland sites there are two main points to be made:

- the educational value from the point of view of developing a relationship to nature
- the educational value for the child of playing in and with nature.

With regard to the first point it is generally accepted that people will only try to protect that which they know and understand. With all the current environmental problems in the world it seems more important than ever that people regard nature and the natural world as something worth conserving and caring for. Areas in which a child can experience nature (for example wilderness areas) thus form the fundament for understanding and recognizing nature for urban children. Breuste (1994) also stresses the importance of experiencing nature for the later recognition of the importance of nature conservation (see also Preuss 1998). Degünther (2006) warns, however, against a dangerous moralization, which is already noticeable in children and teenagers. Many young people see nature as being in need of help and protection which then leads them to avoid any disturbance of natural sites and even to stay away from them for fear of causing damage; thus instead of them becoming closer to nature they end up shying away from it.

Of course some areas need to be protected so that their wildlife can survive, but it is also vital to have natural areas that can be played in, used and enjoyed by people so that they can gain an understanding of nature for themselves or which they can then pass on to others (Figure 7.4). It is then this emotional experience through the provision of natural play areas which provides the basis for the future respect for nature, or similarly, the lack of these which leads to a less caring attitude to nature (Hüther 2008).

Taking into consideration the second point the educational value of playing in natural areas for children is substantiated by various areas of research as well as by studies and observations of children playing in urban wildlife areas or wilderness areas. A study by Blinkert *et al.* (2008) revealed that children tend to play more in groups and play more complex games on nature experience areas (*Naturerfahrungsräume*) – some of which can be found on mature wastelands – than on normal playgrounds. Such areas also help in a child's development

Figure 7.4 Well-established woodland (designated nature conservation area) on a former industrial site in Leipzig, Plaussig. Photo: D. Rink.

as "creativity and self-responsibility are strengthened in children who enjoy experiences through playing in a varied, semi-natural and changeable environment" (Hüther 2008, p.15).

Wastelands also provide a way out of the restrictive corset of the ever more compact city and urban lifestyle. Children need to be able to play on their own, away from the watching eye of parents or neighbors. They need to be able to test their own boundaries, evaluate risks, build up their self-confidence – all this is difficult in a structured, supervised lifestyle (Tschäppeler *et al.* 2007, p.26). This is substantiated by findings on the importance of experiencing nature for psychological and social development of children (see Gebhard 2008).

Although it is important that children learn about nature through their own experiences, there are also arguments for providing educational experiences for children to improve their understanding of the world they live in. The potential for urban wastelands to provide nature areas near schools is recognized by many authors (see Gilbert 1992). Urban wastelands are often used as sites for environmental education, as well as areas where children and adults can experience and be close to nature. There are many examples of nature areas that have been created on former wasteland sites. Usually the previous vegetation is incorporated as far as possible into the site, but depending on the resources available, other wildlife features may also be developed. In Camley Street Nature Park in London a nature area has been established on an old wasteland with an educational center – providing urban residents with the opportunity to learn about the natural world on their doorstep. An interesting example of the use of wasteland is given by Mark Hansen, a teacher in Portland, Oregon who made use of a nearby wasteland to introduce his pupils to the wildlife on their doorstep. Despite initial worries about the smell of the place and getting dirty, he found that after a while the children began to understand the relationships within their environments and to think in terms of whole systems. After a year he found that some, although not all, children had begun to form a different relationship with nature (Hansen 2005), demonstrated by the comment of one of the pupils: "I learned that nature is not just a thing. It is something that you discover" (Hansen 2005).

Thus it seems that urban wastelands play an important role in the fabric of the urban landscape. They can help to counter the restrictive and planned recreational time of today's children as well as providing urban residents with an opportunity to get to know nature on their doorstep. It should however be noted here that dangers do exist on wasteland sites, such as unsafe buildings, contaminated rubbish, glass, etc. If they are to be promoted as wild spaces, or *Naturerfahrungsräume* (nature experience areas, as in Germany) then some degree of management is required. This also helps to make the sites more acceptable to local people. Simple interventions such as mowing a grass verge on the side of the site and removing unsafe structures increase the safety and acceptance of a site without disturbing its wild character.

7.8 CONCLUSIONS

The central theme discussed here has been the existence of wilderness on wasteland sites, its potential and importance in urban areas. In an urban context the main issue with respect to the use of the term "wilderness" lies in its communicative function with respect to the public: it can and should be used to bring attention to wildlife found on wasteland sites. Urban inhabitants need to be helped to perceive this type of wild space differently so that sites can be accepted and valued by local people. For this site development, accessibility and planning designation of the site is of decisive importance: the local population wants to know whether the area is safe and whether it can be used. This type of wilderness experiment in the city – even if it is only small – can help to change the current esthetic perception of urban nature. In particular, wildlife projects for children, which are now found in many European countries, can provide the basis for the later recognition and acceptance of this type of nature. No less important for the future of urban wilderness will also be whether the planning authorities relinquish their demands on design, management and control of urban greenery and open up to new approaches. In particular shrinking cities offer starting points; many of these having already developed a series of interesting projects in which wildlife on wastelands has been incorporated into urban development.

With respect to the city–nature relationship it is also debatable whether urban wilderness is suitable as a utopia for the city of tomorrow, or whether it portrays a warning sign of the perforated city. Urban development is still much too influenced by the paradigm of the managed and designed city green to be inclined towards the influences of wilderness. Urban wildernesses tend to stand out from the "managed" urban green and, up to now, have had a provocative effect; however this is deceptive as their number is limited and they actually make up only a small proportion of the whole of the urban green space. Where does their potential and function then lie? Urban wildernesses are experimental areas, they open up new design possibilities for urban and landscape planning and extend the spectrum of urban open spaces. These wildernesses build modern spaces for perception, which provoke and can change the view of the urban resident. They offer approaches for a new esthetic sense of urban nature through their lack of design and unmanaged appearance, often interspersed with the remains of urban and industrial influences. Urban wildernesses provide experience areas in particular for children, who enjoy the freedom and uncontrolled use of such sites. Adults also sometimes enjoy these informal green spaces and may develop new possibilities of involvement through urban wildernesses. Finally, urban wildernesses also provide another type of protected area, they can serve to test out new management approaches in nature conservation. Up until now these functions and potentials have not nearly been utilized in full, thus urban wilderness still offers exciting perspectives for the future.

REFERENCES

Abs, M. (1992) Die Bedeutung von Industrieflächen aus tierökologischer Sicht. *LÖLF- Mitteilungen* 17(2), 27–31.

Bauer, N. (2005a) *Für und wider Wildnis. Soziale Dimensionen einer aktuellen gesellschaftlichen Debatte*. Haupt, Bern.

Bauer, N. (2005b) Attitudes towards wilderness and public demands on wilderness areas. In: Kowarik, I., Körner, S. (eds) *Wild Urban Woodlands*. Springer, Berlin, pp. 47–66.

Becker, C.W., Giseke, U. (2004) Wildnis als Baustein künftiger Stadtentwicklung ? *Garten und Landschaft* 2, 22–23.

Blinkert, B. (1998) Aktionsräume von Kindern. In: Schemel, H.-J. (ed.) *Naturerfahrungsräume*. Angewandte Landschaftsökologie Heft 19. Bundesamt für Naturschutz. Bonn-Bad Godesburg, pp. 103–114.

Breuste, J. (1994) Urbanisierung des Naturschutzgedankens. Diskussion von gegenwärtigen Problemen des Stadtnaturschutzes. *Naturschutz und Landschaftsplanung* 26(6), 214–220.

Breuste, J., Breuste, I. (2001) Stadtnaturschutz – theoretische Positionen und empirische Befunde zur Nutzung und Akzeptanz von Pflegegrün und Wildnatur in der Stadt Halle/Saale. *Geobotanische Kolloquien* 16, 25–36.

Burkhardt, I., Dietrich, R., Hoffmann, H., *et al.* (2008) *Urbane Wälder*. Abschlussbericht zur Voruntersuchung für das Erprobungs- und Entwicklungsvorhaben, "Ökologische Stadterneuerung durch Anlage urbaner Waldflächen auf innerstädtischen Flächen im Nutzungswandel – ein Beitrag zur Stadtentwicklung". Naturschutz und Biologische Vielfalt, Band 63. Bundesamt für Naturschutz, Bonn.

Chipchase, A. (1999) Ecological characteristics of the flora of urban derelict sites in London. *The London Naturalist* 78, 19–33.

Cordell, K., Tarrant, M.A., McDonald, B., Bergstrom, J. (1998) How the public views wilderness. *International Journal of Wilderness* 3(4), 28–31.

De Sousa, C. (2003) Turning brownfields into green space in the city of Toronto. *Landscape and Urban Planning* 62(4), 181–198.

De Sousa, C. (2008) *Brownfields Redevelopment and the Quest for Sustainability*. Elsevier, Amsterdam.

Degünther, H. (2006) Wieviel Natur braucht der Mensch? Die Bedeutung von Naturerleben für Kinder. *SpielLandschaft-Bremen* 4, 32–36.

Del Tredici, P. (2010) Spontaneous urban vegetation: reflections of change in a globalized world. *Nature and Culture* 5(3), 299–315.

Dettmar, J. (2002) Alternative Wildnis. *Garten und Landschaft* 5, 15–17.

Diemer, M., Held, M., Hofmeister, S. (2003) Urban wilderness in Central Europe. Rewilding at the urban fringe. *International Journal of Wilderness* 9(3), 7–11.

European Parliament (2009) *European Parliament resolution of 3 February 2009 on Wilderness in Europe (2008/2210(INI)*. http://www.europarl.europa.eu/sides/getDoc.do?pubRef=-//EP//TEXT+TA+P6-TA-2009-0034+0+DOC+XML+V0//EN (accessed 11 April, 2011).

Evans, P. (2001) Nature's urban outposts are worth conserving too. *Guardian Weekly* June 14–20, 20.

Ganser, K. (1999) *Liebe auf den zweiten Blick. Internationale Bauausstellung Emscher Park*. Harenberg, Dortmund.

Ganser, K., Dettmar, J. (ed.) (1999) *IndustrieNatur – Ökologie und Gartenkunst im Emscher Park*, Stuttgart.

Gebhard, U. (2008) Die Bedeutung von Naturerfahrungen in der Kindheit. In: Schemel, H.-J., Wilke, T. (eds) *Kinder und Natur in der Stadt*. BfN-Skripten Band 230. Bundesamt für Naturschutz, Bonn, pp. 27–44.

Gilbert, O. (1992) *The Flowering of the Cities. The Natural Flora of "Urban Commons"*. English Nature, Peterborough.

Giseke, U. (2007) Urban redevelopment – enlivening open-space culture? *German Journal of Urban Studies* 46(1), 10–23.

Haase, D. (2008) Urban ecology of shrinking cities: An unrecognized opportunity ? *Nature and Culture* 3(1), 1–8.

Hansen, M. (2005) Exploring our urban wilderness. *Rethinking Schools* (19)3. http://www.rethinkingschools.org (accessed 13 April 2011).

Hard, G. (2001) Natur in der Stadt? *Berichte zur deutschen Landeskunde* 75(2/3) 257–270.

Hard, G., Pirner, J. (1988) Die Lesbarkeit eines Freiraumes. *Garten und Landschaft* 98(1), 24–30.

Harrison, C. (2001) *Wilderness in Cities*. http://www.geog.leeds.ac.uk/courses/level3/geog3320/studentwork/groupc/wildernessword.doc (accessed 22 March 2011).

Hartz, A. (2007) *Stadtlandschaften als Entwicklungspotenzial*. Hauptthesen des Tages, Herbsttagung der DGGL. http://www.dggl.org/pdf/DGGL_T_hesen_A_Hartz.PDF (accessed 22 March 2011).

Harvey, P. (2000) The East Thames Corridor: a nationally important invertebrate fauna under threat. *British Wildlife* December 2000, 91–98.

Herbst, H. (2003) *The Importance of Wastelands as Urban Wildlife Areas – with particular Reference to the Cities Leipzig and Birmingham*. UFZ Bericht Nr. 2/2003. Stadtökologische Forschungen Nr. 35. Umweltforschungszentrum Leipzig-Halle.

Hüther, G. (2008) Das Erleben von Natur aus Sicht moderner Hirnforschung. In: Schemel, H.-J., Wilke, T. (eds) *Kinder und Natur in der Stadt*. BfN-Skripten Band 230. Bundesamt für Naturschutz, Bonn, pp. 15–26.

IUCN (International Union for Conservation of Nature and Natural Resources) (1994) *Guidelines for Protected Area Management Categories*. Gland, Cambridge.

Keil, A. (2002) *Industriebrachen – Innerstädtische Freiräume für die Bevölkerung. Mikrogeografische Studien zur Ermittlung der Nutzung und Wahrnehmung der neuen Industrienatur in der Emscherregion*. Dortmunder Vertrieb für Bau- und Planungsliteratur, Dortmund.

Klausnitzer, B., Klausnitzer, U. (1993) Städtische Brachflächen – potentielle Naturschutzgebiete für Insekten? Ein Literaturüberblick. In: Wittig, R., Zucchi, H. (eds) *Städtische Brachflächen und ihre Bedeutung aus der Sicht von Ökologie, Umwelterziehung und Planung*. Geobotanische Kolloquien 9, Frankfurt. Natur & Wissenschaft, Solingen, pp. 31–44.

Konopka, T., Wüstendörfer, W. (1995) Zur Wertschätzung städtischer Brachen durch die Stadtbevölkerung. *Stadt und Grün* 11, 763–771.

Kowarik, I., Körner, S. (eds) (2005) *Wild Urban Woodlands*. Springer, Berlin.

Körner, S. (2005) *Natur in der urbanisierten Landschaft. Ökologie, Schutz und Gestaltung*. Müller und Busmann, Wuppertal.

Lachmund, J. (1999) Mapping urban nature. Bio-ecological surveys and urban planning in Germany, 1975–1998. In: Gigerenzer, G., Kurtz, E. (eds) *The Expert in Modern Society*. Plenum, Amsterdam, pp. 231–248.

Nicholson Lord, D. (2005) *Challenging Orthodoxy*. UWN Chair. Speech to UWN Greener Cities Conference, Manchester, 24 February 2005. http://www.urbanwildlife.org.uk/green_cities (accessed 22 March 2011).

Oliver, L., Ferber, U., Grimski, D., Millar, K., Nathanail, P. (2005) The scale and nature of European brownfields. In: CABERNET (Concerted Action on Brownfield and Economic Regeneration Network) (ed.) Proceedings of CABERNET 2005 – The International Conference on Managing Urban Land. SAXONIA Standortentwicklungs- und verwaltungsgesellschaft mbH and Land Quality Press, Nottingham, pp. 274–281.

Oswalt, P., Rieniets, T. (2006) *Atlas of Shrinking Cities. Atlas der schrumpfenden Städte*. Hatje Cantz, Ostfildern.

Preuss, S. (1998) Psychologische Zugänge zu Natur und Landschaft. In: Schemel, H.-J. (ed.) *Naturerfahrungsräume*. Angewandte Landschaftsökologie Heft 19. Bundesamt für Naturschutz. Bonn-Bad Godesburg, pp. 125–130.

Rebele, F. (1988) Ergebnisse floristischer Untersuchungen in den Industriegebieten von Berlin (West). *Landschaft und Stadt* 20(2), 49–66.

Rebele, F., Dettmar, J. (1996) *Industriebrachen. Ökologie und Management*. Haupt, Stuttgart.

Reidl, K. (1998) *Ökologische Bedeutung von Brachflächen im Ruhrgebiet*. Tagungsband 18. Jahrestagung, "Biotopkartierung im besiedelten Bereich". Recklinghausen, pp. 9–21.

Rink, D. (2005) Surrogate nature or wilderness? Social perceptions and notions of nature in an urban context. In: Kowarik, I., Körner, S. (eds) *Wild Urban Woodlands*. Springer, Berlin, pp. 67–80.

Schwarz, C. (2004) Extensive Freiraumgestaltung als Strategie im Stadtumbau. *Informationen zur Raumentwicklung* 11/12, 735–747.

Sukopp, H., Wittig, R. (1993) *Stadtökologie*. G. Fischer, Stuttgart.

Tschäppeler, S., Gresch, S., Beutler, M. (2007). *Brachland. Urbane Freiräume neu entdecken*. Haupt, Stuttgart.

Vedder, A., Drecker, P. (1994) Biotop- und Artenschutz auf Industriebrachen des Ruhrgebietes. *BrachFlächenRecycling* 2, 22–28.

Wittig, R., Zucchi, H. (eds) (1993) *Städtische Brachflächen und ihre Bedeutung aus der Sicht von Ökologie, Umwelterziehung und Planung*. Geobotanische Kolloquien 9, Frankfurt. Natur & Wissenschaft, Solingen.

Zucchi, H., Flisse, J. (1993) Städtische Brachen aus tierökologischer Sicht unter besonderer Berücksichtigung der Wirbeltiere (Vertebrata). In: Wittig, R., Zucchi, H. (eds) *Städtische Brachflächen und ihre Bedeutung aus der Sicht von Ökologie, Umwelterziehung und Planung*. Geobotanische Kolloquien 9, Frankfurt. Natur & Wissenschaft, Solingen, pp. 45–57.

C. FLOODING AND CLIMATE ADAPTATION

CHAPTER 8

Multiscale flood risk assessment in urban areas – a geoinformatics approach

Norman Kerle and Dinand Alkema

Faculty of Geo-Information Science and Earth Observation (ITC), University of Twente, Department of Earth Systems Analysis, Enschede, The Netherlands

8.1 INTRODUCTION

Strong similarities exist between global population and urbanization growth rates over the last century (e.g., Grimm *et al.* 2008) and the increase in the number of natural disasters (e.g., Guha-Sapir *et al.* 2004, Figure 8.1). While not correlating linearly all have been showing a strong upward trend, though with disaster numbers showing a stronger increase since about the 1960s. The coinciding growth rates suggest a possible relationship, although they must be viewed with caution. Both demographic and disaster statistics suffer increasing inaccuracies the further we look back. Population numbers, especially for pre-census times (i.e., well into the 1960s for many parts of the world), are often coarse estimates and for many countries considered uncertain (Goldewijk 2005), also affecting urbanization percentages, which have increased to more than 50% globally (Grimm *et al.* 2008). Similarly, part of the rapid increase in disasters since the middle of the 20th century can be attributed to better reporting and growing interest in previously neglected regions of the world.

The exact relationship between disaster numbers and demographic developments is thus uncertain, that is to what extent they might more reflect increasing urban concentration or population growth in general. One reason for this challenge is the concept of disasters, typically defined as events that cause injury and death or economic losses that exceed the coping capacity of the affected people or region, creating a need for outside assistance. According to the Center for Research on the Epidemiology of Disasters (CRED, Belgium; http://www.cred.be/), an event needs to satisfy at least one of the following criteria to be considered a disaster: 10 or more people reported killed, at last 100 people affected, a state of emergency declared, or a call for international assistance issued. While for CRED economic damage is not considered a defining disaster characteristic, in addition to the number of people affected or killed by an event it is the typical third way to rank a disaster. It is also where urbanization becomes important. While drought and floods have affected and killed most people in the past, leading to millions of fatalities or casualties over large areas or entire countries, the costliest events were focused on or included large urban areas, examples being the 1995 Kobe earthquake (losses of USD 131 billion), or the 1994 Los Angeles earthquake (USD 32 billion). This effect is explained by risk theory that sees disasters as a result of vulnerable elements at risk (EaR) spatially intersecting with a hazard. Cities contain the largest

Applied Urban Ecology: A Global Framework, First Edition. Edited by Matthias Richter and Ulrike Weiland.

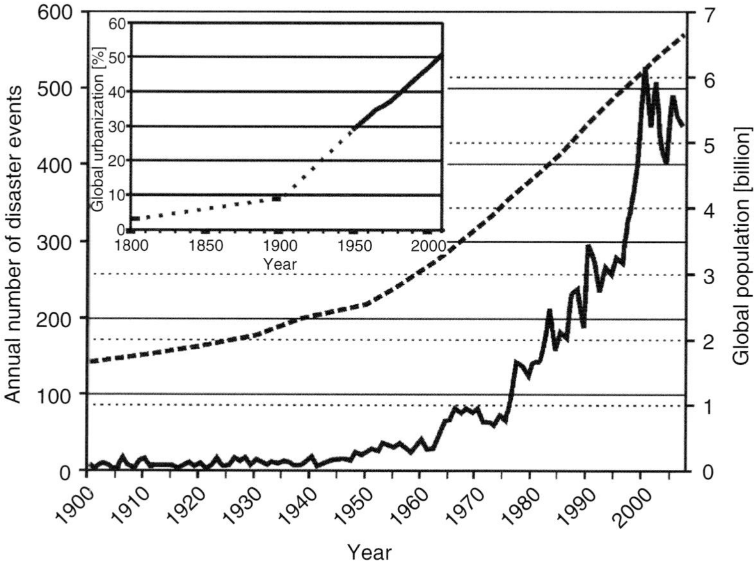

Figure 8.1 Number of annual natural disasters between 1900 and 2007 in bold (primary axis; CRED 2009), and global population for the same period in hatched line (secondary axis; Goldewijk 2005 and US Census Bureau). Inset shows estimated global urbanization rates between 1800 and today (Clark *et al.* 1998, United Nations Population Division (UNPD) 2006). Note that values before 1950 are increasingly uncertain. The Population Reference Bureau gives an urbanization value of 3% for 1800. The value of 9.2% for 1900 is based on the estimation of about 150 million out of approximately 1.6 billion people living in cities (Brown 2006), although the PRB gives a value of 14%.

number of EaR per area unit, be it humans or elements of material value. What then defines the disaster potential is the hazard severity and frequency, and the level of vulnerability of the EaR towards a given hazard type and magnitude. We can thus conclude that population growth in general is related to the rising number of disaster, as it results in a larger number of EaR and assets that can suffer loss, while urbanization has strongly contributed to the rapidly rising overall disaster damage.

In this chapter we conceptualize disaster risk in the context of urban ecology, focusing specifically on flood risk. We show how concepts and theories in both fields display strong parallels, and explain how flood risk can be assessed and managed in ways that correspond to sustainable urban development, i.e., the basis for current urban ecology theory (Weiland and Richter 2009). Similar to this systemic approach that considers complex influencing factors, disaster risk assessment is an integrative procedure built on spatial relationships of relevant parameters, making geoinformatics tools and methods most suitable. In this chapter we introduce the modeling framework used for flood risk assessment, and detail how remote sensing technologies can provide critical input information. We use recent flood modeling work in Naga City (Philippines) as a case study, supported by examples from other urban setting types.

8.2 FLOOD RISK IN THE CONTEXT OF URBAN ECOLOGY

Urban ecology has undergone an extensive conceptual evolution since its descriptive natural history beginnings (Weiland and Richter 2009), showing many links to progress in hazard and risk theory. This can already be seen in the term "disaster management". Although still frequently used it is ill-chosen, as the objective should not be to manage a disaster, but rather to work towards understanding what may lead to a disaster of different magnitudes, and how to avoid or mitigate them. It is thus at best a historically applicable term, where dealing with disasters was reactionary for lack of mitigation and early warning means, but also lack of conceptual understanding of what causes disasters, and how they can be avoided or their impact reduced. It reflects the views of the environmental determinism school, popular in the late 19th and early 20th century, the belief that disasters are caused by a violent nature on helpless society and cannot be avoided (Smith 2004). This view was challenged by the idea that industrialization would lead to technical and scientific progress that would provide a safeguard against hazardous events, and that thus in a post-industrial society nature would be tamed and disasters cease to occur. Central to the theory was that man had the ability to conquer nature. Yet this thinking largely divorced natural events from society, as if they were unrelated. A potentially disastrous event could occur as a result of a natural process, and somehow safeguards, increasingly of an engineering nature, such as straightened rivers and cut-off floodplains, could prevent harm from being done.

A remarkable link between urban ecology and disaster theory lies in the University of Chicago. Led by the Chicago School of Sociology in the 1920s, a departure from the descriptive natural history perspective towards a human–ecological view took place, one that realized the intricate links between society and its built environment (Weiland and Richter 2009). Two decades later the environmental determinism view that had formed the basis of disaster theory was questioned by White (1945), who led what became known as the Chicago School of Natural Hazards Research, based at the same university. The geography-based approach centered on considering both natural and social systems, thus also allowing for solutions other than physical or structural. Central to this approach was a political ecology perspective that tried to consider political, economic and social factors, giving rise to the understanding that not only a hazardous nature is required for calamitous events to occur, but that the specific sociopolitical and economic environment also determines the consequences of an event. This gave rise to the concept of vulnerability that is a person, society or infrastructural element being prone to injury or damage (Blaikie *et al.* 1994).

Many observations resulting from urban ecology research can be used to assess and explain urban disaster risk, and are summarized here.

1 Cities are seen as both the cause and the principal victim of environmental degradation (Richter and Weiland 2008). They are a strong source of pollution, and thus directly (Weiland and Richter 2009), but also via their extensive resource requirements, contribute to global warming. However, by virtue of their concentration of EaR, as well as their disproportionate location in hazardous areas (e.g., along coastlines and rivers), cities are also poised to suffer most from sea level rise or stronger windstorms.

2 Global environmental changes are outpaced by local changes (Grimm *et al.* 2008). For example, urban temperature increases (urban heat islands) are more rapid than global warming rates, leading to rapidly rising secondary hazards (e.g., new disease vectors spreading, or increased ozone concentrations).

3 Urban areas have ecological footprints tens to hundreds of times their size, typically also evidenced by changes in their surrounding land cover and land use (Grimm *et al.* 2008). Urban disaster risk theory also considers cities embedded in a more regional context, with hazard sources being potentially far away, and the characteristics of the area in between strongly affecting the hazard, but also vulnerability and capacity (e.g., wide-spread deforestation or river straightening vs. comprehensive flood plain management). This will become particularly clear in the Naga City example.

4 Urban areas have been found to display an increasing dynamic, as well as fragmentation and diversity in terms of structure and function. This is also relevant from a risk perspective, where the correspondingly variable hazard and vulnerability situation has to be adequately addressed. For example, urbanization dynamics especially typical in less developed countries not only lead to high vulnerability rates on the marginalized urban fringe, but frequently also to rapidly changing informal settlements that constitute non-permanent pockets of deprivation and vulnerability.

5 Grimm *et al.* (2008) pointed out the inefficiency of sectoral approaches in urban ecology. The need for integrative, transdisciplinary analysis of the diverse environmental, social and political aspects is also central to urban disaster risk management (DRM).

Blaikie *et al.* (1994) see disasters as socioeconomic processes interacting with physical exposure. They thus focus on the root causes underlying vulnerability – unsafe conditions and dynamic pressure. They see economic, political and demographic processes as the root causes for vulnerability, as they primarily lead to an uneven distribution of resources. In the following section we show, for Naga City, how these concepts are translated into a geoinformatics-based methodology for flood risk assessment, with modeling providing insight into the hazard situation, and remote sensing delivering input for this assessment, as well as on EaR and their vulnerability.

8.3 COMPREHENSIVE FLOOD RISK ASSESSMENT – NAGA CITY, THE PHILIPPINES

The city of Naga in the Philippines is used as an example for some of the flood problems that many cities all over the world experience. Although Naga is not a mega-city by any means – in 2000 the official number of inhabitants was slightly over 137 000 – it has certain characteristics that makes it worthwhile to study it in greater detail: (i) it is a regional center, offering services to the central and southern parts of Luzon Island; (ii) it is growing rapidly, not only in size but also in economic importance; and (iii) parts of the city are flooded almost every year. Figure 8.2 shows the location of the city in its wider geomorphologic setting. Naga is located on the delta plain of the Bicol River, approximately 20 km upstream from its mouth in the Bay of San Miguel. A tributary originating on Mount Isarog volcano joins the Bicol near the center of Naga. The city center's topography is characterized by flat terrain with elevation differences of only a few meters, and most of the urbanized area was constructed on terrain less than 5 m above sea level, making the lowest parts highly prone to flooding. Although Naga has suffered from floods throughout its history, the recent accelerated growth has made the city expand into areas that used to act as natural water retention zones (marshes and wetlands). Traditionally these areas were used as settlement for the poor due to lack of affordable land in less flood prone parts, but the development policy of the municipality has attracted investors that base their land use decisions more on the potential economic value of the land than on its hydrological characteristics. Massive landfilling without appropriate drainage infrastructures has amplified the flood problems in neighboring areas and has increased flood depths and duration in areas where flooding before was marginal (ADPC 2001).

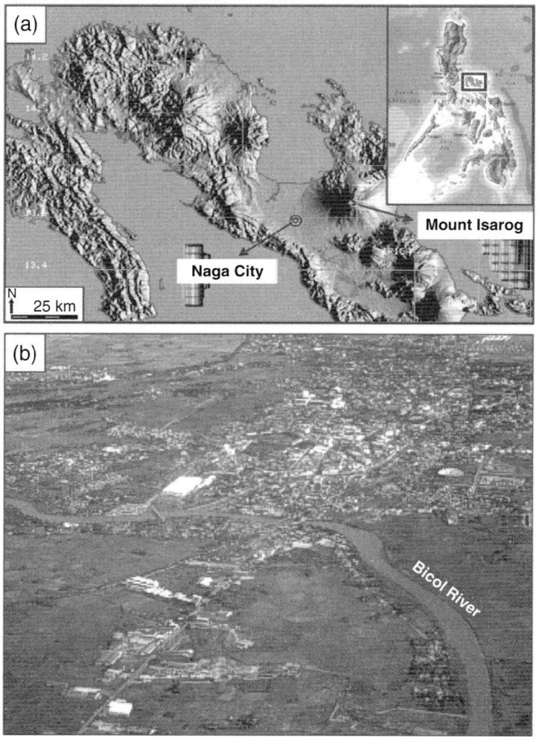

Figure 8.2 (a) Hillshade model showing Naga City and environs (inset shows location of the area), and (b) aerial view of the city and the Bicol River.

8.3.1 Floods in Naga

Floods occur regularly in Naga, especially during the typhoon season from June to December. In this period, the area may be affected by two to six typhoons each year. These storms are accompanied by extremely intensive rainfall, high wind speeds and low atmospheric pressure. The combination of the latter two effects can cause storm surges in the Bay of San Miguel of several meters above mean sea level, especially if the winds come from the north-east. These surges affect directly the coastal areas north of the city, but also hamper the outflow of the Bicol River into the Bay. In this period, the discharge of the Bicol River is close to its bankfull condition, with small increases in water volume resulting in the overflow of the river banks at their lowest parts, and causing inundation of back-swamp areas and other low-lying parts of the delta. Any obstruction to river water discharging into the Bay of San Miguel aggravates the flood hazard in the city surroundings. The high-intensity rainfall adds to these problems in two ways: (i) the Naga River, with its small catchment on the western slopes of Mount Isarog, is highly susceptible to flash floods, and (ii) the drainage of the impermeable surfaces of the city is directed immediately towards the Bicol River. These two systems respond quickly to high-intensity – short-duration events and may add a large volume of water within a very short time period to the river system.

Especially towards the end of the monsoon period, when the capacity of the Bicol basin to buffer more water is severely reduced, events with a relatively small return period may already cause extensive and widespread disruption (ADPC 2001). Some parts of the city are flooded almost every year. A flood with a return period of 5 years can potentially inundate 98% of the low lying parts of the city, affecting directly almost 50 000 persons. In general the depth of the flood can reach up to 2 m, depending on the location and the topography and it can last from 3 hours to several days, but in some parts that are poorly drained, the flood water may remain for more than 2 weeks.

8.3.2 Naga's flood management practices

Under Philippine law, the municipality is the prime responsible for the coordination and delivery of services and effective governance to its citizens. This includes disaster management activities. The city of Naga has three main units that have specified tasks in the response and rehabilitation efforts once a disaster occurs: (i) the Naga Disaster Coordinating Council, in charge of coordinating and managing operations, providing information to the public, and managing available resources; (ii) the Emergency Rescue Naga, including first responders for rescue work, but also information providers such as radio networks and support from municipal bodies; and (iii) the Naga City Disaster Mitigation Council, which advises on disaster mitigation activities and policy making. The city has implemented the following disaster preparedness and mitigation measures: a municipal typhoon warning system, establishment of eight central evacuation centers, stock-piles of relief goods and medicines, improvement of the drainage systems including the rivers, construction of dikes, relocation of squatters and restoration of the vegetation cover. However, most of these measures have been hindered by setbacks, lack of resources, or resistance from the population.

8.3.3 Model-based flood scenario studies

To gain a better understanding of the complex flood system in the Lower Bicol area in general and in Naga city in particular, a sequential series of hydrological and hydraulic models was used to characterize and quantify the floods. These types of models allow an evaluation of future scenarios, to test, for example, the effectiveness of mitigation measures or to assess the consequences of (large-scale) terrain modifications. A significant difficulty in applying such an approach is that it requires a substantial amount of data that is not always readily available in the appropriate format or with the required degree of detail and accuracy. Most of the efforts made in this study were to overcome these difficulties. A one-dimensional (1D) basin model (HEC-RAS) was applied to estimate the discharge in the Bicol and Naga River as a function of the rain and the basin characteristics, such as shape, size, river cross-sections, land cover (to estimate evapotranspiration and surface roughness) and soils (to estimate the infiltration capacity). This resulted in hydrographs at predefined locations along the rivers for rain storms with different return periods (2, 10, and 20 years).

The topography in the Lower Bicol area is relatively flat and is rather complex due to embanked infrastructure, canal dikes and micro relief. In such terrain the flow of water cannot be described satisfactory along predefined (1D) flow lines, hence a full 1D2D hydrodynamic model was used. This model, Sobek (see for example, Dhondia and Stelling 2002), is a combined 1D and 2D flow model where the water flow in the river is calculated in a 1D domain and the flow of water over the dry surface in a 2D domain. These two domains are modeled simultaneously and interactively. The calculation scheme of Sobek ensures a proper water mass and momentum balance (Hesselink *et al.* 2003). For the 1D domain, the model requires cross-section data of the river channels, and for the 2D domain a good representation of the terrain surface, including all flow-affecting structures (a DSM – a digital surface model), a map that represents the resistance to flow primarily due to vegetation cover, and the flow boundary conditions. The upstream boundary conditions are the hydrographs generated for the different return periods by the 1D basin model. The lower boundary is the fluctuating sea level of the San Miguel Bay (obtained from a hydrographic station near the Bicol mouth). The DSM was constructed for this study based on the topographic data available (map scale 1:50 000), combined with elevation data for the embankments from the road and railroad offices and survey data from various engineering works. The final map covered the whole Bicol delta downstream of Naga city. Sobek solves the flow equations using a finite difference solution, which requires surface data to be presented in raster format. In this study a grid spacing of 75 m was

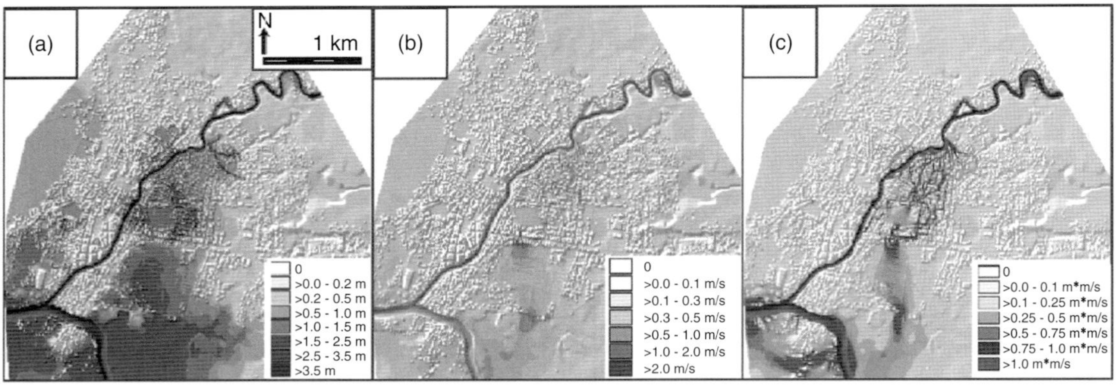

Figure 8.3 Characteristics of the 1 in 50 years flood for Naga City: (a) maximum water depth, (b) maximum flow velocity, and (c) maximum impulse (measured in water depth times velocity).

used. The resistance the flow experiences is strongly related to the land cover which was derived from a Landsat TM image. The values for Manning's coefficient for the land cover units in the area were derived from literature (e.g., from Arcement and Schneider 1989) and internet sources, such as the Engineering ToolBox, see http://www.engineeringtoolbox.com.

The output of the model are hourly water depth and flow velocity maps as well as the maximum inundation depth and maximum flow velocity maps. The hourly maps can be used to derive maps such as the flood propagation (in time), kinetic energy (impulse) of the flowing water mass, speed of rising of the water level, and an estimate of the duration of the inundation for the whole Lower Bicol area.

The actual flood modeling focused on the Naga City municipal area to obtain more detailed flow information at a higher grid resolution (7.5 m). Analysis of the whole Lower Bicol area at this resolution is currently impossible with standard computer hardware. As upstream boundary conditions the same hydrographs were used; however, the downstream boundary now consisted of a discharge–water level relationship (Q_h-relation or rating curve). Due to tidal and storm-surge back-water effects in this part of the river section, the rating curve was derived from the results of the whole Lower Bicol area model to ensure the validity of the curve under these specific model conditions. In total, data from seven sources, including on topography, street levels, and building blocks, were combined and integrated into a DSM that included the buildings as solid blocks. The buildings thus guide the water through the streets, which has pronounced effects on the flow velocity and the kinetic energy of the water flow (see Figure 8.3; also compared with Figure 8.7(d)).

An additional aspect of the study focused on the drainage system within the city to assess its contribution to the flood situation. The detailed results are not presented here; however, they showed that the drainage capacity is insufficient to cope with the extreme amount of rain associated with tropical storms and typhoons (>200 mm/day) and that streets will act as drainage lines funneling the water to the lowest parts of the city where it will cause flooding of buildings and infrastructure. It was also concluded that it is very difficult to alleviate this problem because of the enormous amounts of water that need to be drained and the dimensions of the drainage infrastructure this would require. This is a familiar problem in many typhoon-affected cities in flat terrain. It thus seems more appropriate to look for other solutions, such as local water storage and infiltration rather than increasing the drainage capacity.

8.3.4 Linking flood modeling with disaster management

In collaboration with the Naga City Disaster Mitigation Council the perception of floods of the affected population was further explored. The study of Peters Guarin (2008) showed that the people were very much aware of the flood problem and that they had special coping strategies to deal with minor floods. An interesting aspect of this work was the introduction of the term "manageability" to express the degree to which people

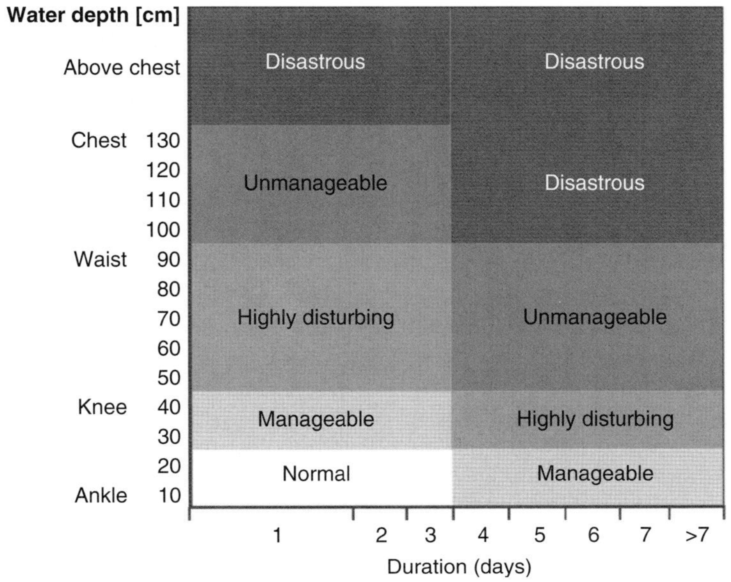

Figure 8.4 Flood perception as a function of water depth and flood duration in terms of manageability (adapted from Peters Guarin 2008).

could manage to live with floods in their neighborhoods. The study applied a participatory approach to obtain primary data from the affected people directly. The two main variables that defined the manageability of flood events were found to be the maximum water depth and the duration of the flood (Figure 8.4).

Combining the flood model results and derivative maps with community-based flood perception information results in maps that indicate which neighborhoods require immediate assistance and which areas can still cope with the situation. The flood model allows the evaluation of possible future scenarios, which will help the city authorities to anticipate relief efforts and to assess the effects of large-scale terrain modifications, such as landfill operations, and to assess the effectiveness of possible mitigation measures.

8.3.5 Naga as example for other flood-prone cities

Although Naga is not a mega-city, its flood-related problems can be seen as a showcase for other coastal cities in the (semi-) tropics, such as Bangkok or Jakarta. The similarities lie in the low elevations with respect to sea level, the small natural variations in elevation, soil and subsoil conditions and complex hydrology. These cities are very sensitive to events such as surface subsidence and sea-level rise. Lower terrain topography in combination with higher sea levels will bring the sea shore closer to the built-up areas and will hinder discharge of river- and drainage water into the sea. Also high-intensity rains in combination with a growing percentage of impermeable surface and insufficient drainage capacity increase the chances for floods. The continuous growth of these cities and increasing complexity of the topography due to embankments, landfills and large structures may aggravate the consequences of floods in terms of flood depths and duration, which in the end may increase the frequency of flood disasters.

8.4 THE ROLE OF REMOTE SENSING IN FLOOD RISK ASSESSMENT AND MANAGEMENT

Comprehensive flood modeling is based on a variety of data sources, some of which are quasi-static, while others are highly dynamic. In Section 8.3 we showed the need for the following information:

Flood hazard:
- surface topography
- surface roughness
- soil type and infiltration capacity

- input hydrograph
- frequency/magnitude information.

Flood vulnerability:

- elements at risk (people, infrastructure [buildings, industrial and transport facilities])
- physical, social, economic, and environmental vulnerability
- capacity.

8.4.1 Quasistatic hazard data

Of the above information types typography can be considered the most static, although depending on the definition of what constitutes the surface this can vary. With its origins more than 100 years ago, topographic information has traditionally been derived from digital elevation models (DEMs), prepared using aerial photographs and photogrammetric techniques (e.g., McGlone *et al.* 2004). DEM here is a generic term that can be separated into models that include all above-ground features (DSMs), or only the actual terrain (digital terrain model – DTM). Depending on the photo scale such DEMs can achieve centimeter vertical accuracies and submeter resolutions, the latter typically exceeding the processing capabilities of hydrodynamic models (see Section 8.3.3). While photogrammetry is a highly accurate technique it has two principal disadvantages, illustrated in Figure 8.5(a): as it is based on point matching in two overlapping (stereo) images, it inherently models not the ground surface, but includes all above-ground features, such as buildings and vegetation. Photogrammetry thus produces DSMs. In addition, observations from two different positions are required to determine the parallax, that is the relative shift of features in the images as a result of their elevation, and the underlying concept for DEM generation. This, however, results in occlusion where ground features block a complete view of the ground (illustrated by black bars next to the building). Point matching for those areas is not possible, resulting in interpolation and DSM errors. A 30 m global DEM based on photogrammetrically processed ASTER satellite data was released in June 2009, offering an interesting potential for flood modeling.

The principle of observing the ground from different positions also underlies DEM generation from radar (*ra*dio *d*etection *a*nd *r*anging) data. As shown in Figure 8.5(b), we can use data from instruments mounted on separate satellites or airplanes, or from observations made at different times. However, radar-based photogrammetry, or radargrammetry, differs from that described for optical data in two important ways: (i) radar instruments are always side-looking, especially in mountainous terrain leading to distortions or radar shadow; (ii) radar is an active technique, meaning that the sensor emits its own energy (radar waves) that are reflected by the ground or above-ground features. This means that radar observations are not limited to daytimes, and are not hindered by atmospheric conditions (clouds, smog, etc.). However, in principle, radar signals are also retuned by the

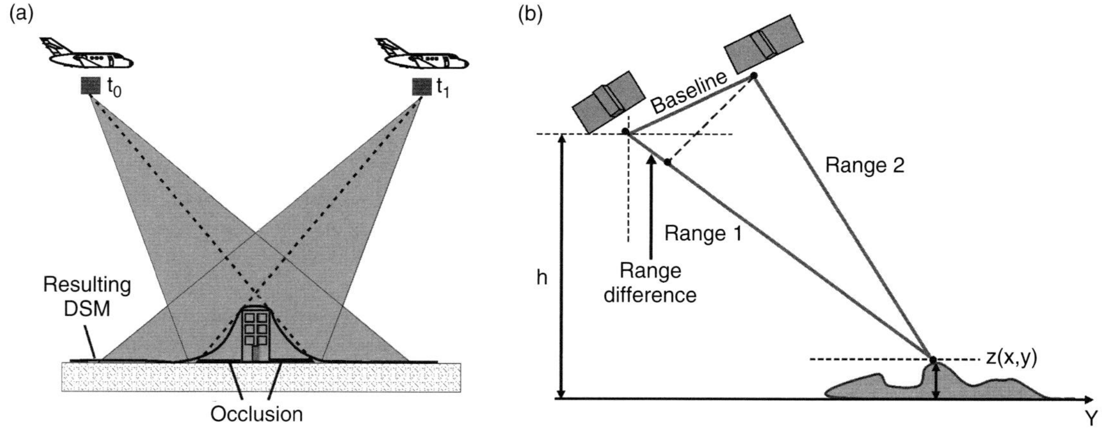

Figure 8.5 (a) Principle of photogrammetry, leading to DSMs that can suffer from occlusion, and (b) principle of radar-based DSM generation. Either two radar instruments (mounted on airplanes or satellites), or observations at different times, separated by a known baseline distance, observe the ground, leading to range differences resulting from distance variations between ground and instrument.

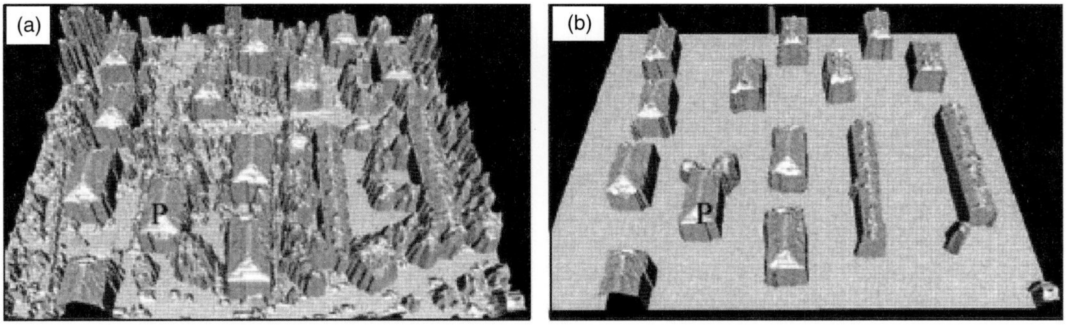

Figure 8.6 (a) DSM derived from laser scanning, with (b) vegetation and other artefacts removed to retain buildings only. Reproduced from Dash *et al.* 2004 with permission from Elsevier.

feature closest to the sensor, such as tree canopies or roof tops rather than the ground, thus also resulting in DSMs.

Although having global coverage, the ASTER DEM resolution is still insufficient for urban area models, while occlusion and oblique incidence for optical and radar data, respectively, also lead to undesirable artefacts. These problems are avoided by laser scanning systems (lidar – *l*ight *d*etection *a*nd *r*anging), increasingly used for detailed topographic assessment, especially in urban areas. Like radar, lidar is an active sensing technique. It is currently only used with airborne platforms or for ground-based sensing, as a satellite-mounted laser would result in a very large footprint and thus low spatial resolution.

For surface mapping lidar has several advantages over the previously described methods. Although a lidar system scans the ground and is thus also liable for occlusion effects in densely built-up areas, appropriate filtering can generate DSMs where also vertical features (e.g., building walls) are accurately modeled. Lidar is particularly useful in mapping vegetation, as part of the laser beam typically reaches the ground through the canopy. This provides sufficient information to detect and identify tress, but also to remove them from the DSM. If roads in between buildings are also accurately mapped in terms of their elevation, morphological modeling can be used to remove buildings from the lidar data as well, effectively leading to a DTM. However, if a DSM was to be used to model water flow in urban areas, it is more appropriate to remove the features that pose no or little obstacle to surface flow (such as trees), while retaining (semi-)solid features (e.g., buildings; see Figure 8.6(b)). Many studies have investigated the effect of erroneous topography on hydrological modeling (e.g., Kenward *et al.* 2000).

8.4.2 Dynamic hazard data

The distinction between dynamic and quasi-static data is somewhat artificial. We here consider topography to be largely static, even though in urban areas, if buildings are considered part of the terrain, this is not typically the case. More variable from a city evolution and system dynamics perspective, however, are land cover and land use, both within and around cities. Land cover refers to the actual surface material, for example, water or grass, while land use specifies its purpose, such as "swimming pool" versus "lake". Land cover assessment is the oldest purpose of remote sensing, and only more recent conceptual and methodological developments have allowed information on land use to be extracted in (semi-)automatic analysis as well. Flood hazard modeling only requires information on land cover, both to correct DSMs (e.g., to remove vegetation, see Section 8.4.1), and to extract surface roughness information. While roughness can also be calculated directly from a DSM, this requires a very detailed and accurate surface model, such as derived from lidar data. It is more typical to extract roughness from image data. In addition to the radar characteristics discussed above it must be noted that, while optical remote sensing data are primarily sensitive to surface chemistry, radar is sensitive to surface moisture and roughness. Detailed radar data can thus be used to estimate the roughness values as input for flood models (e.g., Dobson *et al.* 1995). However, also land cover identified in optical data can be used, by linking land cover types to roughness (King *et al.* 2005), or by considering image texture characteristics (McGarigal *et al.* 2009).

Land cover not only determines surface roughness and thus the effect it has on surface flow; it can also provide some information on infiltration capacity. This is easiest when infiltration is nearly entirely prevented, as is the case for impervious surfaces that are readily detected with remote sensing. In unvegetated areas it is also possible to identify soil types, either in conjunction with ground data (Hahn and Gloaguen 2008), or by using hyperspectral scanners – instruments that record upwards of 200 spectral channels for extremely detailed spectral surface characterization. Assessing infiltration capacity of vegetated areas (such as on crop lands or in forests) is very difficult if no direct observation of the soil surface is possible.

With the exception of storm surges or flooding resulting from excessive precipitation that exceeds the capacity of the drainage system in an urban area, the downstream hydrograph is a function of water accumulation in the upstream catchment, typically the result of precipitation, but also caused by rapid snow melting or dam breaks. Although it is easiest and most accurate to employ a network of rain gauges to measure rainfall amounts, in the absence of those, or for short-term predictions of rainfall rates, remote sensing can also be used to estimate precipitation amounts. For example, the geostationary GOES satellites (two satellites covering the US), Meteosat Second Generation (MSG, positioned over Africa and also covering Europe), or METSAT over Asia, have been used to estimate rainfall in near-real time, using temporal infrared information on cloud top temperatures and their changes. This provides data suitable for flash flood detection. Another instrument specifically designed for rainfall estimation in tropical areas is the Tropical Rainfall Measuring Mission (TRMM). Since 1998 this satellite has been used to measure tropical precipitation (for a mission overview see Kummerow *et al.* 1998), though data accuracy has been found to depend on the precipitation regime. However, in principle remote sensing is well suited to provide critical precipitation input data to model flood potential for a given catchment.

The assessment of natural hazards typically entails a frequency-magnitude assessment of previous events. The extent of standing surface water can be well mapped with radar instruments, as water results in very low radar reflectance (i.e., a near-black signature), but also as it penetrates clouds that typically accompany rainfall-based flood events. There are, however, two important limitations: (i) there are currently only eight operational non-classified radar satellites (compared to >50 optical systems), with their orbital characteristics limiting their temporal resolution (i.e., the frequency with which they can observe the same areas on Earth), to at least several days and (ii) all radar satellites are commercially operated and data thus expensive. In severe flood disaster situations the International Charter "Space and Major Disasters" (Ito 2005) may be activated, which provides for rapid custom satellite data acquisition and processing, with results being made available free of charge to the humanitarian community (see http://www.disasterscharter.org). A detailed overview on the utility of airborne and spaceborne remote sensing in emergency response is given by Kerle *et al.* (2008) and Zhang and Kerle (2008), respectively.

8.4.3 Mapping elements at risk

As explained in Section 8.1, anything of value that can suffer damage or loss constitutes an element at risk, be it a building, a person, or a field with crops, but also less tangible items such as biodiversity or an economic system. Each of those elements should, in principle, be assessed with respect to all existing hazards and their potential magnitudes. With increasing understanding of the diversity of EaR, risk theory has evolved. Where previously only physical vulnerability was considered (e.g., buildings or infrastructure that can be damaged to destroyed in a given event), we now consider social, environmental and economic vulnerability as well (Ebert *et al.* 2009), reflecting the systemic approach to risk assessment.

It is clearly easiest to consider mapping physical elements, such as buildings, from remote sensing data. Especially recent high resolution images, such as from the IKONOS (1 m resolution) and Quickbird (0.66 m) sensors, are very well suited to map infrastructure, especially in densely built-up urban areas. For flood assessment the height of buildings is of particular importance, as it allows an estimation of the number of people living in a given building, but also to calculate partial submersion rates for a given building and flood scenario. Recent research has shown that so-called segmentation- or object-based methods are particularly well suited to create models of urban areas. Figure 8.7 shows part of a Quickbird image of Tegucigalpa, Honduras, and how semantic building groups were extracted using object-oriented analysis (Figure 8.7(b), Shamaoma *et al.* 2006). In addition to the optical image lidar data were used to calculate the height of individual buildings. The combined results were then used to model a 50-year flood scenario for the city, parts of which are shown in Figure 8.7(c, d). It must be noted here that individual physical vulnerability can vary dramatically between individual buildings or infrastructure types, and must always be assessed

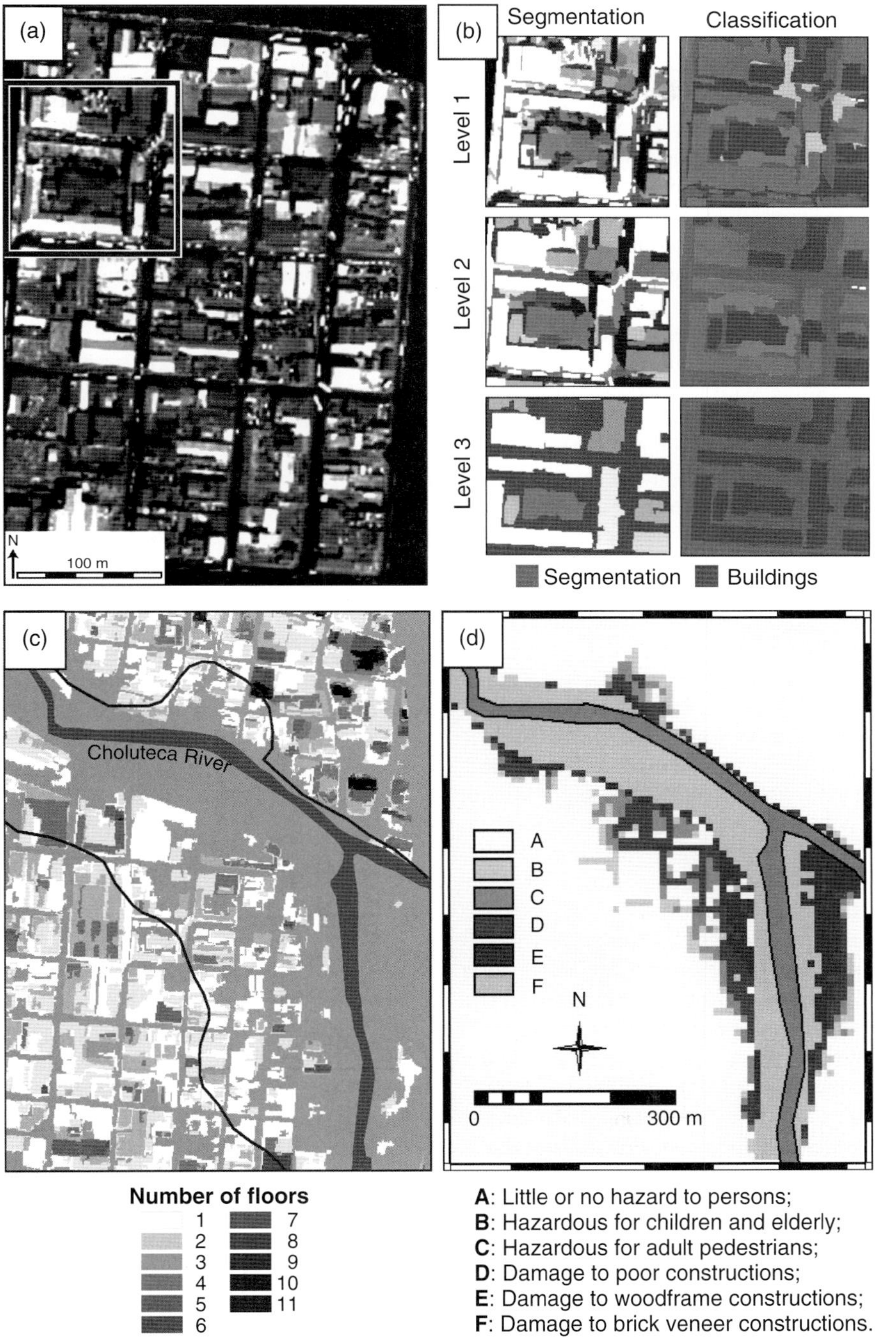

Figure 8.7 (a) Building footprint extraction for parts of Tegucigalpa, Honduras, and (b) three segmentation and classification levels (close-up area indicated by box in (a)). Small objects, such as cars, are gradually removed in the rule-based classification stage, while at level 3 only whole buildings as semantic groups remain. (c) Result of building height extraction for parts of Tegucigalpa, with the 50-year flood boundaries in black, and (d) flood hazard map based on 2D flood modeling of a 50-year flood event (Shamaoma *et al.* 2006).

in the context of a specific hazard and event magnitude. Thus for detailed risk assessment (e.g., for critical infrastructure such as bridges), ground based studies, typically by engineers, need to be included.

The mapping of non-physical EaR and assessment of their respective vulnerabilities is more challenging but not impossible. Environmental vulnerability, for example, can be based on an integration of remote sensing data, for example to identify sensitive areas of particular ecological value, and auxiliary data on vegetation health or biodiversity. Addressing social vulnerability, defined as "people's differential incapacity to deal with hazards, based on the position of the groups and individuals within both the physical and social worlds" (Clark *et al.* 1998), requires a different approach. Ebert *et al.* (2009) used physical proxies for social vulnerability indicators, such as the proportion of built-up areas, building height or road condition (surfaced versus unsurfaced), but also distance of buildings to mapped flood hazard zones and to infrastructure and lifelines. The method worked well to estimate social vulnerability that had previously been based on detailed census data that do not always exist.

As described in the previous sections, hazard and risk theory have undergone substantial developments, and conceptual advances are still occurring. However, many questions remain unanswered that again show parallels to urban ecology. While it is straightforward to calculate risk in monetary terms for a single hazard if information on event probability and severity, but also on infrastructure elements and their value are available, this is proving more difficult for multi-hazard and multivulnerability risk assessment. Often feedback processes between different hazards are poorly understood (e.g., flooding leading to increased slope instability), challenging modeling efforts. Additionally, non-physical vulnerabilities can not be measured in the same terms as building damage, leading researchers to use indices instead (Smit and Pilifosova 2003), although no agreement yet exists on such a comprehensive risk index. There are further problems related to the visualization and communication to decision makers of such multidimensional risk information (Kerle 2008).

8.5 DISASTER RISK IN THE CONTEXT OF URBAN ECOLOGY – AN OUTLOOK

The purpose of this chapter was to conceptualize disaster risk in the context of urban ecology, and to demonstrate the utility of geoinformatics tools in risk assessment and management, focusing specifically on flood risk. Urban ecology shows strong parallels with DRM theory, both in evolution and currently applied concepts: cities are the cause and the victim of environmental degradation; they do not exist in isolation but show a strong reciprocal relationship with the surrounding environment; they experience global changes on an amplified scale, leading to a variety of secondary hazards; the growing environmental footprint and resulting land cover and land use changes surrounding the urban areas can also increase the risk; the increasing trend to urban fragmentation in terms of units and functions that has been described in urban ecology theory mirrors the increasingly variable risk; and the need for integrative, transdisciplinary approaches, which is also echoed in DRM.

The chapter not only showed the strong links between urban ecology and risk concepts and theories, but also detailed how risk can be assessed and managed in ways that correspond to sustainable urban development, which in turn forms the basis for current urban ecology theory. Urban ecology calls for a systemic approach that integrates the complex influencing factors; similarly, disaster risk assessment and management can be seen as an integrative procedure built on spatial relationships of relevant parameters. This also explains why geoinformatics tools and methods are particularly suitable, and suggest that urban planning, development and management based on urban ecology principles can equally employ and benefit from such tools.

In this chapter we introduced a modeling framework for flood risk assessment, and explained how remote sensing technologies can deliver critical input information. Recent flood modeling work in Naga City (Philippines) was used as a case study, but we also showed how flood risk management based on a comprehensive understanding of the underlying hazard and vulnerabilities, and employing geoinformatics methods, forms the key to managing and mitigating flood and other natural risks in other parts of the world as well.

REFERENCES

Arcement, G.J., Schneider, V.R. (1989) *Guide for selecting Manning's roughness coefficients for natural channels and flood plains*. Water Supply Paper 2339, US Government Printing Office, Washington D.C.

Asian Disaster Preparedness Center (ADPC) (2001) *Naga City Disaster Mitigation Plan*. Bangkok, Thailand.

Blaikie, P., Cannon, T., Davis, I., Wisner, B. (1994) *At Risk: Natural Hazards, People's Vulnerability, and Disasters*. Routledge, London.

Brown, L.R. (2006) *Plan B 2.0: Rescuing a Planet under Stress and a Civilization in Trouble*. W.W. Norton, New York.

Clark, G.E., Moser, S.C., Ratick, S.J., *et al.* (1998) Assessing the vulnerability of coastal communities to extreme storms: the case of Revere, MA, USA. *Mitigation and Adaptation Strategies for Global Change* 3, 59–82.

Dash, J., Steinle, E., Singh, R.P., Baehr, H.P. (2004) Automatic building extraction from laser scanning data: an input tool for disaster management. In: Singh, R.P. (ed.) *Monitoring of Changes Related to Natural and Manmade Hazards Using Space Technology*. Pergamon-Elsevier Science LTD, Oxford, pp. 317–322.

Dhondia, F.F., Stelling, G.S. (2002) Application of a one dimensional - two dimensional integrated hydraulic model for flood simulation and damage assessment. *Proceedings of the Fifth International Conference on Hydroinformatics*, Cardiff, UK.

Dobson, M.C., Ulaby, F.T., Pierce, L.E. (1995) Land-cover classification and estimation of terrain attributes using synthetic-aperture radar. *Remote Sensing of Environment* 51, 199–214.

Ebert, A., Kerle, N., Stein, A. (2009) Urban social vulnerability assessment with physical proxies and spatial metrics derived from air- and spaceborne imagery and GIS data. *Natural Hazards* 48, 275–294.

Goldewijk, K.K. (2005) Three centuries of global population growth: A spatial referenced population (density) database for 1700–2000. *Population and Environment* 26, 343–367.

Grimm, N.B., Faeth, S.H., Golubiewski, N.E., *et al.* (2008) Global change and the ecology of cities. *Science* 319, 756–760.

Guha-Sapir, D., Hargitt, D., Hoyois, P. (2004) *Thirty Years of Natural Disasters 1974–2003: The Numbers*. Center on Epidemiology of Disasters (CRED), Louvain-la-Neuve, Belgium.

Hahn, C., Gloaguen, R. (2008) Estimation of soil types by non linear analysis of remote sensing data. *Nonlinear Processes in Geophysics* 15, 115–126.

Hesselink, A.W., Stelling, G.S., Kwadijk, J.C.J., Middelkoop, H. (2003) Inundation of a Dutch river polder, sensitivity analysis of a physically based inundation model using historic data. *Water Resources Research* 39, SWC1.1–SWC1.17.

Ito, A. (2005) Issues in the implementation of the International Charter on Space and Major Disasters. *Space Policy* 21, 141–149.

Kenward, T., Lettenmaier, D.P., Wood, E.F., Fielding, E. (2000) Effects of digital elevation model accuracy on hydrologic predictions. *Remote Sensing of Environment* 74, 432–444.

Kerle, N. (2008) Geodata: limitations in current concepts. *GIM International: the Global Magazine for Geomatics* 22, 13.

Kerle, N., Heuel, S., Pfeifer, N. (2008) Real-time data collection and information generation using airborne sensors. In: Zlatanova, S., Li, J. (eds) *Geospatial Information Technology for Emergency Response*. Taylor & Francis, London, pp. 43–74.

King, C., Lecomte, V., Le Bissonnais, Y., Baghdadi, N., Souchère, V., Cerdan, O. (2005) Remote-sensing data as an alternative input for the 'STREAM' runoff model. *CATENA* 62, 125–135.

Kummerow, C., Barnes, W., Kozu, T., Shiue, J., Simpson, J. (1998) The Tropical Rainfall Measuring Mission (TRMM) sensor package. *Journal of Atmospheric and Oceanic Technology* 15, 809–817.

McGarigal, K., Tagil, S., Cushman, S.A. (2009) Surface metrics: an alternative to patch metrics for the quantification of landscape structure. *Landscape Ecology* 24, 433–450.

McGlone, J.C., Mikhail, E.M., Bethel, J.S. (2004) *Manual of Photogrammetry*. American Society for Photogrammetry and Remote Sensing (ASPRS), Bethesda.

Peters Guarin, G. (2008) *Integrating local knowledge into GIS based flood risk assessment, Naga city, The Philippines*. Unpublished PhD thesis, Wageningen University & ITC, Enschede, The Netherlands.

Richter, M., Weiland, U. (2008) Stadtökologische Traditionslinien. In: Rehberg, K.-S. (ed.): Die Natur der Gesellschaft. Bd. 1. Campus, Frankfurt/Main, pp. 475–488.

Shamaoma, H., Kerle, N., Alkema, D. (2006) Extraction of flood-modelling related base-data from multi-source remote sensing imagery. In: Kerle, N., Skidmore, A.K. (eds) *ISPRS Mid-term Symposium 2006: from Pixels to Processes*. Enschede, The Netherlands, pp. 643–649.

Smit, B., Pilifosova, O. (2003) From adaptation to adaptive capacity and vulnerability reduction. In: Huq, S., Smith, J., Klein, R.T.J. (eds) *Enhancing the Capacity of Developing Countries to Adapt to Climate Change*. Imperial College Press, London, pp. 9–25.

Smith, K. (2004) *Environmental Hazards*. Routledge, London.

United Nations Population Division (UNPD) (2006) *World Urbanization Prospects: The 2005 Revision*. United Nations, New York.

Weiland, U., Richter, M. (2009) Lines of tradition and recent approaches to urban ecology, Focussing on Germany and the USA. *GAIA – Ecological Perspectives for Science and Society* 18, 49–57.

White, G.F. (1945) *Human Adjustment to Floods: a Geographical Approach to the Flood Problem in the United States*. Department of Geography, University of Chicago, Chicago.

Zhang, Y., Kerle, N. (2008) Satellite remote sensing for near-real time data collection. In: Zlatanova, S., Li, J. (eds) *Geospatial Information Technology for Emergency Response*. Taylor & Francis, London, pp. 75–102.

CHAPTER 9

Urban open spaces and adaptation to climate change

Marialena Nikolopoulou

*Kent School of Architecture, University of Kent, Canterbury, United Kingdom**

9.1 CITIES, CLIMATE CHANGE AND THE ROLE OF OPEN SPACES

Cities are integrated systems composed of various components with interconnecting links: individual buildings, urban blocks, settlements, roads, parks and other open spaces, transport and other human activities all affecting the environmental quality. Against this complex background "occupants" – whether inside or outside buildings – are an integral part of the environmental system. In 2007, for the first time in the history of mankind, the urban population exceeded that of rural areas across the globe, while in Europe that figure is over 80%. The occupants' actions not only influence energy use (activities, buildings, transport) but also aggravate environmental quality in cities (increased ambient temperatures and pollution emissions), which subsequently further affects their behavior and activities in such a system.

The World Health Organisation acknowledges that global climate change will have profound effects on the health and well-being of citizens throughout the world. The mean air temperature in Europe is expected to rise by 2 to 6 °C by 2100 (IPCC 2007). Although southern Europe will be affected the most, northern countries will also experience severe problems.

The urban populations are more at risk as a result of the urban heat island. This refers to the phenomenon of higher temperatures in cities than in the surrounding countryside. The intensity of the heat island is determined by the thermal balance of the urban environment and can reach a temperature difference of 10 °C between the city and the surrounding rural environment (Oke 1987). Two of the main reasons contributing to the rising urban air temperatures are the increased heat capacity from the built environment, along with the large impervious surfaces in cities; these are designed to promote rainwater runoff, thus altering the urban moisture content and reducing opportunities for cooling by evaporation in the cities.

Extensive areas of concrete, bricks and asphalt used for walls, roofs, streets, pavements, etc. provide large areas of high heat capacity, which can store great amounts of heat. This heat is stored during the day and re-emitted in the evening. However, as night time temperatures will also increase under climate change, the buildings will give off heat (and thus cool) more slowly, significantly increasing the discomfort for those living in such spaces.

Despite the uncertainties surrounding predictions, one of the potential impacts on health in the United Kingdom by 2050 will be from the increased summer temperatures, with heat-related deaths increasing to 2800 per year (Department of Health 2002). This follows the heat wave of 2003 (which will be considered to be typical by the 2040s) in which excess deaths in northern France alone were 15 000. What was initially estimated to be 22 000 deaths across Europe during the summer of 2003 has escalated to 52 000 (Larsen 2006). Such figures are independent of the increased hospital admissions, affecting the more sensitive

*Formerly at Department of Architecture and Civil Engineering, University of Bath, Bath, United Kingdom (moved 2011).

Applied Urban Ecology: A Global Framework, First Edition. Edited by Matthias Richter and Ulrike Weiland.

groups of the population, with significant financial implications for the public and private sector.

Increased summer temperatures have a significant energy penalty, leading to higher use of air-conditioning, which is rising exponentially as a result of global warming, as well as due to people's expectations for comfort. In Europe air-conditioning sales increased by 22% in the period 2002–2006 (Santamouris 2007). Thus in order to improve thermal comfort conditions in buildings, energy consumption and corresponding CO_2 emissions will increase, while the heat output of these systems in the external environment will further affect the already high ambient temperatures. A vicious circle then begins, as a poor microclimate and high ambient temperatures will further increase the use of air-conditioning, as the external air is not regarded as being appropriate for natural ventilation. Under these conditions it is expected that people will inhabit outdoor areas for longer periods of time in the evening while the buildings are cooling down. This can only take place if the urban open spaces provide beneficial conditions.

It is not only lifestyle changes and the increasingly evident café culture that demand usable open spaces, but the fact that open space can provide enhanced microclimatic conditions, which can help to cool people down. This opportunity needs to be exploited, as cooling people is far more sustainable than cooling buildings, but in order to do that we need to ensure that urban open spaces provide desirable conditions.

However, how do people use outdoor spaces and what is the relationship between the thermal environment and thermal comfort conditions? Such knowledge will enable us to use urban design constructively to enhance resilience in the face of environmental change. Integrating the outdoor context with the built environment would enable us to re-evaluate the need for open spaces and how we could rely on the network of external public spaces and outdoor living to play an important role in cooling cities. These spaces need to adapt to accommodate people and their needs to enable people to use them. Otherwise, people will retreat to indoor air-conditioned spaces, further burdening the environment.

Acknowledging the critical role urban open spaces can play under climate change, this chapter focuses on their thermal environment and outdoor thermal comfort. It examines how environmental stimuli and thermal perception affect their use, and how mechanisms of personal adaptation enable people to adapt in a range of environmental conditions. The last section examines design interventions that can increase our adaptive capacity under conditions of climate change by improving the urban microclimate and reducing the thermal load on pedestrians in open spaces.

9.2 OUTDOOR COMFORT

The great complexity and variability of the outdoor thermal environment, spatially and temporally, means that the theoretical thermoregulatory models, which have been developed for indoors, do not adequately describe outdoor comfort conditions.

Conventional comfort theory relies on a steady-state model where the production of heat is equal to the heat losses to the environment, aiming to keep a constant core body temperature of 37 °C. The environmental parameters, which influence thermal comfort and provide thermal satisfaction, are air temperature, wind speed, humidity, and thermal radiation. Clothing levels and metabolic activities further affect the heat generated and dissipated by the human body. The overall thermal comfort zone thus falls within a narrow band. In outdoor spaces these conditions are encountered within a much wider range and are more variable. This leads to more complex heat exchange mechanisms between the human body and the surrounding environment, rendering such models inadequate in this context. This complexity further increases when the diversity of activities in open spaces is taken into account.

The realization of the lack of understanding of outdoor thermal comfort has led to a number of field surveys aiming to understand these interrelationships (Nikolopoulou *et al.* 2001, Spagnolo and de Dear 2003, Givoni *et al.* 2003, Stathopoulos *et al.* 2004, Eliasson *et al.* 2007). The most extensive work is that which has been carried out within the framework of the EU-funded project RUROS (Nikolopoulou and Lykoudis 2006), which evaluated outdoor thermal comfort conditions in different geographic locations. It encapsulated the large geographic, climatic and cultural variation encountered within Europe through 9189 interviews with users of urban open spaces. The seven cities investigated included Athens and Thessaloniki in Greece, Milan in Italy, Fribourg in Switzerland, Kassel in Germany, and Cambridge and Sheffield in the United Kingdom. Two case study sites in each city provided the platform for detailed environmental and human monitoring. The field surveys took place periodically within a year to obtain the seasonal variation. Comparing objective environmental parameters and attendance in space with subjective responses to questionnaires, the project aimed to establish the impact of environmental stimuli on the use of space and people's thermal sensation and comfort evaluation.

Here we focus on two different aspects of the field surveys: the impact of microclimatic conditions on the use of space, as well as thermal sensation and comfort.

9.3 USE OF SPACE

How does microclimate affect the use of space? Recent research in the United Kingdom (Nikolopoulou *et al.* 2001) has shown that responses to the microclimate may be unconscious, but they often result in a different use of open space under different climatic conditions. In a different climatic context, in Canada, the more extreme microclimatic conditions caused by tall buildings in business districts have been studied in relation to human activities and behavior in one of the first studies to understand the way people respond to microclimatic conditions (Zacharias *et al.* 2001). Another study in the more traditional urban fabric of Göteborg also linked attendance to microclimatic factors. In fact, the researchers demonstrated that 47% of the variation in attendance in an urban square could be attributed to variations in air temperature, clear skies and wind speed, where temperature had the biggest weighting (Eliasson *et al.* 2007).

In a Mediterranean climate, the two case studies in Athens, which were investigated in the RUROS project, provided interesting information on the impact of environmental stimuli on the daily and seasonal patterns of use (Nikolopoulou and Lykoudis 2007). One of the sites, Karaiskaki Square (approximately $80 \times 100\,m^2$), is a neighborhood square with infrastructure for activities including a playground, a coffee shop with outdoor seating, benches, etc. The second site is by the seashore ($200 \times 100\,m^2$), an important area for greater Athens attracting visitors from neighboring municipalities, with areas for promenading, swimming, a big playground, and coffee shops. Both sites have areas with vegetation as well as exposed areas characterized by large heat capacity materials such as concrete pavements and absence of shading.

9.3.1 Seasonal profile

As expected, during the summer visitors prefer to sit in shaded areas, whereas in fall and winter sunlit areas become more popular. Areas that combine large heat capacity materials and lack of shading are not used as much in the summer due to the resulting uncomfortable thermal environment; their use is mostly limited to the evening.

Air temperature and solar radiation were the most dominant parameters in relation to the use of space as respective correlations with wind speed and relative humidity are very weak. Investigating individual locations, it was found that the correlation between presence in the sun and air temperature is – 0.23 (Pearson correlation coefficient, $p < 0.01$), demonstrating that as air temperature increases the number of people found in the sun is reduced.

However, it is interesting to notice how creating favorable microclimates, in what would be described as otherwise harsh conditions, can increase attendance in the space. The coffee shop in Karaiskaki Square provides extensive shading from mature trees at its perimeter, as well as awnings which can be drawn across the top of the sitting area to protect people from the sun. These succeed in improving the microclimate. Hence, although overall increasing air temperature is a reason for staying away from the square, the effective microclimate of the coffee shop renders it a more popular area compared to the rest of the square.

Wind speed is also a contributing factor to the decreasing presence of people, particularly in winter. When the analysis concentrates on people sitting in the sun, wind appears to have a positive effect in the summer – when it is most welcomed, particularly by those sitting directly in the sun. The effect is negative in winter, reducing presence significantly ($R^2 = 0.19$) since a sunlit position is, usually, also exposed to the wind.

However, it is not only the number of people that is influenced by microclimatic parameters. The time spent outside is also affected. A recent study in hot arid climates (Aljawabra and Nikolopoulou 2009) used a similar methodology to investigate comfort conditions in Marrakesh (Morocco) and Phoenix (Arizona, USA). Intensity of solar radiation was an important parameter, and it was found that, as the intensity of mean solar radiation increases, the time people spend outdoors decreases (Figure 9.1).

9.3.2 Diurnal profile

So far we have been discussing the impact of climatic parameters on the use of open spaces on a seasonal

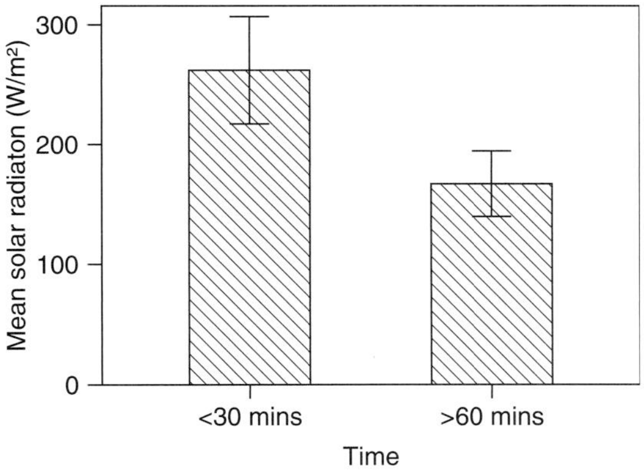

Figure 9.1 Time spent outdoors for different mean solar radiation intensity levels, for Phoenix and Marrakesh (Aljawabra and Nikolopoulou 2009).

basis. Could we expect to see such contrasting profiles diurnally as well? Understanding these, we can then explore whether design could have an effect in influencing such use.

In a previous study in the United Kingdom, Nikolopoulou *et al.* (2001) demonstrated how people enjoy feeling warm in the specific climatic context and thus the maximum attendance of people outside was found at higher air temperatures. They then speculated whether the situation would be reversed in a hotter climate. How dynamic is the relation between the use of space and microclimate? This is examined for a typical hot summer day in Karaiskaki Square in the summer of 2001.

Attendance is significantly lower in the morning than in the evening. In the morning, when air temperature has not yet risen, the use of space is higher than at midday, reducing as the air temperature rises (Figure 9.2). Groups of people found in the space normally comprise mothers with small children, the elderly, etc. The use of the space increases again in the evening, as the temperature drops, and becomes particularly high after 8p.m., when the open spaces are considered to be a much cooler place to escape to from the hot indoor environment by neighborhood residents.

The attendance profile follows a course inverse to the fluctuation of air temperature. It is worth highlighting that the overall range of air temperature monitored for the day is from 31–35 °C. This normally would be regarded as a deterrent for any outdoor activities. When the previous days, however, have been characterized by high air temperatures, people have become used to the heat, they expect it to be hot in

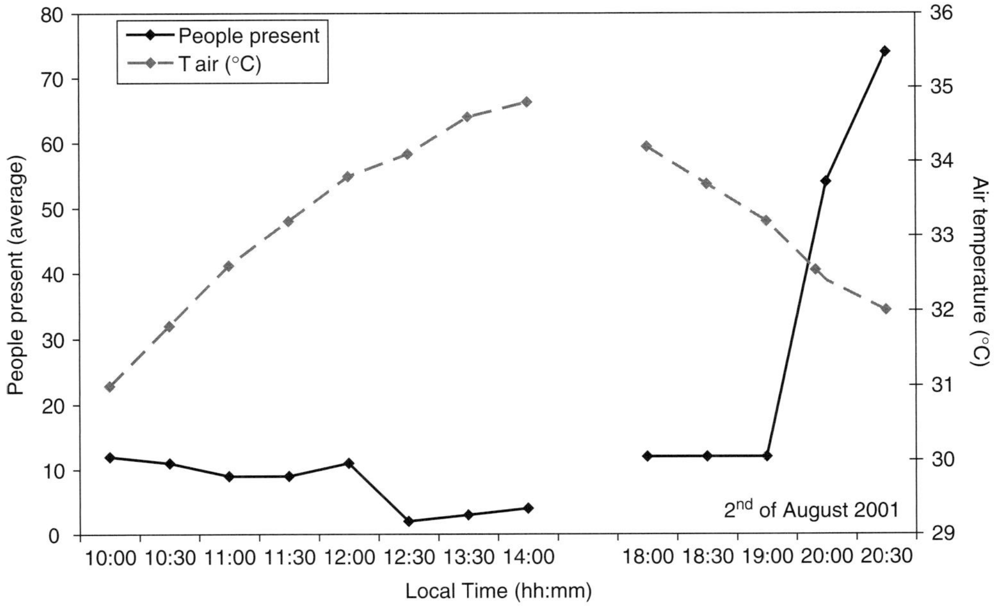

Figure 9.2 Daily course of air temperature and the mean number of people at Karaiskaki Square, Athens (Nikolopoulou and Lykoudis 2007).

the specific climatic context, are aware of what actions to take to improve their comfort state, and thus are not particularly bothered by it. "Past experience" and "expectations" explain why an air temperature of 30 °C is regarded as relatively cool on a day when the maximum air temperature reached 35 °C, with outdoor activity, as evident from the increased presence of people in the square, reaching high levels. This issue of "adaptation" is examined in Section 9.5.

Beyond the profile of a single day, it is also interesting to examine the mean diurnal pattern of use of space for both sites in the different seasons. The seasonal variation of the mean diurnal distribution of the number of people in Karaiskaki Square confirms that there is a relationship between the amount of people present and the local meteorological conditions.

In summer (Figure 9.3) the pattern of attendance is similar to that presented in Figure 9.2. Presence is fairly stable in the morning period and is at relatively low numbers, decreasing after midday. In early evening, presence increases significantly as it is getting dark and the air temperature is dropping.

In fall (Figure 9.3), the highest occurrence of people is normally in the mid-morning period, coinciding with the period when small children would be taken to the playground and elderly people meet up to exchange news, read their papers, etc., while it reduces as the day moves towards the afternoon. It is interesting that throughout the day, absolute numbers in fall are higher than in the summer for most of the day. This can only be attributed to climatic parameters, as other factors are the same.

In winter (Figure 9.3), presence is low at the beginning of the day, increasing rapidly to reach maximum levels at midday, and then decreasing rapidly again to give more stable figures in the afternoon. The rapidly increasing presence in the morning period is due to the low air temperatures at the start of the day, warming up significantly till midday, which then also coincides with people having their lunch break outdoors. With a mean air temperature of 14.4 °C and a maximum of 19.2 °C, outdoor activities are well supported with minimum personal adjustments. Similar seasonal variation was observed at the seashore; attendance also demonstrated strong dependency on the prevailing meteorological conditions.

In summary, air temperature and solar radiation are the most coherent dominant parameters in relation to the use of space, as respective year-round correlations with wind speed and relative humidity are very weak.

This also demonstrates how fluid the relationship between the indoor and outdoor environment can be

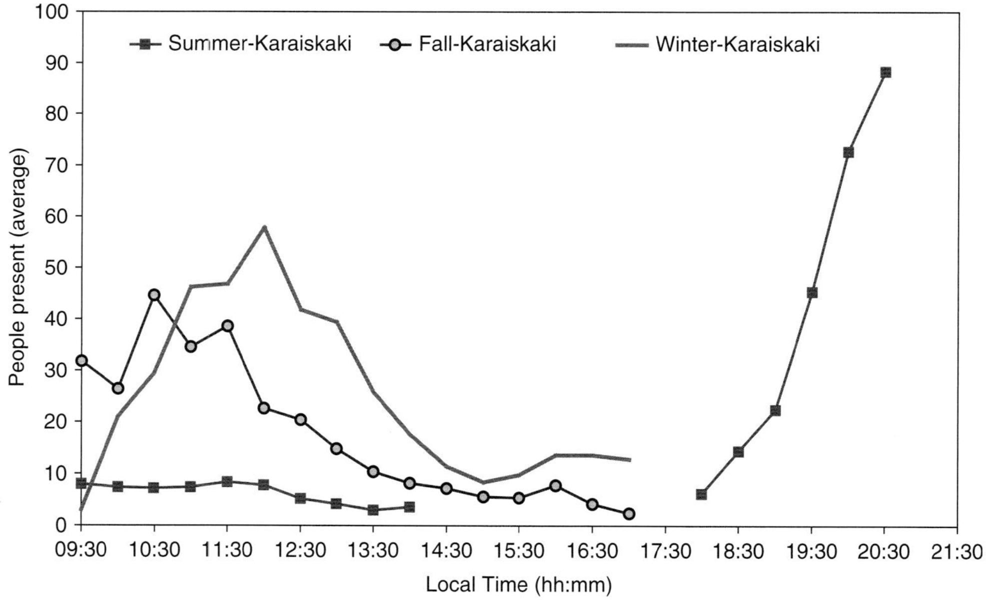

Figure 9.3 Seasonal variation of the mean diurnal distribution of the number of people at Karaiskaki Square, Athens (Nikolopoulou and Lykoudis 2007).

in a well-designed space. The boundaries between the two can be blurred and social life can extend into open spaces. This can be critical when considering our adaptive capacity for conditions of climate change. If building occupants did not have the option of moving outside for a period of time while the building fabric is cooling down, they would have no other option but to take drastic measures to improve the environmental conditions, using energy intensive means such as air-conditioning.

9.4 THERMAL PERCEPTION

Thermal comfort is examined for the 14 sites in the seven cities across Europe. The objective microclimatic parameters investigated included air temperature, solar radiation, wind speed and humidity. Nearly 10 000 questionnaire-guided interviews were carried out in the different cities at different seasons (Nikolopoulou and Lykoudis 2006).

The mean microclimatic data recorded during the interview periods are in accordance with the data recorded at the meteorological station for all cities and seasons, with small differences attributed to the effect of the urban fabric. The highest mean air temperature in the summer was recorded in Athens (30.1 °C) and the lowest in Sheffield (21.3 °C). In winter, the highest mean air temperature was in Athens (16.4 °C) and lowest in Kassel (5.4 °C).

People's thermal sensation was reported on a 5-point scale, varying from "very cold" (−2) to "very hot" (+2), which was defined as the actual sensation vote (ASV). Investigating correlations between the microclimatic variables and ASV revealed that ASV correlates better with globe temperature ($r = 0.53$, $p < 0.01$, Pearson correlation coefficient), than with air temperature ($r = 0.43$, $p < 0.01$), which is attributed to the radiant effect of the sun. The relatively weak correlations between microclimatic variables and ASV indicate that one parameter alone is not sufficient for the assessment of thermal comfort conditions.

The relationship between wind speed and wind vote is moderately weak ($r = 0.26$, $p < 0.01$), implying increasing comfort moving from calm to moderate wind conditions and discomfort as wind increases. This is, however, further affected by the desired cooling effect of wind in different seasons. These influences are consistent with the impact on attendance in the open spaces, examined in the previous section.

Due to the large variations between people, it is impossible to achieve 100% of people feeling thermally comfortable in any given space. This is recognized by the comfort requirements set by international standards (ISO 7730, 2005), where they specify that the thermal environment has to be comfortable for 90% of the people. Given the fact that the indoor environment is, in most cases, strictly controlled, it would be expected that in the outdoor context, where environmental parameters vary significantly annually as well as seasonally, thermal comfort would be significantly lower.

Examining overall comfort levels for the different cities, it is then surprising how high these are for all cities, seasonally as well as annually (Figure 9.4), demonstrating that the vast majority of people are satisfied with the environment. The percentage of overall comfort on a yearly basis is over 75% for all cities, reaching 91% for Cambridge. Even in Athens in the summer, when high air temperature is frequently a source of discomfort, overall comfort is 73%, reaching 93% in winter. This suggests an enormous adaptive capacity across different climatic zones and cultural backgrounds.

This is consistent with the finding of another more recent survey in the United Kingdom, the ASCCUE project (Wilson *et al.* 2008). That project employed a methodology similar to RUROS, with surveys carried out in the cities of Lewes and Manchester, and the results are aligned with those of RUROS.

Differentiating thermal comfort from thermal sensation, we can investigate the distribution of people voting for the different categories. We would expect that thermal comfort normally lies in ASV from −1 "cool" to +1 "warm", assuming that most people voting for the extremes "very cold/hot" are broadly thermally uncomfortable. Around 90% of the population that participated in the surveys for each country voted in the region of $-1 \leq \text{ASV} \leq +1$ (Figure 9.5). This is particularly interesting considering the great variation of the microclimatic conditions across Europe. Only 4% reported to be "very hot" (+2) and another 4% "very cold" (−2), figures regarded as acceptable, even in the tightly controlled indoor environment. The majority of the votes is for neutrality (0), with 44%, with "warm" (+1) and "cool" (−1) votes at a nearly equal split of 24%. Personal adaptation mechanisms are to be attributed to the very small amount of extreme votes under a very wide range of climatic conditions.

Focusing on the conditions where people report to be thermally neutral, that is "neither warm nor cool", it

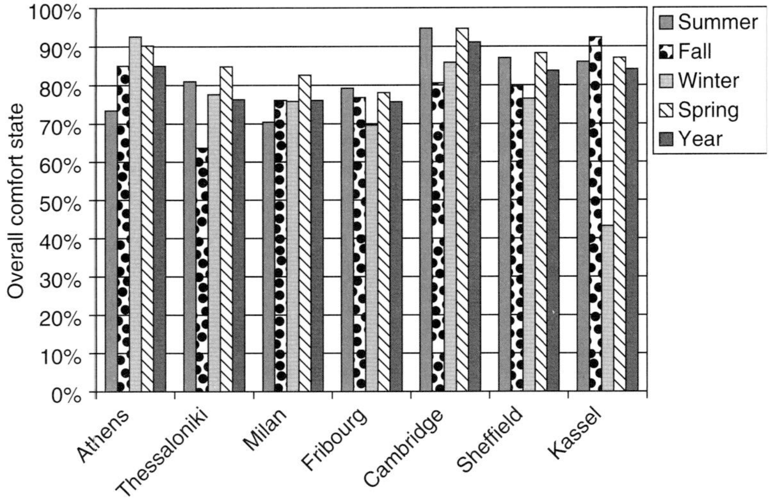

Figure 9.4 Percentage distribution of overall comfort state for the different cities in different seasons (Nikolopoulou and Lykoudis 2006).

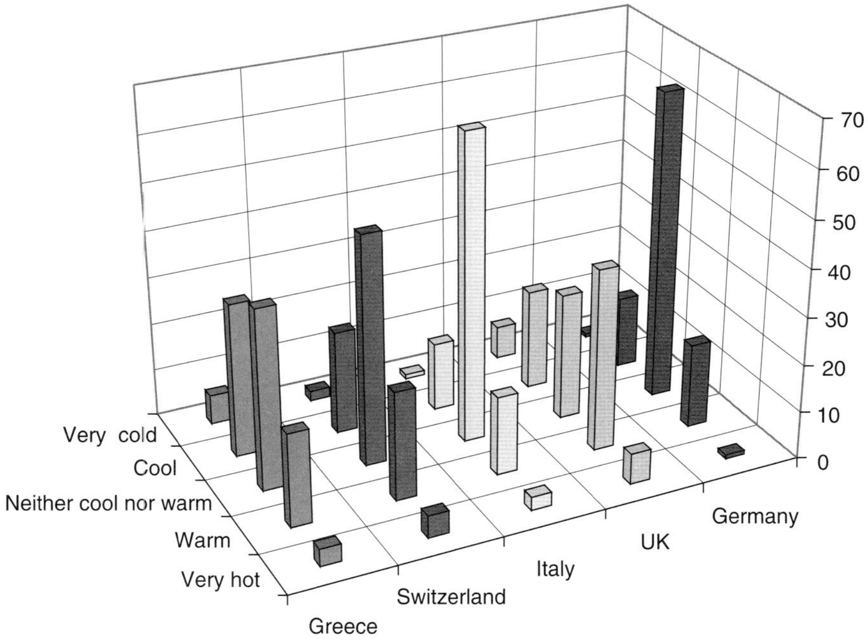

Figure 9.5 Percentage distribution of the actual sensation vote of the interviewees (ASV) throughout the year for the different countries (Nikolopoulou and Lykoudis 2006).

is particularly interesting to see how wide the neutral zone is. The annual neutral temperature across Europe, i.e., the temperature where people feel neither warm nor cool presents a range of 10 K, as for Athens it is 23 °C and for Fribourg 13 °C.

As there are no important physiological differences in the human body between the different parts of Europe, there must be other parameters that become responsible for this wide variation.

9.5 ADAPTATION

Understanding how adaptation works in the context of outdoor thermal comfort can help us not only to design better urban environments, but most importantly to employ these mechanisms to increase our cities' adaptive capacity. Such actions can play an important role under climate change.

We can broadly define "adaptation" as the gradual decrease of the organism's response to repeated exposure to a stimulus, involving all the actions that make them better suited to survive in such an environment. In the framework of thermal comfort this may involve all the processes which people go through to improve the fit between the environment and their requirements. In the past, Nikolopoulou (2004) defined three kinds of adaptation; physiological, physical, and psychological. Physiological adaptation or acclimatization implies changes in the physiological responses resulting from repeated exposure to a stimulus, leading to a gradual decreased strain from such exposure (Clark and Edholm 1985); it is thus not of central importance in this context.

9.5.1 Physical adaptation

Physical adaptation refers to the changes a person makes in order to adjust him/herself to the environment, or alter the environment to his/her needs. In this context two different kinds of adaptation were identified, *reactive* and *interactive*.

Reactive adaptation refers to personal changes taking place, such as altering one's clothing levels, position, etc. The RUROS analysis confirmed the seasonal variation of clothing, with a correlation coefficient between air temperature and clothing insulation levels of −0.61 ($p < 0.01$), where clothing insulation reduces as air temperature rises. This strong relationship was apparent irrespective of geographic location.

Changes to the metabolic heat can also be viewed as an adaptive action, either changing one's metabolic rate (e.g., by moving around as opposed to sitting), or with the consumption of cool drinks to reduce one's metabolic heat. In fact there is a tendency for lower physical activity as air temperature increases, even though the correlation is weak ($r = -0.20$, $p < 0.01$). The consumption of cool drinks has been demonstrated to affect the metabolic heat produced, reducing it by 10% (Baker and Standeven 1996). Thus, although the consumption of cool drinks is not solely a response to thermal conditions, there are increased occurrences of people having cold drinks as air temperature rises ($r = 0.19$, $p < 0.01$).

Spatial variation, by changing one's position, is another way to avoid discomfort, and strongly depends on microclimatic conditions. In a previous study in the United Kingdom, 43% of the variation of people sitting in shade was attributed to air temperature (Nikolopoulou 2004).

In the case of *interactive adaptation*, people interact, making changes to the environment in order to improve their comfort conditions, opening a window or a parasol, etc. Due to the nature of open spaces, such interaction is limited, as it requires the presence of specific elements (e.g., movable shading devices) and clear decisions on their operation. However, such a form of adaptation can be employed extensively in the design of open spaces.

9.5.2 Psychological adaptation

As thermal comfort is defined as a "state of mind that expresses satisfaction with the thermal surroundings" (ASHRAE 2004) understanding psychological adaptation presents an important challenge.

Different people perceive the environment in a different way, hence human response to a physical stimulus is not simply a function of its magnitude, but also depends on the "information" that people have for a particular situation. This goes significantly beyond thermophysiological models. Thermal, emotional and perceptual assessments of a physical place may be intertwined with psychological schema-based and sociocultural processes (Knez and Thorsson 2006), which justifies Knez *et al.* (2008) referring to the psychology of outdoor place and weather assessment. Psychological factors therefore influence the thermal perception of a space. These were initially analyzed by Nikolopoulou and Steemers (2003), but are examined here in view of the wealth of information provided by the RUROS project and other studies.

Naturalness

"Naturalness" as a term was employed by Griffiths *et al.* (1987), describing an environment free from artificiality. People appear to tolerate wide changes of the physical environment, provided they occur naturally, or as McIntyre (1980, p.245) describes it in relation to the increased dissatisfaction of fluctuations of temperatures in air-conditioned buildings: "fluctuations are tolerated if produced by God, but not if produced by the management".

The best example that this is an important parameter in people's perception of outdoor spaces is highlighted in Figure 9.5, where comfort levels are very high for a wide range of air temperatures across Europe. In such places, where all the climatic changes occur naturally, wide changes of the physical environment are accepted.

A consistent and persistent message is also given by a recent international review, which confirmed the importance of a natural environment for human well-being across different cultures (Matsuoka and Kaplan 2008).

Expectations and experience

"Expectations", i.e., what the environment should be like, rather than what it actually is like, greatly influence people's perceptions. This is directly affected by people's experience (Wohlwill 1974), in this case with similar microclimatic conditions.

In the studies in Marrakesh and Phoenix, people who considered the conditions typical for the season tended to spend more time in the space (Aljawabra and Nikolopoulou 2009), than those who did not find it typical or were not familiar with the climatic conditions of the area (Figure 9.6).

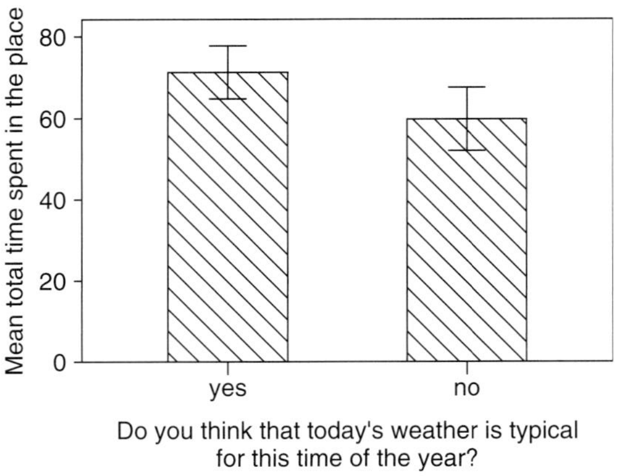

Figure 9.6 Mean time spent outdoors as function of weather expectation (Aljawabra and Nikolopoulou 2009).

The RUROS study has provided extensive evidence to support this argument (Nikolopoulou and Lykoudis 2006). The low percentage of hot discomfort in climates such as Greece and Italy in the summer (Figure 9.4) is, to a great extent, due to the influence of experience and expectations, as in both countries hot summers are expected, people have learned to cope with them and are not seriously affected by them. On the other hand, in Switzerland summers are normally cooler, resulting in thermal discomfort from the heat even though microclimatic conditions are more favorable than in Italy and Greece for the same season.

Such effects become more prominent when looking at the neutral temperatures. Physical adaptation, such as clothing, can only partially justify this extensive range. Comparing neutral temperatures with the respective long-term climatic temperatures for the different cities (Figure 9.7), it is apparent that they follow the profile of the respective climatic temperatures on a seasonal basis in all cities. The difference between neutral and climatic temperature is particularly interesting. In the summer the two sets of temperature lie very close, with the biggest difference noticed in winter. The intermediate seasons lie in between, with spring neutral temperature being closer to the respective climatic air temperature than fall.

This difference between neutral temperature and climatic air temperature appears to be inversely proportional to the mean climatic air temperature of the region; as climatic air temperature increases, the closer the neutral temperature is to it, as is the case in summer. This also appears to have a physical explanation as physiology suggests that in warmer conditions people have the required mechanisms to adapt more easily (Höppe 2002), provided the heat does not become life threatening.

The role of recent experience in influencing expectations is also demonstrated through the relevant shift

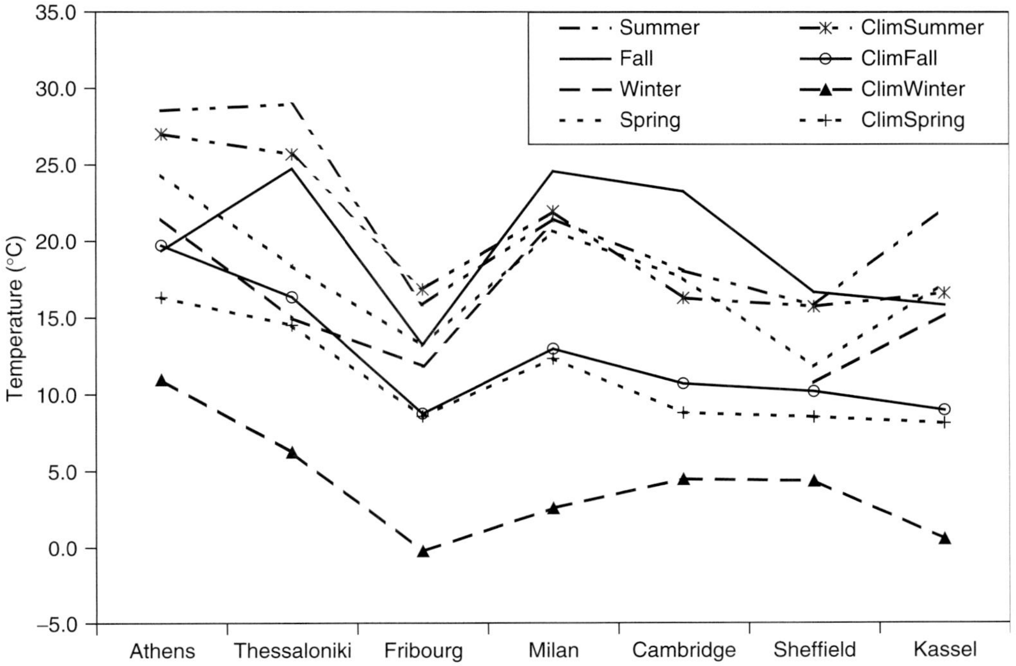

Figure 9.7 Neutral temperatures compared with the relevant climatic air temperature for different seasons for the different cities (Nikolopoulou and Lykoudis 2006).

of the comfort zones for different times of the year, as people seem to prefer temperatures which follow the profile of the preceding season (Figure 9.8). Thus warmer temperatures are expected in fall, following the hot climatic conditions of the summer, whereas in spring, cooler temperatures are regarded as comfortable, following the cold conditions of winter.

Time of exposure

Exposure to discomfort can be tolerated well if people anticipate that it is short lived (Evans 1982, Campbell 1983, Baker 1993). In line with the argument above, unless exposure to discomfort is life-threatening, tolerance to the thermal environment is great.

Perceived control

Perceived control differs from actual control; instead it suggests available choice and it does not imply that this control is exercised.

Existing literature acknowledges that people with a high degree of control over a source of discomfort tolerate wide variations, are less annoyed by it, and the negative emotional responses are greatly reduced. (Evans 1982, Paciuk 1990). Choosing a location to sit in (sun versus shade), presence by free will, and the choice to leave when a microclimatic stressor is no longer welcome, give individuals the feeling of being in control of a situation.

Perceived control is critical in outdoor spaces as actual control over the microclimate is very limited.

Environmental stimulation

"Thermal monotony and the fallacy of neutrality" (Brager and de Dear 2003) best highlight our obsession with thermal comfort being defined as neutrality. International standards are obsessed with uniform conditions to achieve thermal comfort, when in fact it is increasingly accepted that a static environment becomes rapidly intolerable and a variable environment is preferred.

This same definition of comfort, Sennett argues, is associated with rest and passivity. By lowering the amount of stimulation and relaxing its intensity,

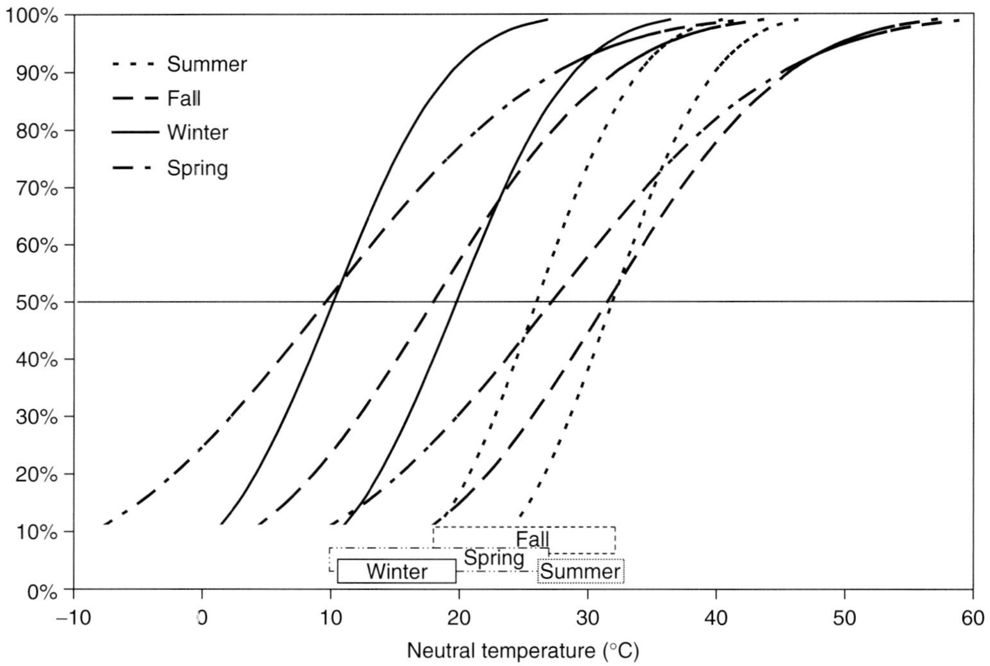

Figure 9.8 Percentage distribution of change of neutral temperature (°C) to shift from cool to neutral and from neutral to warm for Thessaloniki in different seasons. Width of comfort zone for the different seasons is transposed at the bottom of the chart (Nikolopoulou and Lykoudis 2006).

comfort becomes "an essay in monotony" (Sennett 1994, p.364). From a human ecology perspective, change and variation are important to reduce boredom or monotony (Wohlwill 1974).

Even our language and vocabulary has different words and phrases to refer to thermal stimulations and sensations. The Cambridge Dictionary defines a breeze as "a light and pleasant wind" (*Cambridge International Dictionary of English* 1995). It can be warm or cool but not unwanted wind movement. When we want to refer to pleasant thermal conditions, we use phrases such as "nice and warm" or "nice and cool" but never "neutral". Neutral means not noticeable and involves connotations of grayness or blending in with the background, suggesting boredom and indifference, precluding the development of exciting feelings/thoughts. Satisfaction may only be achieved through sensation and not neutrality.

There have been books written celebrating this experience, as we become aware of a space from the impact that it has on our senses. Rasmussen (1962) wrote *Experiencing Architecture*, and Heschong wrote *Thermal Delight in Architecture*. In the latter, she describes how, unlike other senses that can be isolated by closing our eyes, blocking our ears etc., the thermal sense cannot be easily isolated, it is constantly active, bound up with the experience of our bodies (Heschong 1979).

Provision of positive stimulation is important for satisfaction with a place and environmental stimulation becomes of primary importance in outdoor spaces. This is an important asset of these areas and one of the main reasons that people spend time in these spaces.

9.6 DESIGN INTERVENTIONS

But how do we enable design to play a critical role in providing pleasant thermal stimulation? How can we use thermal comfort as a tool and understanding of psychology to inform about the design of open space? How does the urban environment enhance our adaptive opportunity in view of our vulnerability under climate change? These questions need to be considered with practical suggestions for implementation in the context of outdoor thermal comfort.

The urban environment can be regarded as a living laboratory, and by using small or large scale interventions we can use urban design constructively to improve the environmental conditions in these spaces and outdoor thermal comfort in relation to climate change mitigation measures. Urban centers are particularly vulnerable areas as a result of the urban heat island and poor air quality. NGOs and local authorities (CABE Space 2008, Land Use Consultants *et al.* 2006) indeed promote outdoor living and the café culture in urban centers as a way to escape indoor heat. For that to be effective, the outdoor space needs to be considered as an integral part of the urban environment.

Key strategies that can improve outdoor thermal comfort in warm conditions include a range of methods and technologies to improve the urban microclimate in areas people use. All of these have a significant impact on thermal comfort, reducing the thermal load on the human body and the risk of overheating. Most of them also increase the capacity for personal adaptation to take place, both physically as well as psychologically. It is possible for good design to enhance environmental diversity, a sense of place as well as respond to the biggest challenge our century faces – climate change.

9.6.1 Materials

The physical properties of materials used in the urban environment for vertical walls, pavements and streets greatly affect the urban microclimate. Optical properties of materials in particular (i.e., albedo to solar radiation (or solar reflectance)) greatly determine surface temperatures in the urban environment, which influence thermal balance through radiant exchanges. Thus the darker the surface color, the higher is the surface temperature and hence the higher the thermal load on people in outdoor spaces. Traditional Mediterranean cities with their white painted walls and roofs (Figure 9.9) present such an example.

Various studies have been carried out on the thermal performance of materials used in pavements and streets on the urban microclimate. The resulting surface temperature, heat storage and subsequent emission of heat to the atmosphere are significantly greater for dark colored surfaces. Experiments have shown that surfaces with white elastomeric coating with an albedo of 0.72 were 45 K cooler than those with a black coating with an albedo of 0.08, while a white surface with 0.6 albedo was only 5 K warmer than ambient temperature (Taha 1997), thus significantly reducing the thermal load on the urban environment.

Figure 9.9 Traditional white settlements in the Mediterranean region – the island of Santorini. Photo: M. Nikolopoulou.

In a different climatic context, Watkins *et al.* (2007) investigated the resulting surface temperatures from different walls in an urban canyon in London, with reflectivities of 0.5, 0.08 and 0.03. By 1p.m., the surface temperature of the darker walls was very similar, while for the light-colored wall, it was 6–10 K cooler. Further analysis demonstrated that if all the surfaces in that canyon had a similar white matt reflectance, the resulting air temperature on a sunny day at the specific latitude could by reduced by 3–4 K. Once again, this presents a significant reduction of the thermal load on the human body.

Akbari (2007) reports that a 10 K decrease in temperature can be achieved by increasing the albedo of asphalt pavement by 0.25. Doulos *et al.* (2004) compared the thermal performance of 93 pavement materials used outdoors. They found that apart from color, surface texture and the construction material also affect the albedo of the tiles. Tiles made of marble, mosaic and stone were coolest, while tiles with smooth and flat surfaces were cooler than those with rougher surfaces.

It is not always desirable, however, to have light-colored surfaces. For this, pigments with improved solar spectral optical properties have been developed recently, which appear dark but have the properties of light colored surfaces (Levinson *et al.* 2005). These cool non-white surfaces have an innovative paint coating technology that achieves a high albedo and low brightness (i.e., high reflectivity for the near infrared and low reflectivity for the visible spectrum) to reduce the thermal load in the urban environment.

9.6.2 Vegetation

Landscaping and the use of vegetation can affect the thermal environment in different ways, significantly contributing to the cooling of the surrounding environment.

Vegetation reduces air temperature by direct shading of surfaces and by moderating solar heat gain through evapotranspiration of the plants. As the plants transpire, they lose water to the atmosphere by evaporation, converting incident solar radiation to latent heat, a process known as evapotranspiration – a major mechanism by which plants contribute to lowering ambient air temperatures.

The resulting lower ambient temperature leads to reduced long-wave radiation emitted from the ground and leaves, as opposed to the surrounding artificial hard surfaces, thus subjecting people to a reduced radiant load and improved thermal comfort conditions in the summer.

Various studies have demonstrated the beneficial effects of vegetation in reducing ambient air temperatures and the cooling load of the nearby buildings. Givoni (1989) reports differences of summer air temperatures of up to 3 K between the air above exposed pavements and behind shrubs in the same area. Dimoudi and Nikolopoulou (2003), using simulations of vegetation native to the Mediterranean area, found a 0.8 K reduction in ambient air temperature for a 10% increase in the ratio of green to built area for the urban texture under consideration.

The importance of the urban greenscape has also been shown to be critical under the various climate change scenarios for the United Kingdom. Adding 10% vegetation in high density urban areas kept maximum surface temperatures at or below the 1961–1990 baseline levels for the 2020s and 2050s low and high emission scenarios, but not for the 2080s high emissions (Gill *et al.* 2007). More alarmingly, by reducing vegetation by 10% (which frequently happens under intense development of urban areas), surface temperatures are shown to increase by 7–8.2 K by the 2080s high emissions scenario.

Vegetation can be integrated into the urban environment in a number of ways. At the building scale it includes pergolas attached to building walls with climbing plants (Figure 9.10), trees near buildings, even green roofs. At the city scale, vegetation can be included in parks, gardens, squares and playgrounds, pavements, riverbanks, car parks, etc., or any other open space where planting is feasible.

Figure 9.10 Effective use of vegetation for shading. Photo: M. Nikolopoulou.

Appropriate vegetation species have to be carefully selected, taking into consideration issues such as depth required for the roots and potential damage to nearby buildings, crown elevation for pedestrians' circulation, as well as issues such as sensitivity to pollution, irrigation and water scarcity, etc. Furthermore, deciduous or evergreen species should be carefully selected in areas where solar access is needed in winter.

Considering the adaptive capacity under climate change, it is worth remembering that vegetation can also effectively reduce water runoff and flooding.

9.6.3 Shading

Shading from direct solar radiation will not only prevent ambient temperature from rising, but will also reduce the thermal load on pedestrians. Shading whole streets using man-made canopies with the available materials has been used for centuries across the world (Figure 9.11). Whether permanent devices, or movable awnings, Rudofsky (1969) provides many strikingly similar examples found in different parts of the world

Figure 9.11 Canopies made of readily available materials. Photo courtesy of Faisal Aljawabra.

and not unique to specific cultures, from Seville in Spain to Osaka, Japan. The architectural vocabulary provides further examples for effective microclimatic control, such as arcades in Italian towns.

In modern times, proprietors of commercial activities are the first to realize the potential for extending the outdoor season and provide means for improving microclimatic conditions, frequently with the use of movable parasols.

9.6.4 Water

Through the process of evaporation, water can cool the ambient air. Furthermore, water has a very high specific heat capacity (c_p is 4200 J kg^{-1} K, whereas for most of the commonly used building materials c_p is around 1000 J kg^{-1} K). Thus it can absorb solar radiation and store very large amounts of heat without increasing its temperature, thus stabilizing ambient temperatures.

In evaporative cooling, water changes its state from liquid to vapor, with a release of sensible heat from the surrounding air, which is accompanied by a drop in air temperature and an increase in humidity. This has been widely exploited since antiquity in vernacular architecture in hot dry climates. Water features can be employed to cool the ambient environment, with ponds, fountains, water cascades, sprays, ponds, etc. (Figure 9.12).

Figure 9.12 Water fountains providing interaction. Photo: M. Nikolopoulou.

Technological advances provide more innovative solutions for microclimatic control, such as micronizers with nozzles to cool the ambient air through evaporation. These were used in Expo '92, in Seville, where outdoor activities were permitted during summer daytime temperatures of 35 °C. Thirty meter high wind-cooling towers were developed, with the top 25 m covered by a white plastic mast (Figure 9.13). In their interior, a system of micronizers was employed, with nozzles spread in sections of the upper part to cool the air caught at the top of the tower through evaporation (López de Asiain 1997). This is an innovative way of adapting the traditional Middle Eastern wind towers, which were later adapted by passing the air over a water pond, for the modern needs of cooling large areas.

9.6.5 Other measures

In the urban environment there is a range of other measures that can be considered to reduce the excess heat in areas people use, which go beyond the scope of this chapter. Such an example is an integrated network of public means of transport and cycling, which would reduce the excessive reliance on private cars, reducing the anthropogenic heat in outdoor spaces, as well as pollution. Further reductions in outdoor heat can also be achieved by a network of pedestrian streets, where people could circulate freely, without the added burden of the heat generated by vehicular traffic. Similarly, when air-conditioning units are installed great care

Figure 9.13 Wind cooling towers and micronizers at the Expo '92 in Seville. Photo courtesy of Jaime López de Asiain.

should be taken to ensure that the hot air is not expelled into the areas used by the public, as this would increase the thermal load significantly.

Improving the microclimate of an area has many more benefits from increasing biodiversity and carbon sequestration with the increased use of urban green to benefits for the image of the city.

Finally, a good thermal outdoor environment has a positive influence on the indoor environment, allowing natural ventilation and reducing the reliance of air-conditioning, breaking the otherwise unavoidable vicious circle.

9.7 CONCLUSIONS

Open spaces can play an important role in cooling our cities, providing networks of clean air, and helping us to escape the heat of the built environment. However, they are not only climate regulators. They are also an important amenity for the neighboring communities, they support urban life and can ensure social cohesion and well-being.

We already have evidence that sensitivity to heat varies for different age-groups. The RUROS project showed that the presence of the age category "older than 65" accounted for less than 5% of the visitors in the summer in Athens, a near fourfold reduction from the rest of the seasons (Nikolopoulou and Lykoudis 2007). We need to ensure that open spaces remain inviting for all people if we are to reduce the over-reliance on air-conditioning.

Understanding the impact of environmental stimuli and thermal perception on the use of open urban spaces can enable these spaces to fulfill their full potential and provide an oasis in an otherwise sweltering urban fabric. As has been demonstrated in this chapter, both ambient temperature and solar radiation have a strong impact on the use of open spaces, while there is a strong relationship between microclimatic and comfort conditions.

Personal adaptation can play an important role in accommodating the wide range of microclimatic conditions found outdoors, but we should not misinterpret this adaptation. Appropriate design is essential to increase the adaptation capabilities at a personal level, balancing exposure and protection to the different climatic elements.

The impact of expectations and need for environmental stimulation should not be misread. We need to understand the difference between the challenging, fulfilling "eustress" (from the Greek *eu*, meaning "good") and damaging distress (Selye 1985). What differentiates eustress from distress is the way the stressor is perceived. Cognitive appraisal, such as past experiences with the stressor, expectations, duration of exposure or available control, will assist the appraisal of the stressor as threatening or harmless. Careful physical intervention can provide the means for eustress to be achieved.

Environmental diversity can increase spatial variation and available choice for users of the space while improving the local microclimate. Naturalness, through integration of vegetation and water elements can have similar effects while also satisfying the need for environmental stimulation.

Understanding people's thermal experience in urban spaces can inform our vocabulary of independent physical interventions. Such knowledge will enable us to use design constructively to improve environmental quality and increase our adaptive capacity under climate change.

REFERENCES

Akbari, H. (2007) Opportunities for saving energy and improving air quality in urban heat islands. In: Santamouris, M. (ed.) *Advances in Passive Cooling*. Earthscan, London, pp. 30–92.

Aljawabra, F., Nikolopoulou, M. (2009) Outdoor thermal comfort in the hot arid climate: The effect of socio-economic background and cultural differences. *Proc. PLEA 26th Conference on Passive and Low Energy Architecture*, Quebec City, Canada, 22–24 June.

ASHRAE (American Society of Heating, Refrigerating and Air-Conditioning Engineers) (2004) *ANSI/ASHRAE Standard 55–2004: Thermal Environmental Conditions for Human Occupancy*, Atlanta.

Baker, N. (1993) Comfort in passive buildings. *Proc. Solar Energy and Buildings*, Academy of Athens, December.

Baker, N., Standeven, M. (1996) Thermal comfort for free-running buildings. *Energy and Buildings* 23(3), 75–182.

Brager, G.S., de Dear, R.J. (2003) Historical and cultural influences on comfort expectations. In: Cole, R.J., Lorch, R. (eds) *Buildings, Culture and Environment: Informing Local and Global Practices*. Blackwell Publishing, UK, pp. 177–200.

CABE Space (2008) *Public Space Lessons: Adapting Public Space to Climate Change*, London, July.

Campbell, J.M. (1983) Ambient stressors. *Environment and Behavior* 15(3), 355–380.

Clark, R.P., Edholm, O.G. (1985) *Man and his Thermal Environment*. Edward Arnold, London.

Department of Health (2002) *Health Effects of Climate Change in the UK*. The Stationery Office, London.

Dimoudi, A., Nikolopoulou, M. (2003) Vegetation in the urban environment: microclimatic effects and benefits. *Energy and Buildings* 35(1), 69–76.

Doulos, L. Santamouris, M., Livada, I. (2004) Passive cooling of outdoor urban spaces. The role of materials. *Solar Energy* 77, 231–249.

Eliasson, I., Knez, I., Westerberg, U., Thorsson, S., Lindberg, F. (2007) Climate and behaviour in a Nordic city. *Landscape and Urban Planning* 82, 72–84.

Evans, G.W. (1982) *Environmental Stress*. Cambridge University Press, Cambridge.

Gill, S.E., Handley, J.F., Ennos, A.R., Pauleit, S. (2007) Adapting cities for climate change: the role of green infrastructure. *Built Environment* 33(1), 115–133.

Givoni, B. (1989) *Urban Design in Different Climates*. World Meteorological Organization, WMO/TD-No. 346.

Givoni, B., Noguchi, M., Saaroni, H., *et al.* (2003) Outdoor comfort research issues. *Energy and Buildings* 35(1), 77–86.

Griffiths, I.D., Huber, J.W., Baillie, A.P. (1987) Integrating the environment. In: Steemers, T.C., Palz, W. (eds) *Proceedings of the 1987, European Conference on Architecture*. Kluwer Academic Publishers, Dordrecht.

Heschong, L. (1979) *Thermal Delight in Architecture*. The MIT Press, Cambridge, MA.

Höppe, P. (2002) Different aspects if assessing indoor and outdoor thermal comfort. *Energy and Buildings* 34(6), 661–665.

IPCC (2007) *Intergovernmental Panel on Climate Change, Fourth Assessment Report*, Geneva.

ISO 7730 (2005) *Ergonomics of the Thermal Environment - Analytical Determination and Interpretation of Thermal Comfort using Calculation of the PMV and PPD Indices and Local Thermal Comfort Criteria*. International Standards Organisation, Geneva.

Knez, I., Thorsson, S. (2006) Influences of culture and environmental attitude on thermal, emotional and perceptual evaluations of a public square. *International Journal of Biometeorology* 50, 258–268.

Knez, I., Thorsson, S., Eliasson, I., Lindberg, F. (2008) Psychological mechanisms in outdoor place and weather assessment: towards a conceptual model. *International Journal of Biometeorology*, DOI 10.1007/s00484-008-0194-z.

Land Use Consultants in association with Oxford Brookes University, CAG Consultants and Gardiner & Theobald (2006) *Adapting to Climate Change Impacts – a Good Practice Guide for Sustainable Communities: a Case Study Companion to the Checklist for Development*. DEFRA, London.

Larsen, J. (2006) *Setting the Record Straight: More than 52,000 Europeans died from Heat in Summer 2003*. Earth Policy Institute. http://www.earth-policy.org/index.php?/plan_b_updates/2006/update56 (accessed 23 March 2011).

Levinson, R., Berdahl, P., Akbari, H. (2005) *Solar Spectral Properties of Pigments, or how to Design a Cool Non-white Coating*. Presentation, symposium "Cool Roofing: Cutting Through the Glare", Atlanta, GA. http://coolcolors.lbl.gov/assets/docs/OtherTalks/HowToDesignACoolNonwhiteCoating.pdf (acessed 23 March 2011).

López de Asiain, J. (1997) *Open Spaces of EXPO '92. Seville Arquitectura (5)*. The Superior Technical School of Architecture of Seville (ETSAS), Seville.

Matsuoka, R.H., Kaplan, R. (2008) People needs in the urban landscape: Analysis of landscape and urban planning contributions. *Landscape and Urban Planning* 84, 7–19.

McIntyre, D.A. (1980) *Indoor Climate*. Applied Science Publishers, London.

Nikolopoulou, M. (2004). Outdoor comfort. In: Steemers, K., Steane, M.A. (eds) *Architecture and Variety: Environmental Perspectives*. Spon Press, London, pp. 101–119.

Nikolopoulou, M., Lykoudis, S. (2006) Thermal comfort in outdoor urban spaces: analysis across different European countries. *Building and Environment* 41, 1455–1470.

Nikolopoulou, M., Lykoudis, S. (2007) Use of outdoor spaces and microclimate in a Mediterranean urban area. *Building and Environment* 42, 3691–3707.

Nikolopoulou, M., Steemers, K. (2003) Thermal comfort and psychological adaptation as a guide for designing urban spaces. *Energy and Buildings* 35(1), 95–101.

Nikolopoulou, M., Baker, N., Steemers, K. (2001) Thermal comfort in outdoor urban spaces: the human parameter. *Solar Energy* 70(3), 227–235.

Oke, T.R. (1987) *Boundary Layer Climates*, second edition. Methuen, London.

Paciuk, M. (1990) The role of personal control of the environment in thermal comfort and satisfaction at the workplace. In: Selby, R.I., Anthony, K.H., Choi, J., Orland, B. (eds) *Coming of Age*. Environmental Design Research Association, Oklahoma City, OK, pp. 303–312.

Procter, P. (1995) *Cambridge International Dictionary of English*. Cambridge University Press, Cambridge.

Rasmussen, S.E. (1962) *Experiencing Architecture*. The MIT Press, Cambridge, MA.

Rudofsky, B. (1969) *Streets for People: a Primer for Americans*. Garden City, New York.

Santamouris, M. (2007) (ed.) *Advances in Passive Cooling*. Earthscan, London.

Selye, H. (1985) History and the present status of the stress concept. In: Monat, A., Lazarus, R.S. (eds) *Stress and Coping: an Anthology*, second edition. Columbia University Press, New York, pp. 17–29.

Sennett, R. (1994) *Flesh and Stone: the Body and the City in Western Civilisation*. Faber and Faber, London.

Spagnolo, S., de Dear, R. (2003) A field study of thermal comfort in outdoor and semi-outdoor environments in subtropical Sydney Australia. *Building and Environment* 38, 721–738.

Stathopoulos, T., Wu, H., Zacharias, J. (2004) Outdoor human comfort in an urban climate. *Building and Environment* 39, 297–305.

Taha, H. (1997) Urban climates and heat islands: albedo, evapotranspiration, and anthropogenic heat. *Energy and Buildings* 25, 99–103.

Watkins, R., Palmer, J., Kolokotroni, M. (2007) Increased temperature and intensification of the Urban Heat Island: Implications for human comfort and urban design. *Built Environment* 33(1), 85–96.

Wilson, E., Nicol, F., Nanayakkara, L., Ueberjahn-Tritta, A. (2008) Public open space and human thermal comfort: the implications of alternative climate change and socio-economic scenarios. *Journal of Environmental Policy and Planning* 10(1), 31–45.

Wohlwill, J.F. (1974) Human adaptation to levels of environmental stimulation. *Human Ecology* 2(2), 127–147.

Zacharias, J., Stathopoulos, T., Hanqing, W. (2001) Microclimate and downtown open space activity. *Environment and Behaviour* 33(2), 296–315.

D. URBAN BIODIVERSITY

CHAPTER 10

Social aspects of urban ecology in developing countries, with an emphasis on urban domestic gardens

Sarel Cilliers, Stefan Siebert, Elandrie Davoren and Rina Lubbe

School of Environmental Sciences and Development, North-West University, Potchefstroom, South Africa

10.1 INTRODUCTION

Urbanization is one of the most important demographic trends of all times due to the environmental changes that it brings about (Grimm *et al.* 2000, Pickett *et al.* 2001). Transformation and fragmentation of natural habitats, alteration of species composition leading to biotic homogenization, and a depletion of ecosystem services offered by natural ecosystems (McPherson 1998, Whitford *et al.* 2001, McKinney 2006, Tratalos *et al.* 2007) are some of the ecological consequences of urbanization. Hitherto, ecological research in urban areas has focused more on developed countries, but a growing body of research literature is emerging from developing countries. Recently, Pauchard *et al.* (2006) have found that 23% of the original wetlands and 9% of the agricultural, forest and shrub land cover types in Concepcion, Chile, have been lost over a 25 year period due to urbanization. In South Africa it was shown that urban land use has the largest impact of ten different land uses on landscape composition, structure and functioning, as well as on overall biodiversity integrity of the moist subbiome of the grassland biome (O'Connor and Kuyler 2009).

The impacts of urbanization on composition, structure and function of biodiversity have also influenced the socioeconomic world of humans (called the anthroposphere by Marzluff *et al.* 2008). Health in terms of physical, mental and social well-being has also been seriously affected in urban areas (Grahn and Stigsdotter 2003, Tzoulas *et al.* 2007). The ecological impacts of the pre-1994 planning of black settlements in South Africa has been summarized by Donaldson-Selby *et al.* (2007). Health problems are related to close proximity of these settlements to industrial areas and waste dumps and to overcrowding and a loss of access to natural assets for food production and livelihoods. It is quite clear that ecological problems can only be understood or resolved if problems within human societies are dealt with. According to Bookchin (1993), economic, ethnic, cultural and gender issues, among others, "lie at the core of the most serious ecological dislocations we face today." This is specifically true for urban areas and Gobster *et al.* (2004) highlighted the importance of social science research in providing answers to management issues in terms of increasing pressures on urban green areas. It should be recognized though that the urban ecosystem is a complex social–ecological

Applied Urban Ecology: A Global Framework, First Edition. Edited by Matthias Richter and Ulrike Weiland.

system consisting of strongly interacting systems or spheres, all linked to the anthroposphere (Marzluff *et al.* 2008). Integration of social and biogeophysical issues in urban ecological studies have been proposed in several studies providing guidelines in the form of conceptual frameworks (Pickett *et al.* 2001, Alberti *et al.* 2003, Yli-Pelkonen and Niemelä 2005, and Tzoulas *et al.* 2007). These frameworks have focused mainly on developed countries and important issues for developing countries, such as different aspects of health (Cilliers 2010), are often neglected. The framework proposed by Tzoulas *et al.* (2007) is an exception as it focused on the linkage of urban green areas with ecosystem and human health, combining different models.

Urban green areas are extremely important to minimize the mentioned negative consequences of urbanization and, according to Cilliers *et al.* (2004), we need to apply sound conservation orientated planning and management practices in urban areas. Some of the earliest studies on urban green areas were done in Germany (Sukopp and Weiler 1988) and attempted to map the mosaic of different urban biotopes based on the small-scale distribution of land uses. The objectives of these and other studies were to identify habitats of conservation significance in urban areas, focusing mainly on "designated natural areas" (Mathieu *et al.* 2007). In an overview on green infrastructure Tzoulas *et al.* (2007) mentioned that "all natural, seminatural and artificial networks of multifunctional ecological systems, within, around and between urban areas, at all spatial scales" should be included in urban ecological studies. Tzoulas *et al.* (2007) followed a typology of urban green areas widely used in the United Kingdom, which includes green roofs, urban parks, green corridors, encapsulated countryside, derelict land, housing green space and domestic gardens, churchyards, cemeteries, school grounds, and open standing or running water. Tzoulas *et al.* (2007) cited several studies that indicate that all different elements of a green infrastructure preserve and enhance biodiversity in terms of habitat, species and genes in the urban ecosystem and that they could have an influence on urban ecosystem health by contributing to ecosystem resilience, organization, and vigor.

The first objective of this chapter is to give an overview of social aspects of urban green infrastructure in terms of benefits, human perceptions and consequences of socioeconomic aspects on the ecology of urban areas. Second, an emphasis will be placed on urban domestic gardens, which are the least understood element of the urban green infrastructure. Domestic gardens are of special importance in developing countries with high unemployment rates, where products derived from these green areas can support livelihoods. Differences in the focus of social aspects of urban ecology between developed and developing countries will be highlighted throughout the chapter.

10.2 SOCIAL BENEFITS AND HUMAN PERCEPTIONS OF URBAN GREEN AREAS

Urban green areas have a wide spectrum of social, environmental and economic benefits and in terms of social benefits Swanwick *et al.* (2003) mentioned existence and use values. Some of the uses of urban green areas include various recreational opportunities, contributions to child development, numerous environmental education opportunities, promotion of social inclusion, development of a personal and community identity, and the strengthening of community ties in local communities (Swanwick *et al.* 2003, Conner 2005). Personal and community health include some of the most important social issues (Tzoulas *et al.* 2007) and are associated with all the green areas mentioned before, providing ecosystem services such as food, clean water, wood, fibre and fuel, and regulating climate, flooding and disease (Millennium Ecosystem Assessment 2005). Several studies and overviews have indicated that interactions with green areas may have positive psychological, physiological, emotional, mental and spiritual effects on urban residents (Frumkin 2001, Takano *et al.* 2002, Chiesura 2004, Senior and Townsend 2005, Gidlöff-Gunnarsson and Öhrström 2007, Tratalos *et al.* 2007). According to Grahn and Stigsdotter (2003) city planning in Western societies involves a social responsibility towards the health of city residents by developing more green areas such as parks. Accessibility of all residents to these green areas should also be improved to reduce stress-related diseases. In developing countries, on the other hand, there is extreme pressure on the existence of parks and other urban green areas and they are increasingly being replaced by residential and other urban developments (Cilliers *et al.* 2004, McConnachie *et al.* 2008). Cilliers 2010) highlighted therefore the importance of more "informal" and multifunctional green areas inside cities in developing countries by mentioning the social, ecological and economic benefits of urban agricultural areas, which also increase livability, equity and sustainability of cities.

Although there is little doubt about the multiple benefits of urban green areas, human perceptions about urban biodiversity in general are highly variable and will influence the composition and quality of urban green areas in the future. Perceptions, values and activities of politicians, decision-makers, planners and urban residents are also mentioned by Yli-Pelkonen and Niemelä (2005) as being important for the preservation of urban biodiversity. Incorporating these specific social aspects into conservation, planning and management of urban green areas is extremely challenging and begs for a transdisciplinary approach (Cilliers 2010). It is also important for ecologists to recognize research on human perceptions and preferences about nature and landscapes that has been widely discussed in the field of environmental psychology, according to Todorova *et al.* (2004).

There is a large body of research papers studying human perceptions towards nature in urban areas, especially from developed countries. A study of urban residents' attitudes towards flowers in street plantings in Japan have indicated a preference for ordered arrangements of low-growing and brightly colored flowers of a single species (Todorova *et al.* 2004). McGee (2007) studied the perceptions of residents towards management of urban forests in Canada and revealed a preference for management practices which impacted them the least; for example, environmental education was regarded as the most important practice as it does not require residents to act upon information, but the removal of flammable vegetation from their properties which requires them to act, was the least supported. In a study on the valuation of conservation features in residential subdivisions in the United States, it was shown that urban residents have a demand for and are willing to pay (WTP) for conservation orientated design (Bowman *et al.* 2009).

Fewer studies on human perceptions of urban green areas have been carried out in developing countries. In a study on a South African university campus, Faul (2008) indicated that students and personnel seek increased interaction with nature in their everyday environment and they acknowledged that urban greening would improve their work efficiency and willingness to perform tasks. Donaldson-Selby *et al.* (2007) studied photorealistic visualizations of urban greening in a low-cost, high-density housing settlement in South Africa and demonstrated that some forms of urban greening, such as dense river vegetation and flower and vegetable gardens, were regarded as unwanted or unacceptable by many urban residents. Cilliers (2010) reported on a number of case studies from South Africa (Roberts 2001, Roberts *et al.* 2005, Davis 2005, Cilliers *et al.* 2007, 2009) in which a participatory approach was followed in developing and implementing conservation strategies in urban areas and emphasized the importance of a total commitment from all stakeholders, including local government.

10.3 CONSEQUENCES OF SOCIOECONOMIC ASPECTS ON THE URBAN GREEN INFRASTRUCTURE

Several studies from developed and developing countries have shown that the socioeconomic status (SES) of urban residents is an important driver of the existence and ecology of urban green areas. In those studies SES was mainly derived from median household income obtained from census data. Based on these parameters, studies showed that there is a significant increase in bird diversity (Smith 2004, Melles 2005, Kinzig *et al.* 2005), plant diversity (Hope *et al.* 2003, Pedlowski *et al.* 2003, Martin *et al.* 2004) and tree cover (Iverson and Cook 2000, Jensen *et al.* 2004, McConnachie *et al.* 2008) in urban areas occupied by residents with a higher SES. The relationship between wealth and plant diversity is called the "luxury effect" by Hope *et al.* (2003) and is mainly caused by a profusion of human resources and higher levels of residential involvement in urban greening efforts in areas with a higher SES (Hope *et al.* 2003, Melles 2005, McConnachie *et al.* 2008). In a study of 10 small towns in South Africa, McConnachie *et al.* (2008) have shown that more densely populated towns with lower income levels have less proportionate municipal green space, as well as less green space maintenance and are more prone to disturbances such as removal of woody plants. In a study in Brazil, Pedlowski *et al.* (2003) have shown that tree abundance and diversity is not evenly distributed among high-, medium- and low-income neighborhoods. This is the result of policy decisions made by local government which have not recognized the current existence of unequal ecological quality. The social and spatial inequality in environmental resources and services is an extremely complex issue of urban life and needs to be addressed (Pedlowski *et al.* 2003).

According to Kinzig *et al.* (2005) planners and policy makers should find ways in which they can increase access to urban nature for the poor through better management of public spaces and better zoning practices, the so-called top-down factors. Kinzig *et al.* (2005), however, also mentioned the importance of bottom-up factors such as management of private green spaces by urban residents, which is determined by economic constraints and cultural ideas. Urban tree cover is, to a large degree, manipulated by bottom-up factors and Tratalos *et al.* (2007) have subsequently shown that areas with lower SES have lower carbon sequestration levels. Carbon reduction is an extremely important ecosystem service for urban residential areas as described by Jo and McPherson (2001). More emphasis needs to be placed on private green spaces, such as domestic gardens in urban ecological studies, in future.

10.4 URBAN DOMESTIC GARDENS

10.4.1 Literature review

Domestic gardens which could include allotment gardens, as part of urban green infrastructure, have to a large extent been ignored in urban ecological studies. In some studies domestic gardens were included in the research of neighborhood vegetation (Hope *et al.* 2003) but were not studied in detail. The main reasons why ecological studies on domestic gardens are quite rare include problems with access due to multiple ownership, the extremely dynamic nature of the gardens, and the fact that they are unregulated habitats with no recommended planning and management criteria or restrictions (Gaston *et al.* 2005a, Mathieu *et al.* 2007). From a scientific point of view, the biodiversity value of domestic gardens is also questioned as they appear to be too small to be biologically significant (Thompson *et al.* 2003, Gaston *et al.* 2005a, Mathieu *et al.* 2007). Ironically, as a composite the total area of private gardens is too large to be ignored as they represent quite a substantial part of the vegetated area in most cities. Domestic gardens cover 21.8–26.8% of the urban area in six cities in the United Kingdom (Belfast, Cardiff, Edinburgh, Leicester, Oxford, and Sheffield) (Gaston *et al.* 2005a, Loram *et al.* 2007) and 36% of the urban area in Dunedin, New Zealand (Mathieu *et al.* 2007).

According to Gaston *et al.* (2005a), domestic gardens offer a great opportunity to increase biodiversity and maintain the provision of ecosystem services in urban areas. It was further indicated by Loram *et al.* (2007) that the greater the degree of urbanization the more important gardens become as a biodiversity resource, as the total area covered by domestic gardens decreases with an increase in human population density and housing density. Although private gardens are quite small, they differ from most other urban green spaces which are notoriously patchy, in that they are interconnected (Smith *et al.* 2005) and link green areas inside the harsh artificial urban environment with each other (Mathieu *et al.* 2007). Domestic gardens are identified by Gaston *et al.* (2005a) as private spaces adjacent to or surrounding dwellings, comprising of lawns, ornamental and vegetable plots, trees, ponds, paths, patios and temporary buildings, which are quite similar all over the world but variable in terms of form and uses. It is therefore important to study domestic gardens as an additional resource for urban biodiversity and to ensure that they are included in the calculation and mapping of urban green space (Mathieu *et al.* 2007). Another reason for more information on the species composition of urban domestic gardens is that they form an important source for exotic species invasion into other urban and natural areas (Henderson 2006). It was indicated by Thompson *et al.* (2003) that the species occurring most frequently in gardens in Sheffield, United Kingdom were weeds or cultivated plants behaving like weeds, indicating that gardens and derelict land have much in common. Sullivan *et al.* (2005) reported from New Zealand that many attractive environmental weeds are still grown ornamentally in gardens and proper management of new housing subdivisions and the encouragement of responsible gardening are therefore important to reduce weed pressure and weed control costs in natural areas.

Most existing biodiversity studies of domestic gardens occurred in developed countries and to be more specific, the city of Sheffield in the United Kingdom. The main objectives of the Sheffield studies were to describe the size and composition of domestic gardens as a resource for biodiversity (Smith *et al.* 2006a,b,c, Thompson *et al.* 2005) and ecosystem functioning (Gaston *et al.* 2005a). Additionally the factors influencing the levels of biodiversity in different gardens (Thompson *et al.* 2003, 2004), and the effect of management manipulations to enhance native biodiversity in gardens (Gaston *et al.* 2005b) were addressed in the garden studies in the United Kingdom. Results of these studies are relevant to all temperate gardens in the northern hemisphere (Thompson *et al.* 2003) and have shown that gardens provide a complex and heterogeneous mosaic of habitats, which have unique ecological features and support quite a diversity of plants and animals.

Studies on urban domestic gardens in Australia were less focused on biodiversity and ecosystem functioning than the Sheffield studies, but more on the diversity of specific garden types and their relation to cultural, social, demographic and environmental variables (Head *et al.* 2004, Zagorski *et al.* 2004, Daniels and Kirkpatrick 2006, Kirkpatrick *et al.* 2007). The reason for this variation in focus lies in the realization that a "variety of social and economic processes encourage utilitarian and consumerist orderings in gardens" and gardens are therefore, "a significant locale for understanding human-nature relations" according to Bhatti and Church (2004). The study on domestic gardens in Hobart, Australia by Kirkpatrick *et al.* (2007) was the first quantitative attempt to determine whether the relationship between SES of urban residents and biodiversity, described earlier for urban areas in general (Hope *et al.* 2003), is also true for domestic gardens. The socioeconomic factors used in the study by Kirkpatrick *et al.* (2007) were more elaborative than those used in the study by Hope *et al.* (2003). Kirkpatrick *et al.* (2007) studied the following possible socioeconomic predictors for tree frequency and garden types in Hobart, Australia, namely median household income, percentages of people older than 65 years, of Australian-born people, of unemployed, of people renting households, of tertiary educated, and of those in professional or managerial jobs. Household income was the best predictor for the frequency of trees in front gardens, according to Kirkpatrick *et al.* (2007), but specific attributes of the physical environment, such as soil type also played a role in the presence of trees, while garden size was not a significant predictor for tree presence. Kirkpatrick *et al.* (2007) also showed that different variables were driving the occurrence of different garden types. The proportion of people with tertiary qualifications was the most explanatory socioeconomic variable for garden types – tertiary educated residents preferred woodland, flower and vegetable, as well as complex native gardens, while non-tertiary educated people were more inclined to simple native gardens, shrubs and trees (Kirkpatrick *et al.* 2007). Unemployment rate and not the proportion of rented dwellings was the best predictor for the absence of gardens. Exotic gardens with no intensive management inputs were associated more with renters (Kirkpatrick *et al.* 2007).

In a study comparing the characteristics of front and back gardens in Hobart, Australia, Daniels and Kirkpatrick (2006) found that the well-known model of showy front and productive back gardens has been replaced with many different models, with a clear association between low unemployment rates and gardens which are non-showy in the front and showy at the back, reflecting the reluctance of some workers to expend energy outside their backyard living space. The importance of cultural background in selecting a specific garden type was demonstrated by Head *et al.* (2004) who studied backyard gardens in Australia from Macedonian, Vietnamese and British-born residents, as well as first-generation migrants. Traditions from their homelands were carried on in most of these gardens within the social and biophysical constraints of their new environments, and there was a clear difference in emphasis on garden types between the different cultural groups (Head *et al.* 2004). It was also shown that intensive backyard food production has lost its emphasis to a large extent in the first-generation migrants (Head *et al.* 2004). The last observation was confirmed by Kirkpatrick *et al.* (2007) in the Hobart study where only 9 of the 1550 gardens studied were productive vegetable and fruit gardens, and only 78 of the gardens have some vegetables and fruit trees amongst a substantial component of ornamentals. Domestic gardens in developed countries focus more on ornamental species.

Although a vast array of studies exists on domestic gardens in developing countries, the emphasis is more on horticulture, urban agriculture and agroforestry in Africa, Asia, Latin America, and Amazonia and also includes home gardens in rural areas (Rugalema *et al.* 1994, Altieri *et al.* 1999, High and Shackleton 2000, Winklerprins 2002, Pandey *et al.* 2007). In developing countries, especially in the tropics, home gardens are described as multilayered, species-rich agroforestry systems, which include a wide variety of trees, shrubs, herbs and vines used as sources of food, medicine, fodder, firewood, construction materials, market products and ornamentals (Lamont *et al.* 1999). The state of literature on tropical home gardens is reviewed by Kumar and Nair (2004) who considered these gardens as "epitomes of sustainability". Some contributions of tropical home gardens to sustainability include biophysical advantages such as efficient nutrient cycling due to multispecies composition, conservation of biocultural diversity, product diversification, non-market value of products and services, and social and cultural values (Kumar and Nair 2004). High plant diversity in tropical home gardens is only regarded as important in terms of providing subsistence and income to the households, according to Pandey *et al.* (2007), while they are regarded as extremely important in the socioeconomic setting of local communities (Mohan 2004). Mohan (2004) quantified the economic and ecological benefits of home gardens in Kerala, India, in terms of cost–benefit analyses and species diversity indices, respectively, and concluded that these gardens are economically profitable and comparable to natural tropical ecosystems in terms of structure and composition. It was also shown that species richness was not affected by the size of home gardens, implying that urban farmers with small gardens intensify their plantings to include the maximum number of plants, probably indicating that they cultivate these plants to fulfill household needs, rather than for market sale (Mohan 2004). The esthetic and social value of home gardens should also not be underestimated, in that they are helping especially women to maintain a sense of esthetic pride, as well as a social and emotional link and a psychological buffer between urban and rural areas as households often move between these areas (Winklerprins 2002, Kumar and Nair 2004, Zobolo and Mkabela 2006). From a sociocultural point of view, home gardens are fast becoming important study areas for ethnobotanists in developing countries, while those studies are nearly nonexistent in Europe and the United States (Vogl *et al.* 2004). From literature it is clear that the emphasis in gardens of developing countries seems to be more on useful plant species and these gardens can be regarded as subsistence gardens.

10.4.2 Case studies from the North-West Province, South Africa

Background

South African urban areas are regarded as unique study areas for integration of socioeconomic and biophysical aspects in ecological studies, due to a dualistic development characterized by a predominantly Third World sector backed by a relatively strong First World infrastructure, according to Cilliers *et al.* (2009). In this section we will discuss some results on the species diversity of urban home gardens along socioeconomic gradients in two contrasting urban settings in the North-West Province of South Africa, namely a medium-sized city and a rural settlement. The major socioeconomic, ecological, planning, management, and legislative constraints of cities, towns and settlements in the North-West Province are summarized by Cilliers *et al.* (2009). Several research questions were formulated for this study of which only the following three are addressed in this chapter:

- What is the contribution of home gardens towards urban green infrastructure in terms of plant diversity?
- Is the relationship between SES and plant diversity in home gardens similar to findings from studies in developed countries?
- Is there any difference in the use value of plants in home gardens along the SES gradient?

Study areas and methods

The first study area was Potchefstroom, one of the largest cities in the province with a population of about 200 000 and which is situated in the Grassland Biome of South Africa (Mucina and Rutherford 2006) (Figure 10.1). Several urban vegetation studies of different land use areas were completed for the city of Potchefstroom, which formed baseline information for biotope mapping (Cilliers *et al.* 2004) and informed planning and developmental issues through spatial development frameworks (Drewes and Cilliers 2004). The Potchefstroom City Council illustrated their commitment towards greener governance by joining the Cities for Climate Protection Program (CCP) of the International Council for Local Environmental Initiatives (ICLEI) and, among other projects, initiated a vegetable cultivation project in previously disadvantaged communities (Cilliers *et al.* 2007, 2009, Cilliers 2010). No background knowledge of urban home gardens of Potchefstroom exists and the impact of the SES of urban residents on home gardens as a resource for biodiversity has not been established yet. Potchefstroom is characterized by a high cultural and socioeconomic heterogeneity and a clear socioeconomic gradient is present from east to west in the city. Five SES classes based on the unemployment rate of residents in the different election wards could be distinguished in the city. This information was obtained from a socioeconomic survey carried out in 2004 and which refined the census data (WorkWell 2004). Additional socioeconomic information collected during the mentioned survey is also given for each class (Table 10.1). Classes 4 and 5, and to a lesser degree class 3, are probably somewhat comparable to areas

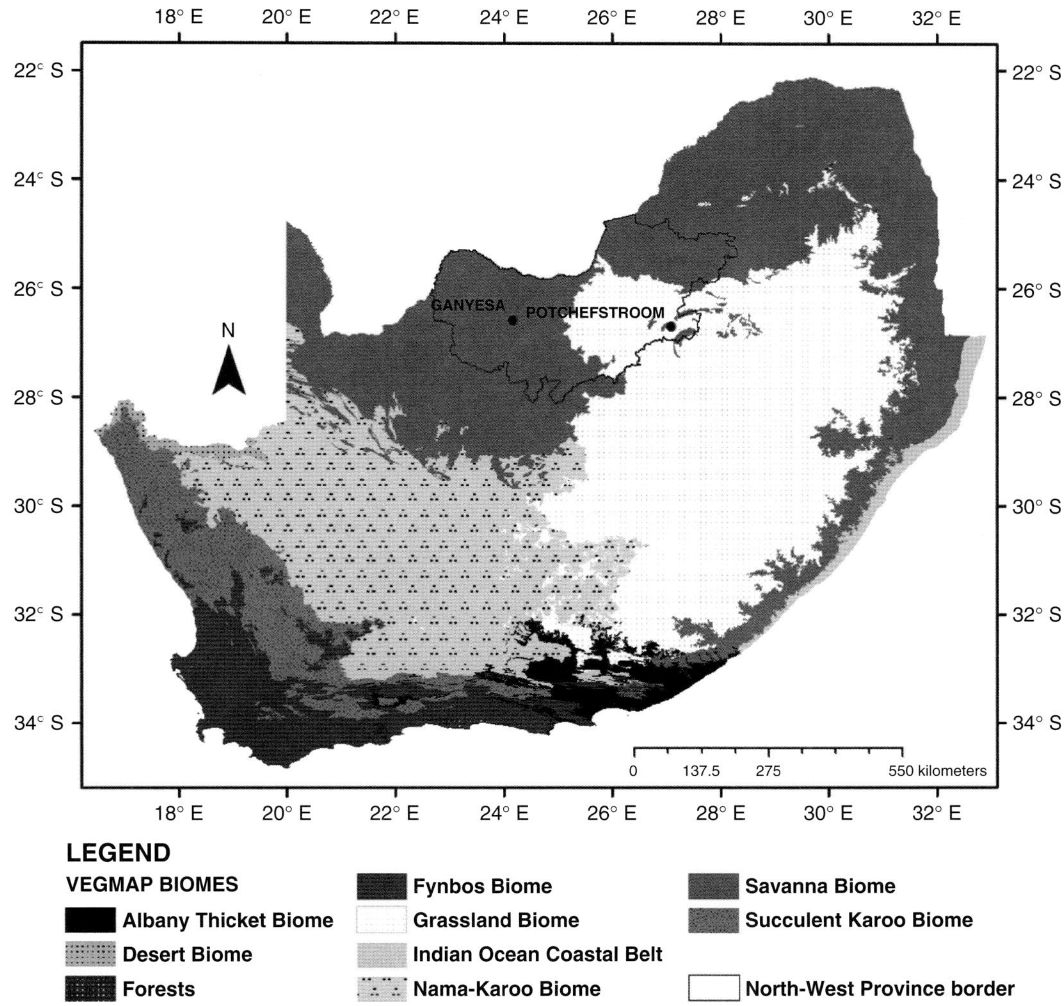

Figure 10.1 Biome map of South Africa indicating the location of the two study areas in the North-West Province.

Table 10.1 SES classes of Potchefstroom, North-West Province, South Africa based on % unemployment with additional information for each class (adopted and extended from Lubbe *et al.* 2010).

SES (class)	% unemployment	Household size (% households with 5 or more persons)	Number of rooms (% with 2 or less rooms)	Schooling status (% no schooling)	Monthly income (% with no income)
1	49.9–56.0	25.3–37.4	19–51.1	6.4–34.9	11.9–27.4
2	21.2–49.8	33–42.2	17.1–34.2	7–11.9	11.4–18.2
3	8.1–21.1	7.4–19	27.8–29.8	17.5–27.5	4–28.9
4	5.7–8.0	12.5–14	9.4–10.3	2.2	9–12.1
5	3.9–5.6	8.8–13.3	7.3–8.4	1.8–4.6	3.6–4.1

Table 10.2 Some of the variables used to determine socioeconomic status (SES) for Ganyesa, North-West Province, South Africa with the different options and scores for each variable.

1. Type of structure	informal house = 1, traditional house = 2, formal house = 3
2. Own or rent house	rent = 1, own = 2
3. Type of sanitation	none = 0, outside home = 1, inside home = 2
4. Water source	natural source = 1, communal water tap = 2, water tap in yard = 3, water tap inside home = 4
5. Distance from nearest water source	more than 1 km = 1, 500 m–1 km = 2, 100 –500 m = 3, less than 100 m = 4
6. Fruit and vegetables	produce/grow = 1, buy and grow = 2, buy = 3
7. Livestock	chicken/geese/ducks/turkey = 1, pigs = 2, goats/sheep = 3, donkey/horse = 4, cattle = 5
8. Method of earning money	support from family members = 1, temporary work = 2, government grant = 3, informal business = 4, own business = 5, permanent job = 6
9. Modes of transport	bicycle = 1, donkey cart = 2, own motor vehicle = 3
10. Monthly income	R 0–500 = 1, R 500–1000 = 2, R 1000–2000 = 3, R 2000–5000 = 4, more than R 5000 = 5

of high- to medium-income in developed countries, while classes 1 and 2 form a more typical scenario for developing countries and, to a certain extent, also for areas of extremely low income in developed countries.

The second study area was the rural settlement of Ganyesa with a population of about 40 000 in the Savannah Biome of South Africa (Mucina and Rutherford 2006) (Figure 10.1), which is situated on the fringe of the Griqualand West Centre of Floristic Endemism (Van Wyk and Smith 2001). From a plant conservation point of view, Ganyesa is therefore extremely important, especially as the area is floristically undersampled (Bester *et al.* 2008). In comparison with Potchefstroom, Ganyesa is culturally and socioeconomically less heterogeneous and is a typical example of a large rural settlement mainly occupied by one ethnic group (Batswana) and which formed part of the Bophuthatswana homeland from the apartheid era. National census data from 2001 was unreliable to determine the SES of the Ganyesa residents, because the settlement is fast expanding and contains large areas for which no data is available. Additionally, many residents do not have a fixed monthly income and therefore median household income alone would not be a reliable measure of SES. A questionnaire was, therefore, formulated based on a selection of socioeconomic aspects of which some are included in Table 10.2. Scores were allocated for each specific answer and, based on the total scores, three SES classes were identified. SES classes 1, 2, and 3 from Ganyesa are comparable to classes 1, 2, and 3 from Potchefstroom, while SES classes 4 and 5 from Potchefstroom (Table 10.1) had no representation in Ganyesa.

Sampling of plant species was done using a 100-point frequency survey along five transects in home gardens along an urban-to-rural gradient, as well as in other land use areas such as natural areas surrounding the urban areas, fragmented natural areas inside the settlement, institutional gardens, wetlands, fallow fields and wetlands. Pearson correlation coefficients were calculated to determine relationships between SES and plant diversity. Exploratory data analyses showed that variables did not deviate from normal.

Results and discussion

In Ganyesa home gardens have the highest total plant diversity (gamma diversity[1]) of all the different land use areas, even more than the natural areas (269 versus 208 species) (Figure 10.2). Natural areas have the highest indigenous plant diversity while home gardens have the highest exotic plant diversity (Figure 10.2). In terms of species richness (number of species per sample plot) natural areas have the highest alpha diversity[2]

[1]Gamma diversity (γ-diversity) is a measure of the total biodiversity over a large area or region. It is the product of α- and β-diversity.

[2]Alpha diversity (α-diversity) is the biodiversity within a particular area, and is usually expressed as the species richness of the area. This can be measured by counting the number of taxa (to correct for area: Simpson's Diversity Index or Shannon Index).

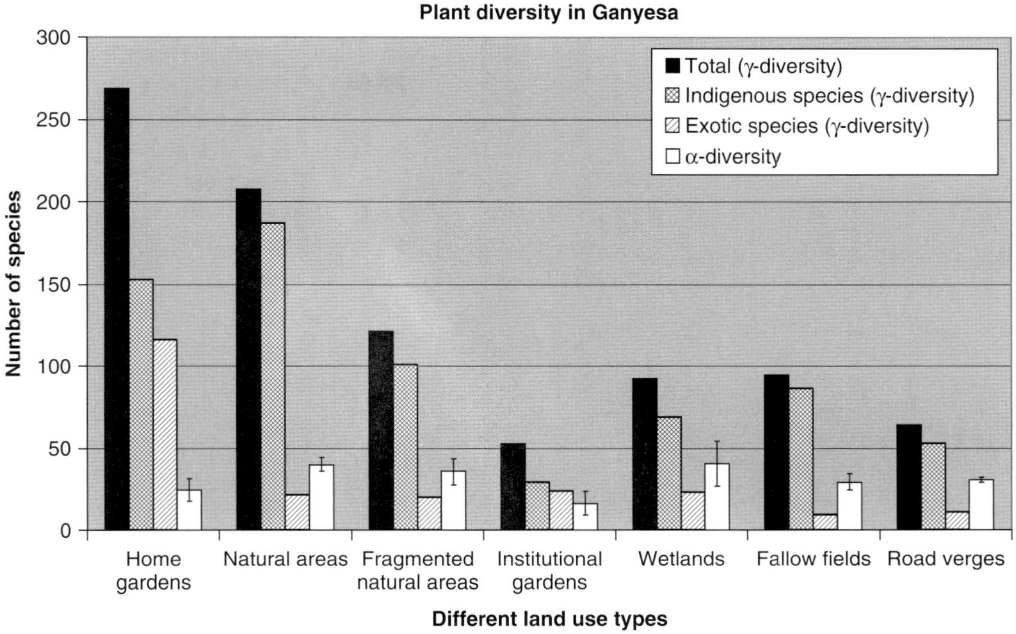

Figure 10.2 Total plant diversity, indigenous species, exotic species and species richness per unit area of different land use areas in the rural settlement of Ganyesa, North-West Province, South Africa.

(mean of 40 species per sample plot) while home gardens have the second lowest alpha diversity (mean of 25 species per plot). As home gardens have a low alpha-diversity, but high gamma diversity, it means there is a high-turnover of species (beta diversity[3]) between gardens. This is an indication of the heterogeneity of cultivation preferences and habitat types in domestic gardens. This is supported by an ordination of exotic species found in gardens, which indicates that sample sites of other land uses, such as the natural areas, are generally grouped together (floristically similar), but home gardens are scattered due to significant variation in species composition per sample site (Figure 10.3). Similar trends in terms of plant diversity were recognized in Potchefstroom (not shown on Figure 10.3), but the total number of species in home gardens (699 of which 73% are exotics) are significantly more than in Ganyesa, most probably due to the higher socio-economic and cultural heterogeneity in Potchefstroom and the presence of residents with higher SES (classes 4 and 5). Home gardens in total therefore contribute more to the overall urban biodiversity than one would expect if gardens are considered on an individual basis.

[3]Beta diversity (β-diversity) is a measure of biodiversity, which works by comparing the species diversity along environmental gradients (Sørenson Similarity Index or Whittaker's Measure).

Considering that beta diversity is high for domestic gardens, it is important to consider what impact social aspects have on this large turnover in species. There are many variables characterizing an urbanization gradient (Hahs and McDonnell 2006), but in the case of home gardens the SES gradient is probably the most pronounced, because unlike the situation in nature, humans determine the plant diversity. The choices of plants are also influenced by uncontrollable environmental factors such as climate, soil, space, etc. However, once this has been recognized in the planning of a garden, the human factors come into play. For instance, a person earning USD 50 per month will think twice before purchasing an ornamental tree of USD 10. Likewise, a college professor, despite earning a livable salary, will probably not spend USD 10 on an affordable ornamental tree if it is a declared invader. Many factors are obviously at play in determining garden plant diversity, such as employment, monthly income, education, availability of water, etc., and this can best be reflected by SES.

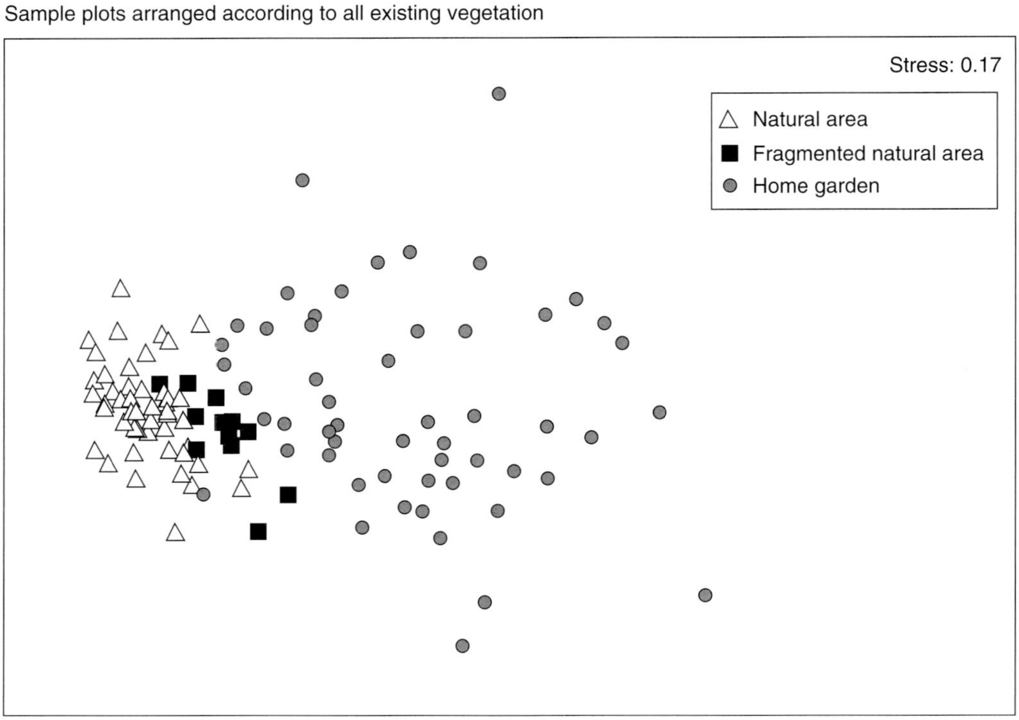

Figure 10.3 NMDS (non-metric multidimensional scaling) ordination of sample plots in natural areas, fragmented natural areas and home gardens in Ganyesa, North-West Province, South Africa.

In Potchefstroom, there is a strong positive correlation between SES and total number of species, and similarly for both the number of indigenous and exotic species (Table 10.3). This confirms results from urban areas in general in the USA (Hope *et al.* 2003) and urban home gardens in Australia (Kirkpatrick *et al.* 2007). There is a weaker positive correlation between highest qualification and total number of species and number of exotic species.

In Ganyesa there is a weak positive correlation between SES and number of exotic species, but a strong negative correlation between SES and number of indigenous species and a weak negative correlation between SES and total number of species (Table 10.3). Correlations between other important components used to determine SES and species numbers were also determined, but only the highest qualification is shown (Table 10.3). Although the correlations were weaker individually, they were quite similar to SES results, indicating that our calculation of SES with combinations of components was reliable. The trends depicted here for Ganyesa, although weak, are different from findings in Potchefstroom and developed countries. It is generally accepted that species diversity increases with increasing SES. However, in developing countries, especially in more rural setups with a lower SES there are specific sociocultural issues that must be considered. When houses are built in cities the land is cleared before construction commences (emphasis on top-down management) and afterwards the cleared land is left to the owner to work and plant. In rural settlements the owner is most often the builder. Restricted resources do not allow for the clearing of the land. Enough surface area is cleared to build the house and the natural vegetation is left intact to be cleared over many years as the need for more surface area arises (emphasis on bottom-up management). These patches of natural vegetation in the domestic garden also have important ecosystem services, as they harbor many edible plants such as leafy vegetables with a weedy nature, called morogo (High and Shackleton 2000). Many are medicinal plants; they provide grazing for

Table 10.3 Correlation coefficient (r) between SES, highest qualification and total number of species, number of indigenous species and number of exotic species for Potchefstroom and Ganyesa, North-West Province, South Africa.

	Potchefstroom	Ganyesa
SES vs. total number of species	0.731*	−0.259
SES vs. number of indigenous species	0.529*	−0.410*
SES vs. number of exotic species	0.721*	0.271
Highest qualification vs. total number of species	0.443*	0.008
Highest qualification vs. number of indigenous species	0.134	−0.190
Highest qualification vs. number of exotic species	0.486*	0.423*

*Indicates statistical significance at $p < 0.001$.

livestock on a temporary basis, and they may act as trap plants for integrated pest management in vegetable gardens following stimulo-deterrent diversion strategies (push–pull system) (Khan *et al.* 1997). Hence, the strong negative correlation between SES and indigenous species diversity, as the latter will be higher with lower SES, whereas communities with higher SES in Ganyesa remove these natural patches rapidly due to more resources and often replace the native vegetation with exotic ornamentals, vegetables or fruit trees.

It is important to determine if any emphasis was placed on specific uses in home gardens in the different SES classes. Gardens of residents with the highest SES (5) in Potchefstroom did not have any vegetable gardens, fruit trees, herb gardens or medicinal gardens, while those with the lowest SES (1) boasted all four of the mentioned groupings of plants (Figure 10.4). Residents with a medium SES (3) have the highest frequency of fruit trees, mainly due to larger plots (small holdings) and an emphasis on the production of vegetables. A transition from ornamental to subsistence gardens was observed along the SES gradient from high to low SES values in Potchefstroom. In Ganyesa there is a strong relationship with the lower SES classes of Potchefstroom when vegetable gardens and fruit trees are considered. However, due to the prevalence of cultural practices in a Batswana dominated community, medicinal gardens are more prominent, and herb gardens to a lesser extent (Figure 10.5). The higher percentage of fruit trees in gardens of medium SES (3) is not due to larger plots as in Potchefstroom, as plot size is not linked to SES in Ganyesa. It can be attributed to age of housing plot or affordability in terms of new houses. Most of the gardens along the SES gradient in Ganyesa could be classified as subsistence gardens, with only some examples of ornamental gardens of residents falling in SES class 3.

10.5 CONCLUSIONS

In this chapter an overview is provided of studies that described various social and other benefits of urban green areas to residents, as well as the variety of human perceptions on urban greening and urban biodiversity. While this is true for developed and developing countries, most of these studies focused on developed countries. It was also shown that socioeconomic aspects are important drivers for urban biodiversity and studies from developed and developing countries have indicated a positive correlation between SES and plant and bird diversity.

From the two South African case studies, it was, however, clear that generalizations in terms of characterization of SES, which in most previous studies was based on median household income from census data, could be misleading. This confirms the suggestion of Hope *et al.* (2003) for more detailed studies on the effect of aspects such as educational level and culture to understand the mechanism of how human choices drive urban plant diversity. An opposite trend in plant diversity patterns of domestic gardens was observed in the settlement (Ganyesa) characterized by residents with a lower SES, than in the medium-sized city (Potchefstroom), which included residents with a higher SES, as well as in developed countries as indicated in the literature. It was only possible to describe this contrasting trend due to the detailed and careful characterization of SES based on multiple social and economic aspects (indicated in Tables 10.1 and 10.2). Additionally, comparison of a more homogeneous settlement with a more heterogeneous city in terms of socioeconomics and culture of their residents contributed to our understanding of conservation of plant

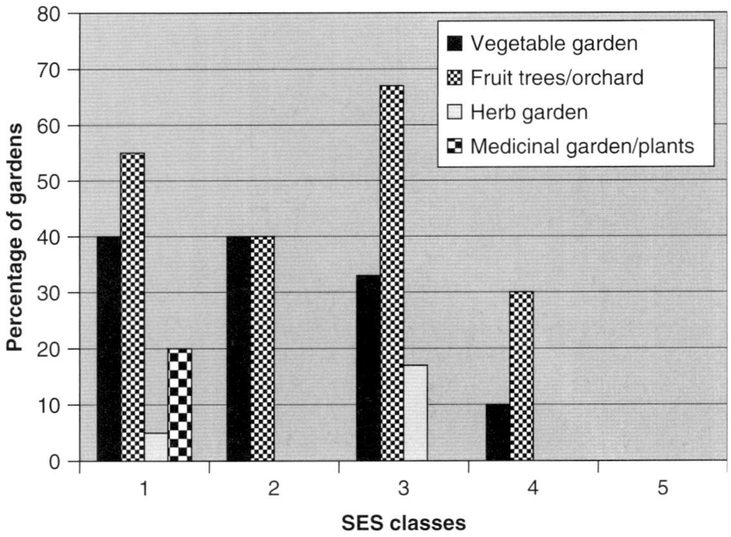

Figure 10.4 Specific uses of home gardens along a socioeconomic gradient in the medium-sized city of Potchefstroom, North-West Province, South Africa.

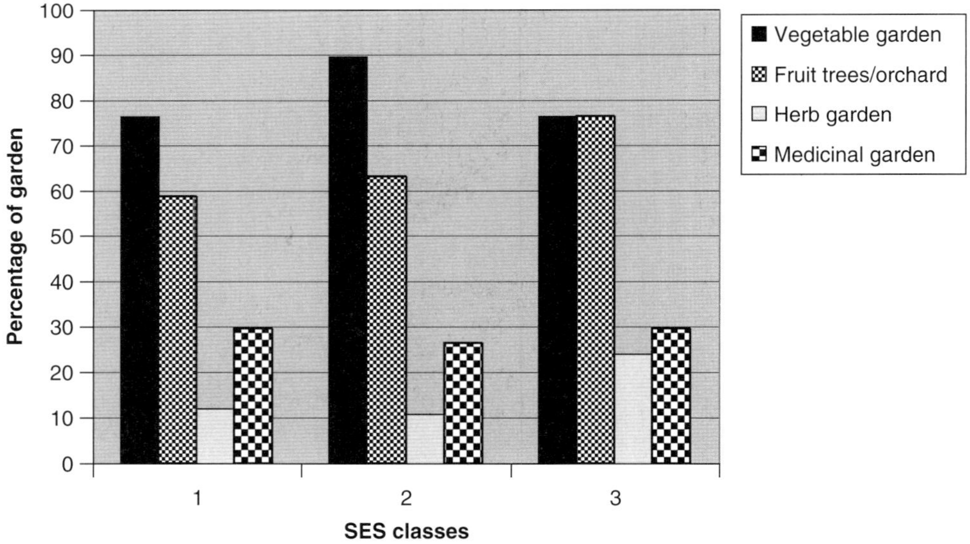

Figure 10.5 Specific uses of home gardens along a socioeconomic gradient in the rural settlement of Ganyesa, North-West Province, South Africa.

diversity and provision of ecosystem services in private green spaces. It is, for example, important in urban home garden research to determine the specific use values of gardens to the residents, namely ornamental garden or subsistence garden. Socioeconomic and cultural aspects are important drivers in determining the garden type and also the specific ecosystem services provided by these gardens (e.g., esthetic value and climate and pollution regulation through carbon reduction only), or also the provision of food and other products. It is a well-known fact that high-income neighborhoods have access to better ecosystem services (e.g., lower temperatures, landscape diversity and ecological dynamism) than middle- and low-income neighborhoods, which are left with more arid landscapes leading to higher annual temperatures and more degraded environments (Pedlowski *et al.* 2002). In future studies on home gardens, specific ecosystem services with supporting, provisioning, regulating and cultural functions (Millennium Ecosystem Assessment 2005) need to be quantified in order to determine their specific effects on human well-being.

Integration of the mentioned social aspects (indicated in Tables 10.1 and 10.2) and their influences on plant diversity in terms of human choices and landscape maintenance is extremely important in the planning, design and management of urban green areas. Care should, however, be taken that top-down management practices as described by Kinzig *et al.* (2005) are not overemphasized and that the described benefits of bottom-up management in terms of the provision of ecosystem services are also taken into consideration. In a management plan proposed for the city of Durban, South Africa, Roberts (2001) emphasized the need for tangible products and poverty reduction through environmental management and expressed the need for urban residents to experience directly the benefits of improved environmental management. A combination of top-down and bottom-up approaches in management, as indicated in the study of home gardens, will address the so-called clashing of perceptions, philosophies and goals between the public, planners, managers, scientists and policy makers in urban areas, as described by McDonnell (2007) in a proactive manner. A removal in the "tension" between the different stakeholders should also lead to the removal of social and spatial inequality in terms of resources and services provided by local government to all the residents in a specific urban setting. Several other studies from South Africa have suggested that any attempt of conservation orientated planning and management of urban areas is doomed without participation and commitment from all stakeholders, including local government (Donaldson-Selby *et al.* 2007, Cilliers 2010); in a study in Brazil, Pedlowski *et al.* (2002) suggested that a participatory approach in urban planning is the solution for achieving social equity.

REFERENCES

Alberti, M., Marzluff, J.M., Shulenberger, E.H., Bradley, G., Ryan, C., ZumBrunnen, C. (2003) Integrating humans into ecology: opportunities and challenges for studying urban ecosystems. *BioScience* 53(12), 1169–1179.

Altieri, M.A., Companioni, N., Canizares, K., *et al.* (1999) The greening of the "barrios": Urban agriculture for food security in Cuba. *Agriculture and Human Values* 16, 131–140.

Bester, S.P., Steyn, H.M., Koekemoer, M. (2008) National plant collecting programme: Is it worth the trouble? *South African Journal of Botany* 74(2), 385.

Bhatti, M., Church, A. (2004) Cultivating natures: homes and gardens in late modernity. *Sociology* 35(2), 365–383.

Bookchin, M. (1993) What is social ecology? In: Zimmerman, M.E., Callicott, J.B., Sessions, G., Warren, K.J., Clark, J. (eds) *Environmental Philosophy: From Animal Rights to Radical Ecology*. Englewood Cliffs, NJ, Prentice Hall, pp. 354–373.

Bowman, T., Thompson, J., Colletti, J. (2009) Valuation of open space and conservation features in residential subdivisions. *Journal of Environmental Management* 90, 321–330.

Chiesura, A. (2004) The role of urban parks for the sustainable city. *Landscape and Urban Planning* 68, 129–138.

Cilliers, S.S. (2010) Social aspects of urban biodiversity – an overview. In: Müller, N., Werner, P., Kelcey, J. (eds) *Urban Biodiversity and Design*. John Wiley & Sons, Ltd, Chichester, pp. 81–100.

Cilliers, S.S., Bouwman, H., Drewes, J.E. (2009) Comparative urban ecological research in developing countries. In: McDonnell, M.J., Hahs, A.K., Breuste, J.H. (eds) *Ecology of Cities and Towns: A Comparative Approach*. Cambridge University Press, Cambridge, pp. 90–111.

Cilliers, S.S., Müller, N., Drewes, J.E. (2004) Overview on urban nature conservation: situation in the western-grassland biome of South Africa. *Urban Forestry and Urban Greening* 3, 49–62.

Cilliers, S.S., Matjila, E.M., Sandham, L. (2007) Urban agriculture utilizing the eco-circle approach in disadvantaged communities in Potchefstroom, South Africa. In: Stewart, G., Ignatieva, M., Bowring, J., Egoz, S., Melnichuk, I. (eds) *Globalisation and Landscape Architecture: Issues for Education and Practice*. Proceedings of conference held at St Petersburg State Forest Technical Academy, 3–6 June 2007, Polytechnic University Publishing House, St Petersburg, pp. 88–91.

Conner, N. (2005) Some benefits of protected areas for urban communities: A view from Sydney, Australia. In: Trzyna, T. (ed.) *The Urban Imperative, Urban Outreach Strategies for Protected Area Agencies*. Published for IUCN – The World Conservation Union, California Institute of Public Affairs, Sacramento, pp. 34–43.

Daniels, G.D., Kirkpatrick, J.B. (2006) Comparing the characteristics of front and back domestic gardens in Hobart, Tasmania, Australia. *Landscape and Urban Planning* 78, 344–352.

Davis, G. (2005) Biodiversity conservation as a social bridge in the urban context. In: Trzyna, T. (ed.) *The Urban Imperative, Urban Outreach Srategies for Potected Area Agencies*. Published for IUCN – The World Conservation Union, California Institute of Public Affairs, Sacramento, pp. 96–104.

Donaldson-Selby, G., Hill, T., Korrubel, J. (2007) Photorealistic visualisation of urban greening in a low-cost high-density housing settlement, Durban, South Africa. *Urban Forestry and Urban Greening* 6, 3–14.

Drewes, J.E., Cilliers, S.S. (2004) Integration of urban biotope mapping in spatial planning. *Town and Regional Planning* 47, 15–29.

Faul, A.K. (2008) Increasing interactions with nature: A survey of expectations on a university campus. *Urban Habitats* 5(1), 58–83.

Frumkin, H. (2001) Beyond toxicity: human health and the natural environment. *American Journal of Preventive Medicine* 20(3), 234–240.

Gaston, K.J., Warren, P.H., Thompson, K., Smith, R.M. (2005a) Urban domestic gardens (IV): The extent of the resource and its associated features. *Biodiversity and Conservation* 14, 3327–3349.

Gaston, K.J., Smith, R.M., Thompson, K., Warren, P.H. (2005b) Urban domestic gardens (II): experimental tests of methods for increasing biodiversity. *Biodiversity and Conservation* 14, 395–413.

Gidlöff-Gunnarsson, A., Öhrström, E. (2007) Noise and well-being in urban residential environments: The potential role of perceived availability to nearby green areas. *Landscape and Urban Planning* 83, 115–126.

Gobster, P.H., Stewart, S.I., Bengston, D.N. (2004) The social aspects of landscape change: protecting open space under the pressure of development. *Landscape and Urban Planning* 69, 149–151.

Grahn, P., Stigsdotter, U.A. (2003) Landscape planning and stress. *Urban Forestry and Urban Greening* 2, 1–18.

Grimm, N.B., Grove, J.M., Pickett, S.T.A., Redman, C.L. (2000) Integrated approaches to long-term studies of urban ecological systems. *BioScience* 50(7), 571–584.

Hahs, A.K., McDonnell, M.J. (2006) Selecting independent measures to quantify Melbourne's urban-rural gradient. *Landscape and Urban Planning* 78(4), 435–448.

Head, L., Muir, P., Hampel, E. (2004) Australian backyard gardens and the journey of migration. *Geographical Review* 94(3), 326–347.

Henderson, L. (2006) Comparisons of invasive plants in southern Africa originating from southern temperate, northern temperate and tropical regions. *Bothalia* 36(2), 201–222.

High, C., Shackleton, C.M. (2000) The comparative value of wild and domestic plants in home gardens of a South African rural village. *Agroforestry Systems* 48, 141–156.

Hope, D., Gries, C., Zhu, W., *et al.* (2003) Socioeconomics drive urban plant diversity. *Proceedings of the National Academy of Sciences of the United States of America* 100(15), 8788–8792.

Iverson, L.R., Cook, E.A. (2000) Urban forest cover of the Chicago region and its relation to household density and income. *Urban Ecosystems* 4, 105–124.

Jensen, R., Gatrell, J., Boulton, J., Harper, B. (2004) Using remote sensing and geographic information systems to study urban quality of life and urban forest amenities. *Ecology and Society* 9(5). http://www.ecologyandsociety.org/vol10/iss2/art10 (accessed 23 March 2011).

Jo, H.-K., McPherson, E.G. (2001) Indirect carbon reduction by residential vegetation and planting strategies in Chicago, USA. *Journal of Environmental Management* 61, 165–177.

Khan, Z.R., Ampong-Nyarko, K., Chiliswa, P., *et al.* (1997) Intercropping increases parasitism of pests. *Nature* 388, 631–632.

Kinzig, A.P., Warren, P.S., Martin, C.A., Hope, D., Katti, M. (2005) The effects of human socioeconomic status and cultural characteristics on urban patterns of biodiversity. *Ecology and Society* 10(1). http://www.ecologyandsociety.org/vol10/iss2/art10 (accessed 23 March 2011).

Kirkpatrick, J.B., Daniels, G.D., Zagorski, T. (2007) Explaining variation in front gardens between suburbs of Hobart, Tasmania, Australia. *Landscape and Urban Planning* 79, 314–322.

Kumar, B.M., Nair, P.K.R. (2004) The enigma of tropical home gardens. *Agroforestry Systems* 61, 135–152.

Lamont, S.R., Eshbaugh, W.H., Greenberg, A.M. (1999) Species composition, diversity and use of home gardens among three Amazonian villages. *Economic Botany* 53(3), 312–326.

Loram, A., Tratalos, J., Warren, P.H., Gaston, K.J. (2007) Urban domestic gardens (X): The extent and structure of the resource in five major cities. *Landscape Ecology* 22, 601–615.

Lubbe, C.S., Siebert, S.J., Cilliers, S.S. (2010) Political legacy of South Africa affects the plant diversity patterns of urban domestic gardens along a socioeconomic gradient. *Scientific Research and Essays* 5(19), 2900–2910.

Martin, C.A., Warren, P.S., Kinzig, A.P. (2004) Neighborhood socioeconomic status is a useful predictor of perennial landscape vegetation in residential neighborhoods and embedded small parks of Phoenix, AZ. *Landscape and Urban Planning* 69, 355–368.

Marzluff, J.M., Shulenberger, E.H., Endlicher, W., *et al.* (eds) (2008) *Urban Ecology: An International Perspective on the Interactions between Human and Nature*. Springer, New York.

Mathieu, R., Freeman, C., Aryal, J. (2007) Mapping private gardens in urban areas using object-oriented techniques and very high-resolution satellite imagery. *Landscape and Urban Planning* 81, 179–192.

McConnachie, M.M., Shackleton, C.M., McGregor, G.K. (2008) The extent of public green space and alien plant species in 10 small towns of the sub-tropical thicket biome, South Africa. *Urban Forestry and Urban Greening* 7, 1–13.

McGee, T.K. (2007) Urban residents' approval of management measures to mitigate wildland-urban interface fire risks in Edmonton, Canada. *Landscape and Urban Planning* 82, 247–256.

McKinney, M.L. (2006) Urbanisation as a major cause of biotic homogenization. *Biological Conservation* 127, 247–260.

McPherson, G.E. (1998) Structure and sustainability of Sacramento's urban forest. *Journal of Arboriculture* 24, 175–190.

Melles, S. (2005) Urban bird diversity as an indicator of human social diversity and economic inequality in Vancouver, British Columbia. *Urban Habitats* 3(1), 25–48.

Millennium Ecosystem Assessment (2005) *Ecosystems and Human Well-being: Synthesis*. Island Press, Washington, DC.

Mohan, S. (2004) *An assessment of the ecological and socio-economic benefits provided by home gardens: a case study of Kerala, India*. Unpublished PhD thesis, University of Florida, USA. http://etd.fcla.edu/UF/UFE0007640/mohan_s.pdf (accessed 23 March 2011).

Mucina, L., Rutherford, M.C. (eds) (2006). *The vegetation of South Africa, Lesotho and Swaziland*. Strelitzia 19. South African National Biodiversity Institute, Pretoria.

O'Connor, T.G., Kuyler, P. (2009) Impact of land-use on the biodiversity integrity of the moist sub-biome of the grassland biome, South Africa. *Journal of Environmental Management* 90, 384–395.

Pandey, C.B., Rai, R.B., Singh, L., Singh, A.K. (2007) Home gardens of Andaman and Nicobar, India. *Agricultural Systems* 92, 1–22.

Pauchard, A., Aguayo, M., Pena, E., Urrutia, R. (2006) Multiple effects of urbanisation on the biodiversity of developing countries: The case of a fast-growing metropolitan area (Concepcion, Chile). *Biological Conservation* 127, 272–281.

Pedlowski, M.A., Da Silva, V.A., Adell, J.J., Heynen, N.C. (2003) Urban forest and environmental inequality in Campos dos Goytacazes, Rio de Janeiro, Brazil. *Urban Ecosystems* 6, 9–20.

Pickett, S.T.A., Cadenasso, M.L., Grove, J.M., *et al.* (2001) Urban ecological systems: linking terrestrial ecological, physical and socioeconomic components of metropolitan areas. *Annual Review of Ecology and Systematics* 32, 127–157.

Roberts, D.C. (2001) Using the development of an environmental management system to develop and promote a more holistic understanding of urban ecosystems in Durban, South Africa. In: Berkowitz, A.R., Nilon, C.H., Hollweg, K. (eds) *Urban Ecosystems: A New Frontier for Science and Education*. Springer, New York, pp. 384–398.

Roberts, D.C., Boon, R., Croucamp, P., Mander, M. (2005) Resource economics as a tool for open space planning in Durban, South Africa. In: Trzyna, T. (ed.) *The Urban Imperative, Urban Outreach Strategies for Protected Area Agencies*. Published for IUCN – The World Conservation Union, California Institute of Public Affairs, Sacramento, pp. 44–48.

Rugalema, G.H., Okting'ati, A., Johnsen, F.H. (1994) The home garden agroforestry system of Bukoba district, North-Western Tanzania. 1. Farming System Analysis. *Agroforestry Systems* 26, 53–64.

Senior, J., Townsend, M. (2005) Healthy parks, healthy people and other social capital initiatives of Parks Victoria, Australia. In: Trzyna, T. (ed.) *The Urban Imperative, Urban Outreach Strategies for Protected Area Agencies*. Published for IUCN – The World Conservation Union, California Institute of Public Affairs, Sacramento, pp. 111–120.

Smith, N.C. (2004) Birds and the urban ecology of Potchefstroom. Unpublished Masters Dissertation, North-West University, Potchefstroom, South Africa.

Smith, R.M., Gaston, K.J., Warren, P.H., Thompson, K. (2005) Urban domestic gardens (V): Relationships between landcover composition, housing and landscape. *Landscape Ecology* 20, 235–253.

Smith, R.M., Warren, P.H., Thompson, K., Gaston, K.J. (2006a) Urban domestic gardens (VI): Environmental correlates on invertebrate species richness. *Biodiversity and Conservation* 15, 2415–2438.

Smith, R.M., Gaston, K.J., Warren, P.H., Thompson, K. (2006b) Urban domestic gardens (VIII): Environmental correlates of invertebrate abundance. *Biodiversity and Conservation* 15, 2515–2545.

Smith, R.M., Thompson, K., Hodgson, J.G., Warren, P.H., Gaston, K.J. (2006c) Urban domestic gardens (IX): Composition and richness of the vascular plant flora, and implications for native biodiversity. *Biological Conservation* 129, 312–322.

Sukopp, H., Weiler, S. (1988) Biotope mapping and nature conservation strategies in urban areas of the Federal Republic of Germany. *Landscape and Urban Planning* 15, 39–58.

Sullivan, J.J., Timmins, S.M., Williams, P.A. (2005) Movement of exotic plants into coastal native forests from gardens in northern New Zealand. *New Zealand Journal of Ecology* 29(1), 1–10.

Swanwick, C., Dunnett, N., Woolley, H. (2003) Nature, role and value of green spaces in towns and cities: an overview. *Built Environment* 29(2), 94–106.

Takano, T., Nakamura, K., Watanabe, M. (2002) Urban residential environments and senior citizens' longevity in megacity areas: the importance of walkable green spaces. *Journal of Epidemiology and Community Health* 56, 913–918.

Thompson, K., Hodgson, J.G., Smith, R.M., Warren, P.H., Gaston, K.J. (2004) Urban domestic gardens (III): Composition and diversity of lawn floras. *Journal of Vegetation Science* 15, 373–378.

Thompson, K., Austin, K.C., Smith, R.M., Warren, P.H., Angold, P.G., Gaston, K.J. (2003) Urban domestic gardens (I): Putting small-scale plant diversity in context. *Journal of Vegetation Science* 14, 71–78.

Thompson, K., Colsell, S., Carpenter, J., Smith, R.M., Warren, P.H., Gaston, K.J. (2005) Urban domestic gardens (VII): A preliminary survey of soil seed banks. *Seed Science Research* 15, 133–141.

Todorova, A., Asakawa, S., Aikoh, T. (2004) Preferences for and attitudes towards street flowers and trees in Sapporo, Japan. *Landscape and Urban Planning* 69, 403–416.

Tratalos, J., Fuller, R.A., Warren, P.H., Davies, R.G., Gaston, K.J. (2007) Urban form, biodiversity potential and ecosystem services. *Landscape and Urban Planning* 83, 308–317.

Tzoulas, K., Korpela, K., Venn, S., *et al.* (2007) Promoting ecosystem and human health in urban areas using green infrastructure: A literature review. *Landscape and Urban Planning* 81, 167–178.

Vogl, C.R., Vogl-Lukasser, B., Puri, R.K. (2004) Tools and methods for data collection in ethnobotanical studies of home gardens. *Field Methods* 16(3), 285–306.

Whitford, V., Ennos, A.R., Handley, J.F. (2001) City form and natural process – indicators for the ecological performance on urban areas and their application to Merseyside, UK. *Landscape and Urban Planning* 57, 91–103.

Winklerprins, A.M.G.A. (2002) House-lot gardens in Santarém, Pará, Brazil: linking rural with urban. *Urban Ecosystems* 6, 43–65.

WorkWell (Research Unit for People, Policy & Performance) (2004) *Potchefstroom Socio-Economic Survey 2004*. Prepared for the Potchefstroom City Council. North-West University (Potchefstroom Campus), South Africa.

van Wyk, A.E., Smith, G.F. (2001) *Regions of Floristic Endemism in Southern Africa*. Umdaus Press, Hatfield, South Africa.

Yli-Pelkonen, V., Niemelä, J. (2005) Linking ecological and social systems in cities: urban planning in Finland as a case. *Biodiversity and Conservation* 14, 1947–1967.

Zagorski, T., Kirkpatrick, J.B., Stratford, E. (2004) Gardens and the bush: Gardener's attitudes, garden types and invasives. *Australian Geographical Studies* 42(2), 207–220.

Zobolo, A.M., Mkabela, Q.N. (2006) Traditional knowledge transfer of activities practised by Zulu women to manage medicinal and food plant gardens. *African Journal of Range and Forage Science* 23(1), 77–80.

CHAPTER 11

Plant material for urban landscapes in the era of globalization: roots, challenges and innovative solutions

Maria Ignatieva

*Department of Urban and Rural Development, Division of Landscape Architecture, Swedish University of Agricultural Sciences, Uppsala, Sweden**

11.1 INTRODUCTION

The globalization process has not only influenced politics and economies but social and cultural life of modern society. Globalization in landscape architecture poses a threat of homogenization, the creation of "placeless" design, and the loss of the uniqueness of the place (Bowring *et al.* 2009). Usually globalization is associated with westernization and an acceptance of the western vision of life and cultural preferences. Today the world has accepted the Euro-American consumer's vision with its "routine modernism" of skyscraper architecture in the central business districts, suburbanization, particular landscape architecture styles, and standardized plant and construction materials (Ignatieva 2010). In this chapter I will concentrate on a discussion of global plant material in urban landscapes.

11.2 THE BEGINNING OF PLANT MATERIAL GLOBALIZATION

The answer to the question "how was the pool of modern global plants formed?" can be found in the most influential European landscape styles of the 18th, 19th and early 20th centuries.

*Formerly at School of Landscape Architecture, Lincoln University, Canterbury, New Zealand (moved 2010).

Formal French gardens are one of the best historical examples of an attempt to "globalize" landscape styles, including planting design. By the end of the 17th and the beginning of the 18th centuries most of Europe and its new colonies accepted the formal French garden, a symbolic representation of the king's power over his people and nature with its straight lines, complicated *parterres de broderie*, bosquets and topiary. Broadleaved European trees such as lime (*Tilia* spp.), oak (*Quercus* spp.), elm (*Ulmus* spp.), hornbeam (*Carpinus betulus*), beech (*Fagus sylvatica*), maple (*Acer platanoides*, and *A. pseudoplatanus)* and horse chestnut (*Aesculus hippocastanum*) were the "core" for creating a "green architecture" for kings and nobility throughout Europe. Only a few conifer and evergreen species were used (*Taxus baccata* and *Buxus sempervirens*). The "core" species were directly exported to all European courts. For example, hundreds of lime trees were transported from Holland and chestnuts, hornbeams and yews from France to La Granja gardens in Spain in the 1720s. The English Hampton Court garden imported lime trees from Holland as well (Hobhouse 1992). Only climate affected this global French planting "pool". For example, in Scandinavian countries and Russia severe winters killed box and yew and forced local gardeners to find alternative solutions in species from local forests. In Sweden and Russia at the beginning of the 18th century *Vaccinium vitis-idaea* and *Juniperus communis* were successfully substituted for box and yew

Applied Urban Ecology: A Global Framework, First Edition. Edited by Matthias Richter and Ulrike Weiland.

in the parterres (Ignatieva 1982). In St. Petersburg, Russia, the locally obtained *Alnus glutinosa* was successfully pruned for Peterhof's maze. In Italy and Spain traditional evergreen plants such as *Cupressus sempervirens* and *Quercus ilex* were widely used for bosquets and topiary. The French style was also transported to the New World colonies (United States, Canada, Mexico and Argentina) and actively used in private gardening and design of open public spaces (Faggi and Ignatieva 2009).

Landscape or "natural style" (later known as the Picturesque style) was the "major contribution to European aesthetics" (Watkin 1982, p.1) in the second part of the 18th and at the beginning of the 19th century. The principal attributes of this "natural" style were irregularity, roughness and sudden variation, the opposite of the main principles (linear and clearness) of the French formal gardens. Serpentine and meandering paths, green openings like meadows, groves, forests and shrubberies of free configurations should form "stereotyped wilderness" (Hobhouse 1992). Picturesque parks were born in Britain and reflected the features of existing natural ecosystems. Open grasslands (most of them secondary, as a result of deforestation), broadleaved forests, groves and tree groups began to be the dominant feature of these picturesque parks. Flowering plants were only allowed if they grew naturally in the fields and woods. For example, the planting of violets, primroses, daffodils, lily of the valley, and daisies was encouraged in woodlands and meadows to make them look like "real wilderness". Plants in the picturesque parks were a tool, just like a painter's pigments, to create very "natural", often monotone, green landscapes.

Different flowering plants were grown mostly in walled kitchen gardens, which were placed at some distance from the house. Most trees for parks were obtained from local sources such as woodlands and hedges. There were only a very limited number of exotic trees used in the picturesque parks. Some introduced plants (e.g., *Cedrus libani* ssp. *libani* was introduced to Britain in 1638) started to have a very important emphasis in picturesque compositions. *Taxodium distichum* and *Robinia pseudoacacia* were introduced after 1742. Plants were amassed into groups of very natural forms and special attention was given to specimen plants. Later American *Liriodendron tulipifera* and *Magnolia grandiflora* arrived in Britain.

The British climate was always very favorable for growing pasture and lawn grasses. Meadows and lawns were the essential elements of open spaces in the picturesque parks and worked as a connection, which allowed Britain to be one large landscape garden. Landscape parks were used for grazing (productive landscapes) but were specially arranged to be beautiful. Leading designers of the picturesque style tried to adopt the *genius loci* ("the genius of the place") and to understand natural scenery perfection.

Grasses such as *Lolium perenne, Poa pratensis, Festuca pratensis, F. rubra,* and *Arrhenatherum elatius*, all native to Britain and its mild climate, had excellent opportunities for growing almost all year round. Derived from English picturesque parks, the lawn started to be a very influential landscape design element for the rest of the world and a subject of admiration and imitation.

English landscape parks were adopted in Europe and also used in the development of public parks in the United States and in British colonies (Australia, New Zealand, and South Africa) and later in other countries (Figure 11.1).

European trees and lawn species successfully arrived in these new countries. For example, in Christchurch, New Zealand all public park plantings were based on broadleaved deciduous trees (*Tilia* spp., *Quercus robur, Aesculus hippocastanum, Ulmus procera, Fraxinus excelsior, Acer pseudoplatanus* and *A. platanoides*). Broadleaved trees and lawn species were the essential contribution of Britain's landscape park movement to the global planting "luggage".

11.3 VICTORIAN GARDENESQUE (1820–1880)

The Industrial Revolution in Europe in the 19th century opened up new opportunities for the development of landscape art. New geographical discoveries and the conquest of new lands by the British Empire introduced a tremendous array of new plants to the garden market. The beginning of mass glass production (sheet glass was invented in 1833 and glass tax was removed in 1845) and the lawn mower (1830) introduced even more opportunities for public and private gardening.

After the dominance of perfectly "green" picturesque landscapes, people were "hungry" for color in the garden. This new period demanded a new garden. The main design principles of Victorian gardens were eclecticism: a mixture of formal elements (for example, French parterres, topiaries), picturesque imagery (winding paths, groves and lawns) and even Chinese

Figure 11.1 Global picturesque. One of the public parks in Adelaide, Australia (left) and the Imperial Palace East Gardens, Tokyo (right).

motifs. In planting design Victorian gardens introduced exoticism, the wide use of new plant species that were introduced from different parts of the world (Ignatieva and Stewart 2009). Victorian gardens always emphasized plant collecting and were a perfect reflection of the plant discovery era.

Victorian gardens had always been based on Christian belief and philosophy. Man was near to God in the garden (reference to Eden and the Gethsemane gardens in the Bible). For wealthy people gardening was "a source of agreeable domestic recreation" and for the poor "beneficial to physical, mental and spiritual well-being". The Victorian era was a triumph of art and horticultural skills over nature. "[...] Nature must acknowledge the supremacy of Art" said the famous gardener Loudon (Carter 1984).

Victorian gardens actually used the picturesque approach of arranging open and closed spaces but its use was primarily for displaying exotic trees and shrubs. Individual plants on lawns were shown to best present their horticultural quality. The appearance of a garden looked more like a collection of different plants that had arrived from overseas (Leszczynski 1999). Loudon recommended the arrangement of plants in the way of showing museum-like specimens, that is to be most easily admired.

Around the 1840s carpet bedding was introduced in the gardenesque gardens. Very elaborate displays based on colorful annuals were grown in greenhouses and moved outdoors in the warm season. Such carpet beds were numerous in public and private gardens. At the beginning plants from Central and South America and South Africa (*Tagetes erecta, T. patula, Begonia semperflorens, Ageratum houstonianum, Zinnia elegans, Heliotropium peruvianum, Petunia × hybrida, Pelargonium zonale, Salvia splendens, Lobelia erinus*, and verbenas) were among the most common. Such flowerbeds could display plants from June to the middle of October. Interestingly the spring bulbs (tulip, narcissus, and hyacinth) and biennial flowers, such as forget-me-not, appeared in such displays only in the 1860s. Hybridization of different annuals, first of all pelargoniums, petunias, verbenas, salvias, and lobelias, gave incredible variety on one hand but also led to standardization of main colors such as yellow, purple, scarlet, blue, pink, and white. It resulted in the reduction of plant stock species' variety in nurseries to be able to accommodate hybrids and cultivars that were in greatest demand. This point was one of the essential turning points towards globalization of plant material in urban environments.

Gardening books and the first gardening periodicals, such as Loudon's *Gardener's Magazine*, which describes different contemporary gardens and plant species, should "improve" urban and suburban agricultural landscapes for "the good of mankind" according to Victorian ethics. This journal can be regarded as the first influential garden magazine, "the father" of all contemporary western garden magazines.

The professional British plant collectors enriched the British garden market tremendously and introduced the modern practice of nursery culture. New plants, commercial nurseries, media magazines, skillful botanists and horticulturalists succeeded in promoting gardening as a popular pastime of the entire British nation.

11.4 INFLUENCE OF THE VICTORIAN GARDEN ON THE GLOBAL PLANTING PATTERN

The British Empire suggested not only a new "model" garden for the rest of Europe and colonial countries. Britain introduced a whole range of garden related attributes and plant material which later, in the 20th century, became symbols of the entire western civilization and an important feature of "global" landscapes.

The introduction of garden (flower) shows was another Victorian "gift" for western civilization. The first flower show was held in Kensington Gardens in 1852 and later became known as the Chelsea Flower Show. Antipodean, North American, and many European countries established their own mini "Chelsea Flower Shows" in the 20th century (Ellerslie Flower Show in New Zealand, Melbourne Flower Show in Australia, Floriade (The Netherlands) and Ottawa Tulip Festival in Canada).

British Victorian botanical gardens and public parks were another very important source of contemporary inspiration. The layout of botanical gardens and public parks was based on a combination of free lines, pathways, plant groupings and single specimens borrowed from the picturesque approach as well as on the gardenesque "clumps of shrubs", individual flowering shrubs and free standing flower beds arranged on the lawn. Some formal elements such as straight avenues and geometric flower parterres could also be included in the botanical garden design. The majority of plants in botanical gardens (as in all other gardenesque gardens) should be viewed individually as if in a museum. Randomly scattered beds, herbaceous plants, and shrubs were widely spaced for such individual viewing.

Among other elements from the Victorian era that were borrowed for "global" planting design were rockeries. They were favorites in private and public garden with a wide range of alpine perennials (from different mountainous regions such as the European Alps, the Himalayas, and the New Zealand Southern Alps) and dwarf conifers (pines and junipers).

Rockwork was a very fashionable motif in baroque and rococo gardens and later in the picturesque parks. By the 1840s alpine plants were widely recommended for English rockeries. Loudon advocated using rockeries in gardens. He wrote about "the contrast between the smooth flat surface of the lawn, with the uniformity of the circular beds, and the great irregularity of the surrounding rockwork" (Carter 1984). Together with lawns and colorful flowerbeds, rockeries also started to be a very strong attribute of Victorian gardens. With discoveries of new alpine plants in North America, the Caucasus and Japan, the variety of alpine plants increased dramatically by the beginning of the 20th century.

The Victorian gardenesque approach was very popular during the 20th century and was slowly transformed into a very simplified version: "tidy", "pretty", and "colorful". The rich choice of Victorian plant materials was simplified and declined with mass nurseries' commercial activity. The most common annual "global" flower bedding plants that are grown today in nurseries around the world are: *Begonia, Tagetes, Petunia, Salvia, Pelargonium, Viola, Solenostemon (Coleus)*, and *Lobelia* (Figure 11.2).

Victorian gardens appropriated and popularized quite a few exotic and native trees, shrubs and herbaceous plants, which contributed to the creation of global plant pools in commercial nurseries around

Figure 11.2 Simplified version of the modern gardenesque flowerbeds: Moscow, Russia (left), Adelaide, Australia (middle) and Athens, Greece (right).

Europe, in the United States and the rest of the world. Among them are rhododendrons, heath (*Erica* ssp.), *Berberis vulgaris, Arbutus unedo, Taxus baccata, Hedera helix*, elms, pines, and sycamores. From the 1860s masses of naturalized bulbs (daffodils, bluebells), primroses and violets became very popular (Elliott 1986).

The use of many exotic coniferous species (in woodlands, avenues and as single trees) was another very important attribute of Victorian landscapes. From that time on *Cedrus deodara, Araucaria araucana, Cryptomeria japonica, Chamaecyparis lawsoniana, Sequoiadendron giganteum*, and *Sequoia sempervirens* started to be important species in all nurseries.

The herbaceous borders, borders planted with herbaceous perennial flowers (achillea, aconitum, adonis, astromeria, aquilegia, aster, astilbe, hosta, gypsophila, hemerocallis, hollyhocks, phlox, lilies, delphiniums, chrysanthemums, anemone, iris, digitalis, peonies, silene, primula, *Potentilla, Oenothera, Symphytum*, and of course roses) were recommended by Loudon and became very popular by the last quarter of the 19th century. William Robinson, creator of the "wild garden" further promoted the idea of the herbaceous border. The most influential advocate of "the border" was Gertrude Jekyll who planted colorful plants in large masses and arranged them in informal drifts. Her seasonal "schemes" provided continuous flowering from early spring to late autumn including perennials as well as biennials, annuals, bulbs and even small trees and shrubs.

In his "wild garden" William Robinson advocated "the natural" garden with hardy native and exotic plants which "will thrive without further care" (Jellicoe *et al.* 1991). Among his favorites were rhododendrons, camellias, *Kalmia latifolia, Acer saccharum, Crataegus crus-galli, Rubus laciniatus, Viburnum opulus, Cornus alba*, and *Pinus nigra*. Robinson was actively naturalizing perennials and annuals such as bluebells, *Anchusa, Aconitum, Onopordon acanthium*, crocus, chionodoxa, scilla, fritillaria, tulip, and narcissus. "Robinsonian" gardening was very fashionable in North America and Europe. In the United States it was enriched by native woody and herbaceous plants.

The elements of Victorian and Edwardian gardens and plants can be seen in numerous public and private gardens around the world. Not all countries are capable of growing the variety of plant species available in "authentic" English gardens. Some local nurseries added some of their native species to the "English" pool of plants.

Figure 11.3 Japanese garden in Hillwood Gardens (Washington, DC).

Victorian and especially Edwardian Britain discovered Japanese and East Asian gardens through travelers and artists. "Japaneserie" brought to England new plants such as *Magnolia parviflora, Clerodendron, Cornus macrophylla, Caesalpinia japonica*, and *Camellia sasanqua*. Nurserymen John Gould Veitch and Robert Fortune especially traveled to Japan to collect plants and later advertised them in gardener's magazines. The fashion for Japanese magnolias, conifers (cryptomerias, larches, different pines and junipers), maples (first of all *Acer palmatum*), bamboos and of course flowering cherries arrived in the Western nurseries and persists to the present day. Some owners even employed Japanese gardeners to complete "truly" Japanese rock gardens and create "hybrid" gardens (Tachibana *et al.* 2004). This fashion of craving for the "Japanese garden" spread throughout Europe and the United States and can be seen even today in global landscapes in the form of numerous Japanese gardens in botanical collections around the world. Japanese gardens and their elements were packaged for consumption in the global horticultural market (Figure 11.3).

11.5 VICTORIAN TROPICAL AND SUBTROPICAL PARADISE

Botanical voyages of the 17th, 18th, and 19th centuries brought to Europe many new tropical and subtropical plants. Admiration of flowering orchids in Britain, for example, dated back to as early as the 17th century. Tropical plants have been imported in Europe to be grown in special conditions, such as greenhouses, heated conservatories, orangeries, and as indoor plants in "winter" gardens. The arrival of exotic plants depended on exploration of the tropical colonies.

Due to the lack of knowledge about the native habitats of these plants, most people assumed that all of the tropics were hot and steamy jungles or hot and dry desert (Jellicoe *et al.* 1991). This attitude later resulted in "standardization" of collections in greenhouses of most botanical gardens around the world. There were always tropical, cacti, and orchid greenhouses in European botanical gardens. Tropical plants were brought, grown and then propagated first of all in botanical garden greenhouses and then sent to other private collectors and horticulturists. Some nurserymen played a significant role in the spread of certain tropical plants. The most influential botanical garden, the "real" creator of tropical botanical collections, was the Royal Botanic Gardens at Kew and also the closely related Royal Horticultural Society. Step-by-step collections of "chosen" tropical and subtropical plants were grown in palm houses and conservatories in England.

The choice of exotic plants which appeared in greenhouses depended on many factors. First of all it was due to an interest and admiration of new unusual plants as part of God's "paradise", "the Wonders of Creation" concept (Victorian vogue for plant collecting) and also curiosity (empirical mystery of scientists and plant collectors) typical for the age of the Industrial Revolution (Elliott 1986). The number of exotic plants introduced into Britain skyrocketed with the invention of the Wardian case (which is closely glazed) that enabled plants to survive long marine journeys.

Some of the most popular plants for tropical greenhouses were palms. Due to their straight stems "striving to reach for the skies", unusual leaf shape and in most cases significant height, palms were viewed as "godheads" (Reynolds 1997). Palm collections grew rapidly in Kew Botanic Gardens: 6 palm species in 1768, 10 by 1787, 20 in 1813, 40 in 1830 and 420 in 1882. The most popular palms of the Victorian age were *Phoenix canariensis, Trachycarpus fortunei, Washingtonia filifera,* and *Livistona chinensis*. Canarian and Chinese Windmill Palm were the most "hardy" and could also grow outdoors in Britain. Very soon these two palms started to be "classical" symbols of Victorian gardens and became widespread not only around Mediterranean gardens but in the United States and all British colonies (Figure 11.4).

Figure 11.4 Palms as a symbol of the classical Victorian garden: Adelaide, Australia.

Figure 11.5 Tropical "paradise" in Rarotonga (Cook Islands).

Among other "landmarks" of tropical "paradise" are species such as *Bougainvillea*, and *Plumeria* from South America, *Hibiscus rosa-sinensis* from South-East Asia, *Strelitzia* (Bird of paradise) from South Africa, *Heliconia* (tropical Americas and Pacific Ocean), banana tree (herbaceous plants of the genus *Musa)*, *Cordyline terminalis, Codiaem* (Croton), tree ferns and cycads (Figure 11.5). Tropical orchids and bromeliads are another very important feature of Victorian "tropicana".

Among aquatic plants the most popular in Victorian times were *Victoria amazonica, Lotus nelumbo*, and tropical *Nymphaea* and *Cyperus papyrus*.

Colonial botanical gardens followed picturesque and gardenesque traditions of the Royal Botanic Gardens at Kew. From Kew many new subtropical and tropical plants were sent to colonial botanical gardens and to their local commercial nurseries for design of public and private gardens. Interestingly, new botanic gardens in the tropics (the tropical world's Kew Gardens) were established by the directors of the Royal Kew Botanic Gardens or by trained staff such as in botanical gardens in St. Vincent, the West Indies, in Peradeniya (Sri Lanka), at Bagor in West Java, in Singapore, Brisbane, and Kenya (Wijaya 1999). By the beginning of the 20th century the global tropical planting design pool based on British Victorian tradition crystallized and led to the creation of similar landscapes (Ignatieva 2010). This "exotic" tendency of British Victorian landscapes had dramatic consequences for the biodiversity of many tropical countries, especially for island ecosystems. Since only a few native plants from each tropical country were chosen as the tropical "elite", the majority of plant material used for landscape design in tropical countries was based on exotic plants. Even the traditional Pacific island lei (flower arrangement) consists to 99% of introduced plants, which appeared in these countries only by the end of the 19th century. For example, in Hawaii not only plant material of resorts and cities are based on the exotic tropical "pool", but even traditional garlands. *Plumeria*, hibiscus and orchids, bougainvillea, carnation are the most common and only two native species (*Metrosideros polymorpha* and *Alyxia oliviformis* native to the Hawaii Islands) are used in the Hawaiian lei (Wageman and Wageman 2004). Visitors to Akaka Falls State Park in Hawaii (widely advertised as a "tropical paradise") could not even imagine that almost all the plants that they see there today are non-native.

The popularization of tropical and subtropical plants was also reinforced by "subtropical gardening" in England in the1860s to1870s, where some tender subtropical plants were grown outdoors. William Robinson was one of the leading advisors of this style. Plants with large decorative green or colored leaves were the favorites such as tree ferns (*Dicksonia, Cyathea*), banana (*Ensete ventricosum* the False banana, or Abyssinian banana), yuccas, agaves, dracaenas, New Zealand *Cordyline australis*, cannas, New Zealand phormiums, some palms, bamboo and pampas grass.

The arrival of dwarf foliage plants such as *Iresine* and *Alternanthera* from South America introduced new opportunities for ground cover and "carpet-bedding". *Iresine* is one of the most popular ground covers in all subtropical countries. Among the widely used carpet bedding plants were also compact succulent *Sempervirum* and *Echeveria*, which are now also a very important part of the global planting culture.

In most tropical and subtropical countries the local indigenous versions of natural plant communities are largely suppressed. Local nurseries are oriented towards growing "traditional" tropical and subtropical non-native plants. The borrowed gardening traditions of Victorian and Edwardian England resulted not only in unification of plant material but also the unification of styles. Today tropical gardening around the world is a variation of neat lawns, colorful flowerbeds, herbaceous borders and palm groves consisting of the same plant species. This lack of ecological awareness, together with "globalized" tropical plants has led to an identity crisis. Visitors to the Cook Islands, Hawaii, or Vanuatu will not see any sign of local flora in the urban landscape and will feel the sameness of these places.

11.6 MODERN NURSERIES' DIRECTION: GLOBAL POOL OF PLANTS

An unprecedented number of exotic plants arrived in Britain as a result of scientific discoveries and special nurserymen collecting overseas. Exotic species are non-native plants or animals that have been introduced into areas where they do not naturally occur, mostly by human actions. Most British nurseries progressively developed their plant species pool. The opportunity for exchange between colonial nurseries increased with industrial development and the introduction of new and faster transportation. Scientific and technological progress in the field of horticulture allowed the introduction of a lot of new plant varieties and hybrids. It was a very common practice as well to exchange plant material between European countries, for example, between the United Kingdom, Germany, France, and Italy.

Numerous books and garden manuals, television gardening programs, gardening magazines and horticultural shows propagandized a very eclectic range of different garden styles (from old English cottage garden, to Mediterranean tropical or subtropical to alpine, to Californian or Japanese, to dry (xerophytic) garden) with associated plants.

Innovative modernist and post-modernist examples of 20th century garden design developed in all countries around the globe. The choice of plant material in such gardens had some limitations due to minimalist principles of design. Public parks and gardens in most countries around the world followed picturesque–gardenesque traditions of arranging plants: the lawn as an essential element of open park space with scattered groups or individual specimens of deciduous or evergreen trees (depending on the climate) with flowering shrubs, flowerbeds and rock gardens. Average private gardens of middle class suburbia in English-speaking countries explored the Victorian and Edwardian tradition in its simplified version – manicured front lawn with some decorative shrubs, perennial borders and flowerbeds. Most plants in such gardens came as colonial "luggage" from England. For example, many New Zealand gardens have been furnished with "heated glasshouses for winter floral displays, duck ponds, bandstands, rose gardens, rockeries, woodland dells, arboreta (tree collections), herbaceous borders, expansive lawns cut by flowerbeds, mazes, sunken gardens, Japanese gardens, pergolas, and playgrounds" (Leach 2002, p.216). Due to the very favorable climate, New Zealand gardens were also especially embellished with subtropical vegetation, rock gardens and ferneries. Among typical British "signature plants" such as yuccas, agaves, palms, bamboo, and pampas grass, some New Zealand plants were also used in displays (*Phormium* species, *Cordyline australis*, and *Cortaderia richardii*).

Design styles and the choice of plant material play crucial roles in the globalization process. Europe, the United States, Australia and New Zealand share architectural, landscape design and plant material traditions. New countries, who recently joined the Western free market, for example, China, Russia, and other ex-Soviet Union republics, are trying to catch up with western lifestyles, cultural preferences, urban planning and architectural preferences (suburbanization and redevelopment of downtowns with skyscrapers) and styles of gardening and horticulture (Ignatieva and Smertin 2007).

An analysis of existing plant nursery catalogs from different countries around the world confirms the unification of plant material on a global scale and the direct connection of plant material lists in modern nurseries with planting design history creating a pool of "chosen plants".

Randomly chosen plant material catalogues of nurseries from temperate zones in the United States (Northern Nurseries, Cicero, NY 2001), Germany (Lorberg 2005) New Zealand (Southern Woods Nursery in Christchurch, and Ashton Glen Nursery in Otago (Northern Nurseries, Cicero, NY 2001; Ashton Glen Nursery http://www.cfgphoto.com/photos-perennials.html; Southern Woods Nursery http://www.southernwoods.co.nz/) and Russia (Dinos PARK in St. Petersburg (Kompaniya Dinas PARK http://arum.spb.ru/green_pr/dinas_park.htm) were analyzed.

Despite differences in size, history and climatic conditions all these nurseries offered very similar plant material. Among coniferous species *Abies concolor*, several *Picea*, *Cedrus deodara*, *C. libanii* ssp. *atlantica*, different *Chamaecyparis* species and its numerous cultivars, different pines, *Thuja plicata* and *T. occidentalis*, junipers (e.g., numerous cultivars of *Juniperus chinensis*), and *Gingko biloba* were common. More coniferous species were offered in the east coast of the United States, Germany and Russia compared to New Zealand because of the more severe climate in the former countries.

Deciduous species that are common for all nurseries are *Acer platanoides*, *Tilia cordata*, different species of *Fraxinus* (European and North American), *Fagus sylvatica*, *Ulmus glabra*, and *U. parvifolia*, *Carpinus betulus*, different oaks (*Quercus robur*, *Q. palustris*, and *Q. rubra*). These trees reflect the picturesque tradition of using broadleaved trees as a foundation for park design. Among others some "iconic" species of European landscapes are *Salix alba* and *Betula pendula*.

The influence of Japaneserie fashion resulted in more than a century of Japanese species in nurseries throughout the world such as *Acer palmatum*, *Berberis thunbergii*, *Cornus kousa*, *Pieris japonica*, and *Prunus* × *yedoensis*. By 1900 Asian plants were very well established in many European gardens.

Late Victorian and Edwardian garden traditions are reflected in growing decorative flowering shrubs, such as rhododendrons and azaleas (numerous species from North America, Caucasus and Asia), *Rosa*, *Syringa vulgaris*, different species of *Viburnum*, *Weigela*, *Magnolia*, *Hydrangea*, *Forsythia*, *Crataegus*, *Cotoneaster*, and vines such as *Wisteria sinensis*.

Traditional Victorian and Edwardian plants, which were widely used in perennial borders and rock gardens, have very stable positions in most temperate garden nurseries. The favourites are: *Achillea tomentosa*, *Ajuga reptans*, *Anemone* spp., different *Aster*, *Astilbe*, *Chrysanthemum*, *Hosta*, *Iris*, *Sedum*, *Saxifraga*, *Solidago*, *Viola*, peony, lilies, *Hemerocallis*, *Delphinium*, *Aconitum napellus*, *Aquilegia*, *Lupinus polyphyllus*, and *Campanula carpatica*. From the tremendous variety in Victorian times species like *Begonia semperflorens*, *Petunia*, *Salvia splendens, Tagetes* and *Lobelia erinus* were chosen for flowerbeds. Among universal decorative grasses the tall *Cortaderia selloana* has been the most popular ever since the discovery of it for its very interesting foliage in Victorian "subtropical" outdoor gardening.

From the beginning of the 20th century plants arrived in countless numbers in different Western nurseries. American and Asian species appeared in European nurseries and "improved" traditional European perennial and mixed borders, and European and Asian plants reached gardening markets in the United States, Australia and New Zealand. In some nurseries, the global "core" was embellished by some native plants of some more unusual horticultural varieties or by special groups of plants (e.g., topiaries). Nurseries of tropical and subtropical countries also offered the "core" global tropical non-native plant material, which originally had appeared in glasshouses and tropical colonial botanical gardens.

The success of plant breeding in the 20th century satisfied a very broad range of consumers' fashion requirements. Numerous cultivars and hybrids of traditional Victorian plants were also popularized via garden magazines, journals and TV programs. With commercialization and the development of wholesale plant collections even some "natural" landscapes and whole countrysides have been altered by the introduction of exotic species. The most dramatic movement towards completely foreign exotic landscapes can be observed in the antipodes, for example, in New Zealand where there was a very concerted attempt to create a "slice of England". Homesick colonists brought to these new colonies familiar plants and animals. In their new environments, which were very fragile since after thousands of years of isolation native flora was suppressed by much more competitive non-native plants. In New Zealand, for example, today the number of completely naturalized non-native plants is higher than the number of indigenous vascular plants (2500), plus there are over 20 000 additional exotic species in outdoor cultivation (Ignatieva 2010). Most New Zealand urban and agricultural landscapes today are completely dominated by exotic plant material.

Among the strongest recipients of landscapes of consumerism, based on Western garden and nursery traditions of global plants, are countries that have only recently entered the free market world, for example, Russia, China and the United Arab Emirates. In Russia all recent examples of landscape architecture design – from new pedestrian zones to private villas for "New Russians" – clearly demonstrate the dominance of Western examples from design schemes and plant nurseries to construction materials and technologies and even garden magazines. The main elements of all private Russian gardens are a manicured lawn, evergreen junipers and cypresses, alpinarium (rock garden) and mixed borders. All vegetation as a rule is based mainly on exotic species, which arrive from European nurseries. The centuries of garden traditions based on productive landscapes and native vegetation (all the Tsar's palace parks and later public parks were based mainly on native species) are being abandoned and turned towards Anglo-American civilization (Ignatieva and Smertin 2007).

The City State of Dubai in the United Arab Emirates illustrates how the original Muslim paradisal approach towards "greening the desert" has been transformed into a very global urban landscape using the Western languages of picturesque, gardenesque and modernism (Bolletter 2009). Dubai architects tried very hard to create very distinguished, unique and recognizable "Para-Scape", "Para-City", "Logo-Scape", "Theme-Scape", "Pleasure-Scape" (including golf courses), "Muni-Scapes" (green spaces) typologies in urban landscapes. Nevertheless by using this paradisal "green" landscape and "classical" global plant material as well as Western traditional landscape design principles ("generic pastoral settings") they have failed by creating landscapes where "people are living, working and playing in neighborhoods, residential and commercial precincts that could be anywhere in the world" (Donald 2007). For example, lawn covers 70% of Dubai's Safa Park. The main plant species are different palms and bougainvillea. In flower arrangements *Canna* and *Tagetes* are dominant. The only native tree, the date palm (*Phoenix dactylifera*), is quite permanent in all types of planting design in Dubai. Creation and maintenance of such a "paradise-oasis" consumes enormous amounts of energy and water, and has led to the destruction of local desert and marine ecosystems and a social and ecological crisis. Dubai landscapes today have suppressed and rejected any local desert reality context and ecological wisdom in favor of a general familiar Western model that can attract global capital (Bolletter 2009). The reality is that Dubai today is a totally unsustainable city, but is very influential as a model for inspiration within the Middle East region.

11.7 INNOVATIVE SOLUTIONS: SEARCHING FOR NEW ECOLOGICAL PLANTING DESIGN

11.7.1 Europe: United Kingdom

The history of Western planting design shows quite a few attempts at breaking with traditional "fashion" in planting design and trying to design with nature. Many authors see William Robinson as one of the pioneers of ecological gardening. His calls to use native and exotic trees and shrubs in "wild gardens" as long as "they behaved like natives and do need special techniques of growing in English climate" (Hobhouse 1992) at a time of dominancy of Victorian and Edwardian traditions with mainly tender plants and exotic annuals massed in flowerbeds, was a revolutionary approach.

By the first quarter of the 20th century garden design faced a new problem. Escaped non-native species were replacing native species and threatened the environment. One of the classical cases of such "escapees" in the United Kingdom and Ireland is *Rhododendron ponticum* (Cross 1981). This species was made hardier by artificial and natural selection and by hybridization with Appalachian and other rhododendron species. It is easy to propagate and became cheap and popular in the mid and late nineteenth century as an ornamental plant in numerous British gardens. The lowest price was in about 1880, by which time it had escaped widely from cultivation sites. The escapees were ignored by botanical recorders for over 50 years. It was scarcely recognized as a problem until between the two world wars. Only in the second part of the 20th century did Britain introduce major projects to control this garden escapee.

The end of the 20th and the beginning of the 21st centuries in the United Kingdom can be characterized as an era of understanding and appreciation of urban biodiversity and the creation of sustainable urban plant communities: a different typology of "naturalistic herbaceous" plant communities is offered today for urban settings in the United Kingdom. Researchers from the University of Sheffield have been working with different natural and semi-natural vegetation to improve visual attractiveness for citizens and also to enhance urban biodiversity. For example, "pictorial meadows" offer seed mixes of native and non-native colorful species as wildlife-friendly and inexpensive replacements for traditional turfs (Dunnett 2008). Hitchmough (2009) has enhanced grasslands with planted, nursery-grown forbs, which can increase visual and ecological values of urban green spaces, as an alternative to traditional regularly mown lawns. Hitchmough (2009) suggested a wide range of species that can be valuable for visual displays and for wildlife (e.g., as a source of nectar and pollen for insects and birds).

11.7.2 Europe: The Netherlands

Some Dutch garden authors at the end of the 19th and beginning of the 20th centuries strongly advocated the use of native plants. For example, Leonard Springer had concepts similar to Robinson "the artful nature had to be well maintained, well arranged, neat and tidy" and encouraged the planting of native as well as exotic species

The Dutchman Jacob P. Thijsse (the father of the ecological design movement in the Netherlands) proposed the use of a plant communities approach in public park design (Amsterdam Boschplan). He advocated different plant communities from northeast European woods as the most suitable choice for Dutch planting design. Thijsse encouraged the use and propagation of wildflowers for garden design as an important mechanism for creating a sense of place and also for protecting rare species. Thijsse's efforts to protect indigenous vegetation (dunes, heath, moorland, woodland and wetland) are well known in Europe. His natural planting, grouping according to plant communities, grouping of wild plants, flowering meadows were based on natural indigenous plants. This particular emphasis on scientific research differentiates his approach from Robinson's "wild garden", which was based on the naturalization of ornamental plants (native and exotic) with some elements of natural ecological process and not on scientific experimentation with different natural associations (Woudstra 1997).

11.7.3 Europe: Germany

At the beginning of the 20th century German landscape architects also used native plants in their gardens. Unfortunately even very ecological experiments at that time were over-shadowed by the connection of the idea of using native plants with National Socialist ideology. After the Second World War Germany actively developed the field of urban ecology. The work of Herbert Sukopp, "the father of urban ecology" highlighted a research direction on urban biotopes, spontaneous vegetation, and biotope mapping. The "Go Spontaneous" (design with spontaneous vegetation) approach is dated back to Karl-Heinrich Hülbusch and the "Kasseler Schule der Landschafts- und Freiraumplanung" (Kassel School of landscape and open space planning) (Arbeitsgemeinschaft Freiraum und Vegetation 1986) and now is a popular planting design paradigm in Germany. This vision is based on the knowledge of plant community processes (first of all succession) and the ecological as well as the design qualities of plant material (Kuhn 2006). This approach is based on one of the most crucial ideas of modern urban ecology: find the opportunities for increasing biodiversity in each (even very small) biotopes and find a way of integrating nature into urban design. The German respect of "spontaneous vegetation" highlights a shift towards a new naturalistic approach in planting design which could, with time, change people's "traditional" preference from "tidy", and in many cases sterile cultural urban landscapes based on "traditional" garden plants, towards more spontaneous and biodiverse habitats.

11.7.4 United States

In the United States attempts were made to search for sustainable planting design practices from the beginning of the 20th century. Jen Jensen, based in the Midwest, established the "Prairie Style" where indigenous plant material played an essential role. He strongly believed in the ecological essence of existing sites and tried to encourage people to use unique prairie plants before they completely disappeared under cultivation. Jen Jensen's work was very influential in raising an interest in gardening with native plants. The "new style" of perennial gardening by Wolfgang Oehme and James van Sweden suggested "natural" schemes for a very wide range of public and private gardens. The Lady Bird Johnson National Wildflower Centre in Texas encouraged Americans to protect wild flowers and use indigenous plant material. Numerous guides to Natural Gardening (Knopf *et al.* 2002) and the activity of botanical gardens such as Brooklyn Botanic Garden (Marinelli *et al.* 1999) navigate Americans towards the use of indigenous plants in their gardens as an important tool for urban biodiversity protection and the creation of sustainable landscapes.

Today the Midwest continues Jen Jensen's "prairie" style in restoration projects and the introduction of native plants in new subdivisions. The most recent design of the Millennium Park in Chicago (completed in 2004) also highlighted a native prairie theme in their choice of plants. Chicago's green roofs, including the most famous City Hall roof, display native prairie plants and demonstrate potential for biodiversity and truly sustainable design (Arvidson 2009).

11.7.5 New Zealand: modern approach to planting design

The 1970s' environmental movement in New Zealand highlighted the desperate need for the conservation and restoration of lost native ecosystems as well as the need for plantings of natives. Due to dramatic loss of New Zealand native biota to introduced weeds, researchers should use terms such as "native biodiversity" or "indigenous urban biodiversity" to be able to address the problem. This approach is especially acute in urban environments where the most common biotopes are based almost entirely on non-native species.

The New Zealand view of planting design today is very different from European and North American ways of designing with nature. In New Zealand the main strategy is to increase indigenous plants and suppress the number of non-native plants. For New Zealand it takes much more effort to re-establish native plants due to very high competition from weeds and seed banks, which also contain non-indigenous plants (Ignatieva 2010). New Zealand has no opportunity to use many of the well-established Northern Hemisphere approaches such as "go wild" or "go spontaneously". The vision of a new planting design esthetic in New Zealand is strongly based on ecological knowledge of indigenous plant communities. The "plant signature" concept was introduced by Nick Robinson in1993 and further developed by Di Lucas and Colin Meurk (Lucas Associates 1995) who suggested different options of using native plant materials in private gardens based on the existing environmental conditions prior to urban development. Native trees such as *Plagianthus regius*, *Sophora microphylla* and *Podocarpus totara* are used in street plantings of New Zealand cities. Native tussock grasses, shrubs and ground covers such as *Pittosporum*, *Hebe*, *Coprosma*, *Carex*, *Juncus*, *Chionochloa*, and *Leptinella* are used for traffic islands, hedges, swales, ponds, rain gardens, river margins and pervious parking surfaces. There has been quite a strong movement, especially over the last decade, to "go native", not only in gardening but also on a broader scale, for example, in urban and even agricultural landscapes (Spellerberg and Given 2004). To encourage local citizens to use indigenous plants in private gardens special

Figure 11.6 The Formal Garden, Demonstration Gardens "Design with Indigenous Plants", Christchurch Botanic Gardens, New Zealand.

demonstration gardens ("Designing with Indigenous Plants") have been established in some New Zealand Botanical Gardens (Ignatieva *et al.* 2009). A new planting design strategy in New Zealand is directly connected to national identity (Figure 11.6).

There has also been a significant shift in the policies of local nurseries. If in the 1950s only ten native genera were included in tree and shrub lists for ornamental gardening (Leach 2002), now almost every New Zealand nursery provides quite a wide range of trees, shrubs, vines, herbaceous groundcovers and New Zealand alpine native plants. Nurseries such as Motukarara Conservation Nursery grow local native Canterbury plants exclusively, with an emphasis on genetic purity. The nursery propagates over 120 000 plants each year from two hundred different species which are collected from selected 20 ecological sites (Motukarara Conservation Nursery, no date).

11.8 DISCUSSION AND CONCLUSION

Each style in the history of planting design has offered a special palette of plants that contributed to the Western generic "pool" of fashion plants. English picturesque, Victorian and Edwardian gardens presented Western civilization with several clichés such as pastoral landscapes of lawn with scattered groves and specimen trees, rock gardens, flower and carpet bedding, "wild gardens" and perennial borders with an associated wide range of plant material. This plant "pool" reflected garden traditions and discoveries of new colonies as well as an interest in the plant world. Victorian botanical gardens were responsible for creating the list of purposely "chosen" plants from different historical and geographical origins. Edwardian gardens used a blend of different styles (Japanese, for example) for consumption in a growing and demanding horticultural market. By the end of the Edwardian era the "core" of plants which are grown today in commercial nurseries all around the globe was almost completed. Some authors believe that the United Kingdom pushed gardens to be "the main contact zones for cultural exchange between Britain and the world". The 20th century proved the process of transculturation in gardens at all levels of plant material as well as design styles and even hard materials (Tachibana *et al.* 2004).

Plant material plays an important role in modern globalized consumer culture. Countries which have recently entered the market system such as Russia, China, and the United Arab Emirates use global plant material as an important symbol of Western landscapes and for creating "familiar" comfortable landscapes, which can attract global investments in their economies. Globalized plant material contributes to the homogenization of our cities and towns.

Western countries (Europe, the United States, Australia, and New Zealand) are the main contributors to the global plant material and general vision of consumerism and global culture. Nevertheless these countries have experienced ecological and social crisis and have to search for new alternatives to existing unsustainable landscape practice based on global cultural preferences. Today these countries are suggesting new planting design directions based on the knowledge of plant communities, hence respecting local nature and culture. At the beginning of the 21st century Western countries should understand the importance of their mission in offering a new vision to planting design that demonstrates ecological responsibility and integrity.

ACKNOWLEDGMENTS

I gratefully acknowledge the assistance of Dr. Glenn Stewart for editing this manuscript.

REFERENCES

Arbeitsgemeinschaft Freiraum und Vegetation (ed.) (1986) *Krautern mit Unkraut*. Notizbuch der Kasseler Schule Bd. 2. Kassel.

Arvidson, A. (2009) Make no little plans. *Landscape Architecture* 7, 90–107.

Ashton Glen Nursery (No date) http://www.cfgphoto.com/photos-perennials.html (accessed 23 March 2011).

Bolleter, J. (2009) Para-Scape: landscape architecture in Dubai. *Journal of Landscape Architecture*. Spring 2009, 28–41.

Bowring, J., Egoz, S., Ignatieva, M. (2009) "As good as the West": two paradoxes of globalisation and landscape architecture in St. Petersburg. *Journal of Landscape Architecture* 90, 6–15.

Carter, T. (1984) *The Victorian Garden*. Bell & Hyman Limited, London.

Cross, J.R. (1981) The establishment of *Rhododendron ponticum* in the Killarney oakwoods, S.W. Ireland. *Journal of Ecology* 69, 807–824.

Dinos PARK, St. Petersburg (2009) http://arum.spb.ru/green_pr/dinas_park.htm (accessed 23 March 2011).

Donald, G. (2007) From paradise found to identity lost. In: Stewart, M., Ignatieva, M., Bowring, J., Egoz, S., Melnichuk, I. (eds) *Globalisation of Landscape Architecture: Issues for Education and Practice* St. Petersburg's State Polytechnic University Publishing House, St. Petersburg, p. 13.

Dunnett. N. (2008) Pictorial meadows. In: Müller, N., Knight, D., Werner, P. (eds) *Urban Biodiversity & Design: implementing the convention on biological diversity in towns and cities. Third conference of the Completence Network Urban Ecology, Erfurt, 21–24 May 2008. Book of Abstracts*. BfN, Bonn, p. 64.

Elliott, B. (1986) *Victorian Gardens*. B.T. Batsford Ltd., London.

Faggi, A., Ignatieva, M. (2009) Urban green spaces in Buenos Aires and Christchurch. *Municipal Engineer* (UK) Volume 162, Issue 4, 241–250.

Hitchmough, J. (2009) Diversification of grassland in urban greenspace with planted, nursery-grown forbs. *Journal of Landscape Architecture*. Spring 2009, 16–27.

Hobhouse, P. (1992) *Plants in Garden History*. Pavilion, London.

Ignatieva, M. (1982) Whortleberry (*Vaccinium vitis-idaea* L.) in parks. *Leningrad Panorama Journal* 9, 36–38.

Ignatieva, M. (2010) Design and future of urban biodiversity. In: Müller, N., Werner, P., Kelcey, J. (eds) *Urban Biodiversity and Design*. John Wiley & Sons, Ltd, Chichester, pp. 118–144.

Ignatieva, M., Smertin, V. (2007) Globalisation trends in Russian landscape architecture. In: Stewart, M., Ignatieva, M., Bowring, J., Egoz, S., Melnichuk, I. (eds) *Globalisation of Landscape Architecture: issues for education and practice*. St. Petersburg's State Polytechnic University Publishing House, St. Petersburg, pp. 111–115.

Ignatieva M., Stewart, G. (2009) Homogeneity of landscape design language in the urban environment: Searching for ecological identity in Europe, USA, and New Zealand. In: McDonnell, M.J., Hahs, A., Breuste, J. (eds) *Comparative Ecology of Cities and Towns*. Cambridge University Press, Cambridge, pp. 399–421.

Ignatieva, M., Meurk, C., Stewart, G. (2009) The new demonstration native gardens in the Christchurch Botanic Gardens. *Newsletter of the Christchurch Botanic Gardens Inc*. 76, 4–8.

Jellicoe, G., Jellicoe, S., Goode, P., Lancaster, M. (eds) (1991) *The Oxford Companion to Gardens*. Oxford University Press, Oxford, New York.

Knopf, J., Wasowski, S., Boring, J., Keator, G., Scott, J., Glasener E. (2002) *A Guide to Natural Gardening*. Fog City Press, San Francisco.

Kuhn, N. (2006) Intentions for the unintentional spontaneous vegetation as the basis for innovative planting design in urban areas. *Journal of Landscape Architecture* 2, 46–53.

Leach, H. (2002) Exotic natives and contrived wild gardens: the twentieth-century home garden. In: Pawson, E., Brooking, T. (eds) *Environmental Histories of New Zealand*. Oxford University Press. Oxford, New York, pp. 214–229.

Leszczynski, N. (1991) *Planting the Landscape*. John Wiley & Sons, New York.

Lorberg, S. (2005) *Woody Nursery. Catalogue*. Edition 80.

Lucas Associates (1995) *Indigenous Ecosystems of Otautahi Christchurch*. Lucas Associates, Christchurch, New Zealand. Sets 1 to 4. 1995–1997.

Marinelli, J., Ludlam, J., Lingstrom, B., Tebbitt, M. (eds) (1999) *Easy Lawn*. Brooklyn Botanic Garden, Brooklyn, NY.

Motukarara Conservation Nursery (No date) http://www.doc.govt.nz/conservation/native-plants/motukarara-conservation-nursery (accessed 23 March 2011).

Northern Nurseries, Inc. (2001) Cicero, NY. Spring 2001 Wholesale Offering.

Reynolds, J. (1997) "Palm trees shivering in a surrey shrubbery" – A history of subtropical gardening. *Principles* Vol. 41. http://www.palms.org/principes/1997/surrey.htm (accessed 23 March 2011).

Spellerberg, I., Given, D. (eds) (2004) *Going Native. Making use of New Zealand Plants*. Canterbury University Press, Christchurch.

Southern Woods. http://www.southernwoods.co.nz/ (accessed 23 March 2011).

Tachibana, S., Daniels, S., Watkins, C. (2004) Japanese gardens in Edwardian Britain: landscape and transculturation. *Journal of Historical Geography*. Vol. 30, Issue, April 2004, 364–394.

Tritenbach, P. (1987) *Botanic Garden Sand Arks in New Zealand*. Excellence Press, Auckland.

Wageman, J., Wageman, V. (2004) *A Literary Lei: Flowers and Plants of Hawaii*. Watermark Publishing, Corby.

Watkin, D. (1982) *The English Vision. The Picturesque in Architecture, Landscape and Garden Design*. John Murray, London.

Wijaya, M. (1999) *Tropical Garden Design*. Thames & Hudson, London.

Woodstra, J. (1997) Jacobus Thijsse's influence on Dutch landscape architecture. In: Wolschke-Bulmahn, J. (ed.) *Nature and Ideology: Natural Garden in the Twentieth Century*. Dumbarton Oaks Research Library and Collection, Washington DC, pp. 155–185.

E. ENVIRONMENTAL URBAN DESIGN

CHAPTER 12

Ecological infrastructure leads the way: the negative approach and landscape urbanism for smart preservation and smart growth

Kongjian Yu

Graduate School of Landscape Architecture, Peking University, Beijing, China

12.1 INTRODUCTION

12.1.1 Urbanization in China challenges survival

China's urbanization has been considered to be an unprecedented phenomenon in Chinese history as it is in the world. According to the UN report (UN 2007), at the end of 2006 China's total population was 1.3 billion, 44% of it residing in urban areas; this number will reach 70% by 2035. Three facts make this projection more meaningful:

1 The population is spatially unbalanced: about 94% of the population lives on approximately 46% of the land, mainly in the east part of China;

2 The scarcity of natural resources: China possesses 21% of the world's population but only 7% of the world's fresh water and other natural resources. The water volume consumption per capita in China is one fourth of the world average;

3 A landscape with rich cultural heritages: China has a continuous history of civilization of more than five thousand years, and virtually every inch of the land is part of the cultural heritage.

These facts have imposed two challenges to landscape and urban planning in China:

1 Sustainability: The first challenge is about ecological and environmental sustainability. While thousands of dams lie across almost all rivers in this country, a broader population is exposed more than ever to disastrous natural forces, as demonstrated by China's numerous floods and droughts each year. In the north desertification is in a crisis situation. Each year 3436 km^2 of land is being turned into desert, and this figure is increasing every year. At present, the total area of desertification accounts for about 20% of the whole country, furthermore, about 5 billion tonnes of soil erode yearly into the ocean (Jiang and Liu 2004, Zhao *et al.* 2004). Statistics show that in the past 50 years 50% of China's wetlands have disappeared and 40% of the surviving wetlands have been polluted (Chen *et al.* 2004).

The underground water level drops every day. In Beijing for example, the underground water overuse is 110%, and each year the underground water level drops by one meter. Two-thirds of the 662 cities now lack sufficient water and not a single river in the urban and suburban areas remains unpolluted. If present

Applied Urban Ecology: A Global Framework, First Edition. Edited by Matthias Richter and Ulrike Weiland.

trends are not reversed, the World Bank forecasts that by 2020 there will be 30 million environmental refugees in China due to water stress; more than half of China's 662 cities will suffer from water shortages, affecting 160 million people; 90% of cities' groundwater deposits and 75% of rivers and lakes will be polluted. While the GDP growth rate in the past 20 years is impressive in most Chinese cities, the annual loss caused by the environmental and ecological degradation is now between seven and twenty percent of the GDP. This is equal to or even higher than annual GDP growth (Guo 2004).

These are all by-products of China's double-digit GDP growth rate and the nation's swift process of urbanization. One can only ask: Is this sustainable? Can we survive the fast deterioration of the environment and ecology? How can landscape and urban planners play a role to meet these unprecedented challenges?

2 Cultural and spiritual integrity: The second challenge is the loss of our spiritual homeland. Spiritual integrity has to consider the believe that every piece of land and all elements in the landscape are inhabited by various spirits where the ancestors were buried. The trend toward materialism is taking over China at a rapid rate. The Dragon Hills (sacred hills) that housed numerous villages in rural China have been bulldozed. Meaningful and sacred streams and ponds in front of the villages have been filled or channeled in the name of flood control. Landscapes have become commercialized. Gradually, we have lost our spiritual connection to our land and to the Earth.

How can landscape and urban planners assume the role to protect and rebuild such spiritual connections through the design of our physical environment?

In order to meet these two challenges, the methodology of landscape and urban planning has to be reversed.

12.1.2 The failure of the conventional approach in urban development planning

Aside from other social and political reasons, the physical planning approach in the urban development planning field is responsible for the ecological and environmental degradation as well as the loss of cultural and spiritual landscape. The Chinese system of urban planning was inherited from the former socialist Soviet Union. The physical planning of urban development was an extension of the social and economic development planning exercise. The single most important basis for urban development planning is population projection, upon which urban land use, resources allocation, functional zoning and built infrastructure plan are based. In most of the cases the population projection for the long term (20 years) and even the short term (5 years) is never correct, or even nearly correct, partly due to changing government policy.

For example, in 1986 the population of Shenzhen was estimated to become 1.1 million in 2000, but the actual number turned out to be 7 million by the end of 2000. The same is true for Beijing and Shanghai (Yu *et al.* 2005a,b). This model of urban development planning initiated large-scale technocratic projects and mono-functional simplification. Nature was not considered as a system and zoning was directed by economic principles. As noticed in other Soviet-influenced countries: "Destruction of traditional landscape systems and ecological destabilization of landscapes, with extensive erosion, soil salinization, forest dieback and water and soil pollution, were visible results of this process," (Jongman 2001). As a result, planning of land use, as well as facilities and infrastructure can never catch up with the ever expanding population, and the urban boundary, following the extension of infrastructure, just keeps moving across the unplanned rural land indiscriminately. Critical cultural heritage sites and natural habitats are getting lost before any conservation plans are carried out. In this conventional approach any environmental conservation planning, historical protection planning and green system planning, or even flood control planning was subordinate to the comprehensive master planning, which does not cover the rural area beyond the urban development boundary. This shortcoming of the conventional planning system has recently been widely noticed (Yu *et al.* 2005a,b), but changes occur slowly.

In addition to the technical failure of this economic development oriented planning system, the shift of systems from a planned central economy to a market one weakens urban planning power. Physical planning is being subordinated to a socialist reference, as results are observed in a "fried egg" city form. After the reform in the 1980s, the "Leap Forward" type of city was able to show new development zones, a science center and new university cities, which virtually leap out of the master plan area and beyond the former greenbelt, or any comprehensive plan.

It is well recognized that urban planning is playing a less important role in development control and it has been widely recognized that the conventional economic oriented approach to urban development planning failed (Gaubatz 1999, Yeh and Wu 1999, Cheng and Masser 2003, Jim and Chen 2003). For many years the major driving force of urban development has been road infrastructure and the developed urban area.

As a result Chinese urban planners are facing a huge challenge and are required to modify the urban planning system from a centrally planned, to a transitional economy reform. Action for a revolutionary change in planning methods was called upon in the past decade (Yeh and Wu 1999, Wu and Wu, 2003, Yu *et al.* 2005a,b, 2004, Yang 2003, Zhang 2003), Among them, the Negative Approach was proposed (Yu *et al.* 2005a,b, Yu and Padua 2006), as a counterpart to the conventional "positive" approach in the sense of economic urban development. The negative approach is argued in the following three aspects:

1 Rethinking the conventional urban planning model: It is argued that the "population speculation–land use–infrastructure layout" has been proven invalid in dealing with the swiftness of urban development issues and is largely responsible for the degradation of ecological conditions and the chaotic situation of the current Chinese cities. Therefore a revolutionary change to the planning methodology is necessary. The conventional planning approach will simply point out where to develop but will not be able to show efficiently where development shall not take place for the sake of environmental sustainability and cultural heritage preservation.

2 A reversed approach to create ecological infrastructure (EI): As opposed to the conventional approach, which gives priority to economic development, the architectural development and built infrastructure the negative approach to urban planning is a landscape urbanism approach. This gives priority to the identification and planning of EI, which integrates landscape security patterns (Yu 1996) in safeguarding critical natural processes and biodiversity conservation, as well as securing the integrity and authenticity of the cultural heritage across landscapes (Yu *et al.* 2005a,b).

3 A negative form of a city pattern by positively defining EI: As opposed to the conventional ("positive") approach to urban development, in which urban growth is defined by built, gray infrastructure made up of roads and pipes that provide services for urban development, the negative approach acts as the green and unbuilt EI, and functions as a framework for urban growth and definition of urban forms. EI is used not only as a tool for smart preservation (effectively preserving the natural, biological and cultural processes), but also as a tool for smart growth and creation of good urban form, guiding the pedestrian open space system in a compact city. It is argued that the first and most important principle for better urban development is to determine where not to develop (Yu *et al.* 2005a,b, Benedict and McMahon 2006, p.12).

The key in the negative approach is planning the EI and its subsequent application to shape the urban growth.

12.1.3 Green infrastructure leads the way: the negative approach and landscape urbanism

The idea and practice of the negative approach towards urban development planning, in which landscape is first planned as infrastructure as well as an urban form maker, is certainly not a whimsical concept. It has become deeply rooted in both Eastern and Western planning theories and practice, belonging to an ecological planning approach typically referred to under the flag of McHarg's design with nature theory (1969), trying to provide land use and urban planning with a solid ecological basis. The negative approach can be more fashionably and partially described in the context of "landscape urbanism", which Waldheim described as: "a disciplinary realignment in which landscape replaces architecture as the basic building block of urbanism" (Waldheim 2007). Here landscape was considered as an infrastructure of processes and as a field of operation (Corner 1999, 2006).

Feng-shui: landscape as spiritual linkage and carrier of *genius loci*

The first tradition, or the prescientific model, of the negative planning theory is the Chinese ancient art of geomancy, or *Feng-shui*, which always give priority to the natural pattern and processes of Qi (or breath) (Lip 1979, Skinner 1982, Rossbach 1983, Yu 1994, 1996). Placed in order from large to small, the whole national landscape (mountains and water courses) was considered as an interconnected dragon vein and a network of Qi movement. A sacred landscape infrastructure in the fractal form (Yu 1994) is a given and forbidden pattern that any human actions have to come to terms with, such as establishment and construction of villages and cities, roads, bridges and even tombs. In this sense the protected sacred landscape formed the spiritual backbone or network of the live sustaining environment and then became the infrastructure that bears *genius loci* (Yu 1996). This tradition of landscape thinking still flourishes in rural China and has, for thousands of years, created the Chinese landscape full of survival implications and spiritual bearings.

Greenways: landscape as infrastructure of recreation and esthetic experience

In the United States parks and green spaces have served as the fundamental infrastructure to solve urban problems such as congestion and pollution since the late 19th century (Zube 1986, 1995, Little 1990, Ahern 1991, 2002, Fabos, 1995, 2004), the well cited examples include Boston's Emerald Necklace by the Olmsted brothers and the Minneapolis parkway system by Cleveland. At the regional scale green spaces are systematically planned as a metropolitan infrastructure, such as the one shown in Eliot's plan for Massachusetts. A similar idea of natural system protection and green space planning was even implemented at the national scale in mining. This tradition of park systems and parkways, with its function mainly focused on recreation, has recently been picked up by the greenway movement in the United States, but much enriched and integrated with more comprehensive functions, including the protection of natural resources and natural processes, protection of cultural heritage and recreation (Fabos 1995, 2004, Ahern 2002.

Greenbelt: landscape as urban form maker

The third tradition of landscape urbanism and the notion of landscape as infrastructure, originally in the Western world, can be traced back to the European practice of greenbelt, green heart and green wedge concepts, which are used by urban designers as stoppers, separators and connecters of urban development to create a good urban form. According to Kühn (2003) greenbelts between cities and the countryside have been established as a planning notion since the breaking-up of most European city walls in the 18th and 19th centuries. The greening of former walled areas created promenades for recreational uses and city beautification, but also continues to serve as a separator of city and countryside, as the walls did in the city of the Middle Ages. This notion was also proposed by the Chinese pioneer architect Liang Sicheng, who planned the old Beijing City wall as a ring park in the 1950s (Wang 2003). At the end of the 19th century, the idea of the greenbelt as the city stopper was taken up by E. Howard and became a fundamental element of his Garden City model (Howard 1946). Since Howard's time green spaces of a certain pattern have been planned for structuring and defining "good urban form", such as the greenbelt for a compact city (e.g., London or Berlin, see Toft 1995), green hearts for conurbations, and green wedges for development control (Blumenfeld 1949, Moughtin 1996, Jim and Chen 2003, Frey 2000). Similar ideas have been applied in Chinese city planning since the 1950s and still prevail today, such as the two greenbelts planned for Beijing (Mao 2003, Ming and Yang 2003, Ouyang 2003). Current evidence, based on European countries (e.g., Kühn 2003, Amati and Yokohari 2006), the United States (the Washington DC region), Canada's capital city of Ottawa (Taylor *et al.* 1995), and also the Asian and Chinese cities such as Beijing (Yokohari *et al.* 2000, Ming and Yang 2003), show that these greenbelt and wedge dreams have more or less failed. Some of the major reasons for the failure of the greenbelts and green wedges to stop urban development include the following:

1 They are usually planned artificially and arbitrarily and lack the intrinsic relationship between the green elements and the living earth system (Kühn 2003).

2 They lack usage by the residents, due to their poor accessibility and the lack of connectivity between green spaces and housing projects, etc.

3 They usually only serve as barriers to stop the urban sprawl processes and lack integration of various functions, such as flood control, recreational use, heritage protection, and habitat protection.

4 They quickly become development opportunities when peripheral pressure increases; or, as noticed by Ahern (2002, p.121), since they were based on political boundaries and not on natural features, they became vulnerable to land use change and effectively became a "bank" in which undeveloped land was held until development pressure demanded its use.

5 They are impossible to administer and keep safeguarded in a metropolitan region that is fragmented into a myriad of local governments, cutting across greenbelt and wedge jurisdictions.

Evidence shows that negative definitions of landscape forms (derived as an urban containment) are hardly successful in protecting open spaces in growing city regions. Instead landscape has to have a positive definition, based on its uses and perception by people (Kühn 2003). This issue strongly supports the notion that, in order to have a sustainable landscape and urban form, the conventional planning approaches of architectural urbanism and economic development urbanism have to be reversed.

Partly for this reason in the past decades, especially in the United States, the greenway concept has more or less replaced the concepts of greenbelt, green wedge and greenheart as urban form makers. Based on multiple case studies, Taylor *et al.* (1995) demonstrated that greenway evolved from an urban design approach that attempted to impose both landscape form and land use function onto an ecologically based planning approach that addresses natural factors, connections between natural and urban systems, public participation and support, and innovative government involvement.

The concept of greenways was recently further developed into the more comprehensive and interconnected landscape called green infrastructure (GI), which was considered as the "maker of urban form" within urbanizing and metropolitan regions (Searns 1995, Walmsley 1995, Benedict and McMahon 2006). "Greenways have offered a way to provide a counterbalance to urbanization in an affordable way". It is affordable because linear parks and greenways require small amounts of land, when compared with large non-linear open spaces, to accomplish their objectives. Also they can be woven into the urban fabric with minimal disruption. The new generation of greenways takes the adaptive response notion a step further to address broader habitat and infrastructure needs (Searns 1995). Therefore greenways were considered to be a tool for both smart conservation and smart growth (Benedict 2001, Benedict and McMahon 2006, Walmsley 2006).

Landscape as infrastructure for biological conservation

The fourth tradition of landscape as infrastructure is rooted in biological conservation. The biologist E. Wilson commented: "In the expanding enterprise, landscape design will play a decisive role. Where environments have been mostly humanized, the biological diversity can still be sustained at high levels by the ingenious placement of woodlots, hedgerows, watersheds, reservoirs, and artificial ponds and lakes. Master plans will meld not just economic efficiency and beauty but also the preservation of species and races" (Wilson, 1992, p.317). Concepts such as Ecological Infrastructure (Mander *et al.* 1988, van Selm 1988), ecological framework (Kerkstra and Vrijlandt 1990, Buuren and Kerkstra 1993), ecological network (Bishoff and Jongman 1993), extensive open space systems (Ahern 1991), multiple use modules (Noss and Harris 1986), habitat network and wildlife corridors (Noss 1993), landscape restoration framework (Fedorowick 1993), ecological corridor, environmental corridors, framework landscape and ecostructure, etc., are made in different places with different emphases for the preservation of biodiversity in the context of a stressed landscape (for a comprehensive review see Ahern, 1995, 2002, Jongman and Pungetti 2004). Although all these concepts came into the landscape being based on biological and environmental conservation perspectives, differences among them exist. Jack Ahern (1995) has carried out a very comprehensive comparison and analysis of them. These concepts, though slightly different, all indicate that the philosophy of nature conservation is changing from the species-centered and site protection approaches in the early phases, into ecosystem-oriented ones in the present time, emphasizing the significance of integrated conservation infrastructure.

Among them, EI gathers the most comprehensive meaning and is further related to what is to be explored in the context of this chapter. It is argued that the EI concept is an important strategy to move built landscapes, metropolitan regions and cities towards a more sustainable condition (Ahern 1995).

The concept of EI originally emerged in the 1980s in two fields: ecocity study and conservation biology. According to available documents, the term EI first appeared in the MAB program (Man and Biosphere) of UNESCO. In the 1984 report of MAB five principles were put forward: (1) Ecological Conservation; (2) Ecological Infrastructure; (3) Living Standard of Residence; (4) Cultural and Historical Conservation; (5) Induction of Nature into Cities.

The principle of EI refers to the natural landscape and hinterland of the city but is not clearly defined and overlaps with other concepts such as ecological conservation. In biological conservation studies, the term was first used to represent the habitat network and emphasized the biodiversity conservation function of landscape components such as core zone and corridors from practical perspectives (Mander *et al.* 1988, van Selm 1988). The practices on EI in The Netherlands have shown to be good examples, such as the Dutch Ecological Main Infrastructure, which is made up of: (a) natural core areas, (b) natural development areas, (c) corridors or connections, (d) buffer zones (Bohemen 2002).

Similar to EI is the concept of ecological networks, which is an important concept in nature conservation policy in Europe. Authorities in different levels of management are increasingly accepting the ecological network concept as a tool on a regional scale and even on a continental scale (Jongman 1995). The ecological network is clearly defined and includes several or all of the following elements: core areas, corridors and restoration areas and nature development areas (Bohemen 2002). Important parameters include size, shape, proportion, and distribution of biotopes, their continuity in time and space, variation in internal structure, and the ecological contrast to adjacent biotopes (Seiler and Eriksson 1995, pp.253–264). All these have been planned to be parts of the European Ecological Network (EECONET), which combines the concept of an ecological network with a cooperative, operational framework for guiding and coordinating European action (Bennett 1994).

Natural capital and ecological infrastructure: landscape as integrated infrastructure for a sustainable city

Landscape ecology since its emergence in 1939 (Troll 1939), and especially through its rapid development since the 1980s, has become the single most important discipline that provides a sound scientific basis for the planning and design of landscape (Naveh and Lieberman. 1984, Forman and Godron 1986, Forman 1995, Turner 1989, Turner *et al.* 2001). It was argued that like hardly any other discipline, the landscape approach offers holistic assessment and planning tools to define and develop the interface between nature and culture. The definition of landscape as a heterogeneous land area composed of a cluster of interacting ecosystems (Forman and Godron 1986) is fundamental as it brings the discipline of landscape into a field of science dramatically different from its poetic and picturesque past. While scientific research provides a great amount of knowledge about the processes, patterns and changes, a new shift is called upon to bridge the gap between scientific knowledge and its application (Haines-Young 2000, Potschin and Haines-Young 2006), where landscape sustainability becomes the key concept.

The concept of natural capital and ecosystems' services and goods (Costanza and Daily 1992, De Groot *et al.* 2002, Constanza *et al.* 1997, Daily 1997, 2000, Daily *et al.* 2000, De Groot 2006, Millennium Ecosystem Assessment (MEA) 2005) have shed a bright light on the understanding of sustainable landscape and EI. Natural capital is considered to be the stock of natural ecosystems that yields a flow of valuable ecosystem goods or services into the future for society. Since the flow of services from ecosystems requires that they function as whole systems, the structure and diversity of the system are important components of natural capital (Constanza *et al.* 1997). Four categories of service are identified: provisioning, related to production of food and clean water; regulating, related to the control of climate and disease, mediation of flood and draught; life supporting, related to nutrient cycles and providing habitat (suitable living space) for wild plant and animal species; cultural, related to spiritual, esthetic and recreational benefits.

In this sense, EI can be understood as the necessary structure of a sustainable landscape (or ecosystems) in which the output of the goods and services is maintained, and the capacity of those systems to deliver the same goods and services for future generations is not undermined.

The Ecological Infrastructure is what some scholars sought one decade ago (Forman and Collinge 1997, p.129), and "the spatial solution is a pattern of ecosystems or land uses that will conserve the bulk of, and the most important attributes of, biodiversity and natural processes in any region, landscape or major portion thereof". In their ASLA (American Society of Landscape Architects) Award Winning Project "The Growth Pattern of Taizhou City Based on Ecological Infrastructure" Yu defined Ecological Infrastructure as "the structural landscape network that is composed of the critical landscape elements and spatial patterns that are of strategic significance in preserving the integrity and identity of the natural and cultural landscapes and securing sustainable ecosystem services, protecting cultural heritages and recreational experience." (ASLA 2005, Yu and Padua 2006).

A similar concept to EI (and probably more commonly used in the United States) is green infrastructure (GI), the descendant of greenways and defined as an: "interconnected network of green space (including natural areas and features, public and private conversation lands, working lands with conservation values, and other protected open spaces) that is planned and managed for its natural resource values and for the associated benefits it confers to human populations" (Benedict and McMahon 2006, p.3). GI differs from greenways as it emphasizes ecology, not recreation; it includes large, ecologically important "hubs" as well as key landscape linkages; it can shape urban form and provide a framework for growth. It was argued that like other "gray" and "built" infrastructure, GI should be designed holistically, laid out strategically, planned and implemented publicly, funded up front as a primary public investment, and should be the framework of conservation (Walmsley 1995, 2006, Benedict and McMahon 2006, p.35). GI is considered to be a smart solution to our conservation challenge and it promotes both smart conservation and smart growth.

Through decades of meetings and intellectual exchange, and referring to the above comments about GI, one can only come to the conclusion that EI and GI are exchangeable terms, and the above argument about GI is also valid for EI.

These four traditions and ideas about landscape as infrastructure and landscape urbanism finally come together on the basis of the understanding of natural capital and ecosystems services, and are merged into the concept of EI. Other landscape elements such as cultural heritage corridors, riparian buffers and storm water management system, can also be integrated into EI.

Landscape security patterns as a tool to define ecological infrastructure

Here, the EI is defined as the structural landscape network of which essential landscape elements are conformed. Both existing and potential spatial patterns are of strategic significance in preserving natural, biological and cultural processes, which are critical in securing the integrity and identity of the natural and cultural landscapes, as well as securing natural capital that supports sustainable ecosystem services.

It is important to note that EI is process-oriented, not just a visible spatial pattern. Regarding the identification and planning of ecological infrastructure, the processes oriented model of spatial analysis, namely security patterns (SP) approach, might be useful (Yu 1995, 1996). SP are "composed of strategic portions and positions of the landscape that have critical significance in safeguarding and controlling certain ecological processes." Components of the SP have the quality of initiative, coordination and efficiency and are, therefore, strategically important in controlling ecological processes and landscape change. SP can be identified according to the properties on a general surface model of flows and processes. Potential surfaces (accessibility surfaces) are developed using landscape resistance to represent the dynamics of horizontal ecological processes (e.g., species movement, the spread of urban development and water flow). Four strategic landscape portions and positions are commonly identified on the potential surfaces: buffer zones, intersource linkages, radiating routes and strategic points. These components, specified by certain quantitative and qualitative parameters, together with the identified sources (e.g., native habitats), compose the ecological SP at various security levels. These SP could be integrated into an overall ecological infrastructure, and can be used by defenders of ecological processes as defensive frontiers and strategies of spatial bargaining in landscape changes. The landscape SP approach has been applied in multiple cases (Yu 1996, Guo *et al.* 2005, Yu and Padua 2006, Ahern 2007).

12.2 THE NEGATIVE APPROACH: METHODOLOGY

The goal of the negative approach to landscape and urban planning is to meet the challenges of sustainability, including to have a sustainable landscape and sustainable urban form, and also to allow land to be developed without losing its ecological, cultural and spiritual integrity. Instead of using population projection as the basis of economic development planning, the negative approach looks for the ecological and cultural sustainability and integrity as the solid base for development planning. Instead of planning civil infrastructure (gray infrastructure) as the framework for urban development and architectural urbanism, the negative approach considers EI to shape urban form and frame urban development. The overall goals and objectives are smart preservation and smart growth.

With these objectives in mind, the negative approach to planning is composed of the following steps:

12.2.1 Process analysis

Processes associated with critical ecosystems' functions or services are targeted to be safeguarded by EI; thus systematic analysis shall be carried out using Geographical Information Systems (GIS), which are an efficient tool to simulate natural and cultural processes across the landscape. These processes include:

1 Abiotic processes, associated with the regulation and life supporting services of ecosystems.

2 Biotic processes, associated with life support services of ecosystems in providing habitat (suitable living space) for native wild plant and animal species and biodiversity conservation.

3 Cultural processes, associated with information functions including visual perception, heritage protection and recreational activities.

12.2.2 Defining landscape security patterns

Landscape SP are identified for the individually targeted processes. SPs are composed of elements and spatial positions that are strategically important in safeguarding the different processes across the landscape. Models including suitability analysis, minimum cost distance and surface analysis were used in the identification of SP for the individual processes (Yu 1995, 1996). Alternative security levels – low, medium and high – are used to define the attributes of the SP in safeguarding each of the targeted processes.

12.2.3 Defining ecological infrastructure

The overlaying technique is used to integrate the SP for individual processes. Alternatives of EI are developed at various quality levels: high, medium and low. Green lines were drawn to define and protect the EI.

The EI is planned at three scales:

1 The regional scale (hundreds to thousands of square kilometers): at the regional scale green lines are drawn to define the structural elements as corridors and restricted areas for construction.

2 The intermediate scale (tens of square kilometers): at this scale the overall design and management guidelines are developed for EI, and especially for the green corridors that work as critical EI elements in water management, biodiversity conservation, heritage protection and recreation.

3 The small scale (less than 10 square kilometers): at a specific site an urban green network is designed to allow ecosystem services to be delivered into the urban fabric.

12.2.4 Defining urban form at the large scale: urban growth alternatives based on regional EI

Urban development patterns are defined at regional scales based on regional EI. Using the multiple EI alternatives as framing structures, scenarios of regional urban growth patterns are developed. A planning committee, represented by decision makers of the city, planning experts and stakeholders, makes impact assessment for these scenarios. The decision makers can finally select one of the scenarios based on a balanced evaluation of economic, ecological and social benefits.

12.2.5 Defining urban form at the intermediate scale: urban open spaces system based on EI

At the intermediate scale the regional EI is to be integrated inside the urban structure and becomes the urban green space system, which integrates various functions of ecological conservation, heritage protection and recreational activities, commuting and cycling (Figure 12.1).

12.2.6 Defining urban form at the small scale: site-specific urban development alternatives based on EI

Alternative urban development models are designed at an individual site and neighborhood, following EI guidelines developed above for the site. In these urban development alternatives, ecosystem services safeguarded by EI are delivered into the urban fabric so that the conventional urban sprawl can be avoided. Building the EI into the site specific land use schemes, as a new way of development, will not only help the whole city, but will also benefit the site specific development ecologically and economically (Figure 12.1).

12.3 URBAN GROWTH BASED ON EI: A CASE OF NEGATIVE PLANNING FOR TAIZHOU CITY

Taizhou is located on the South-east coast of China, with a total area of 9411 km^2 and a population of 5.5 million. Out of the population, only 0.7 million live

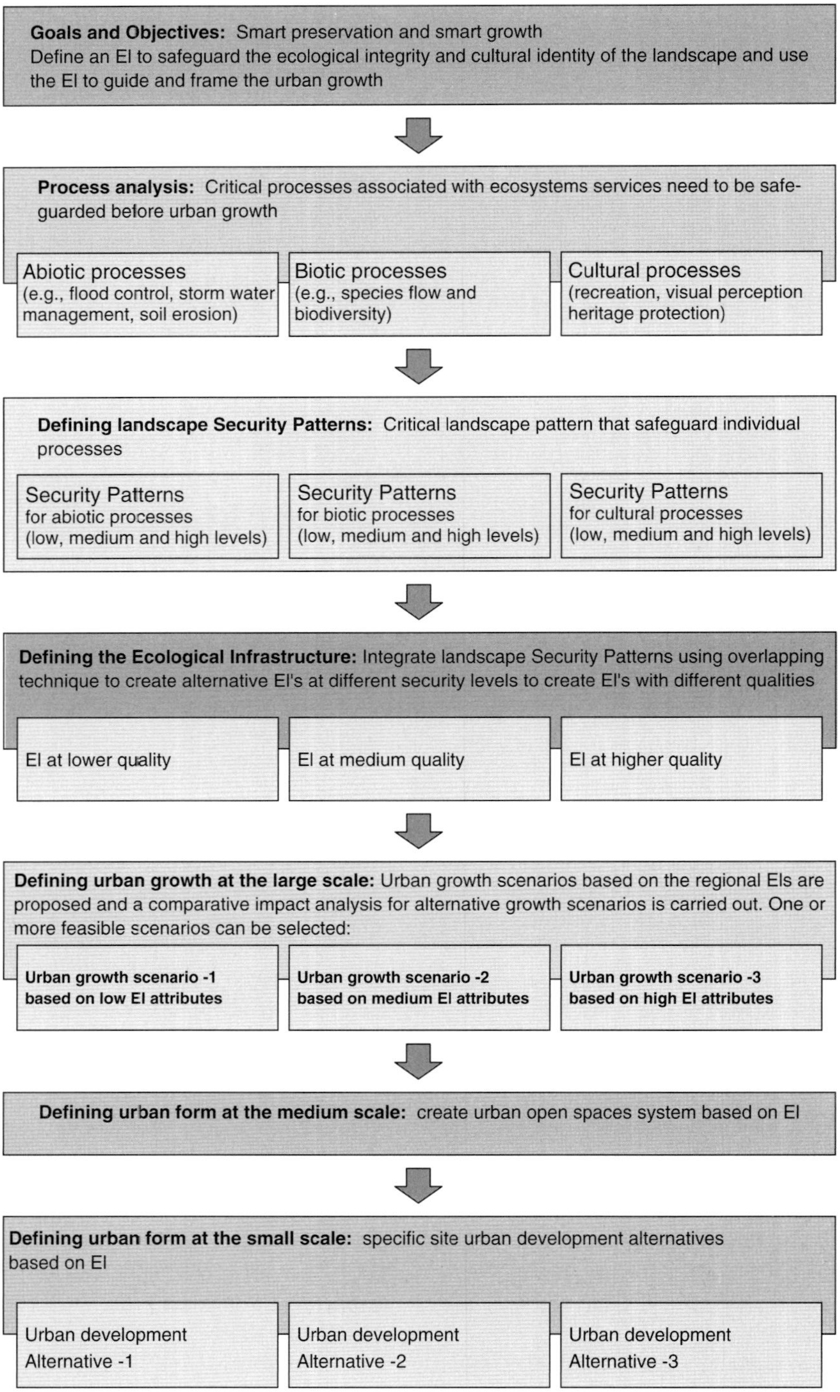

Figure 12.1 The framework of the negative approach of urban development planning based on ecological infrastructure.

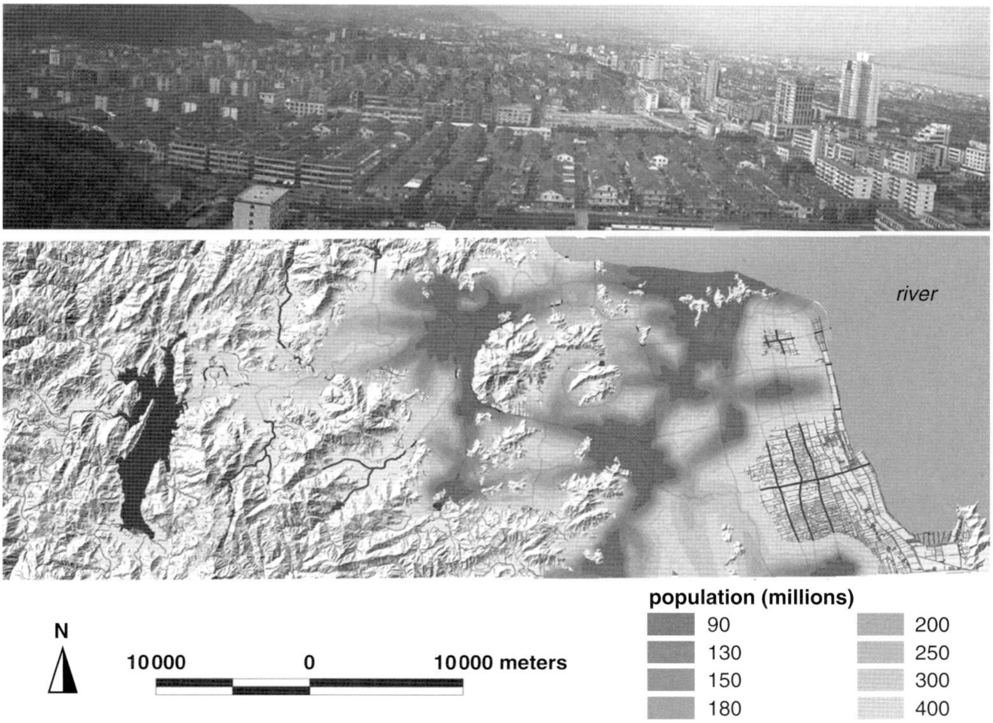

Figure 12.2 The simulation of urban sprawl of Taizhou City based on economic and development oriented model: urban sprawl indiscriminately takes over and destroys the integrity and identity of the landscape.

in urban areas; the urban population is expected to increase to 0.9 million in 2010, 1.3 million in 2020, and 1.5 million in 2030. Although it has quite a rural and agricultural character, it is now one of the fastest growing areas in China due to a boom of small private industries (Figure 12.2).

Under the influence of the monsoon climate and being adjacent to the sea, flooding has been a major hazard. As an adaptation to the storm water and flood problem, the landscape has been shaped into a unique form featuring a network of water courses that integrate natural water systems, wetlands and man-made ditches, as well as cultural heritage features such as bridges, dikes, dams, and vernacular landscapes. This area has long been famous for its rice, fishery, and citrus production. There is a very limited amount of arable or other land available for development.

This water network landscape, which has been effective in preserving the agricultural processes for thousands of years, is now facing the challenge of being destroyed by the swift urbanization process that began in the early 1990s. The wetlands have been filled, rivers have been straightened and channeled, cultural heritage features (not listed as protected historical relics), have been destroyed and visual and recreational experiences have been totally ignored.

Addressing the above situations, the negative approach to urban growth planning was taken to safeguard sustainability of the landscape.

12.3.1 Critical landscape processes

Three categories of processes are targeted:

1 The abiotic processes: in Taizhou area the monsoon storms frequently cause floods and waterlogging, so the main focus of the process analysis is flood control and storm water management.

2 Biotic processes: native biodiversity conservation is the focus of the biotic analysis. Birds are the main concern for this area due to the rich fluvial and intertidal wetland ecosystems. This area has been listed as

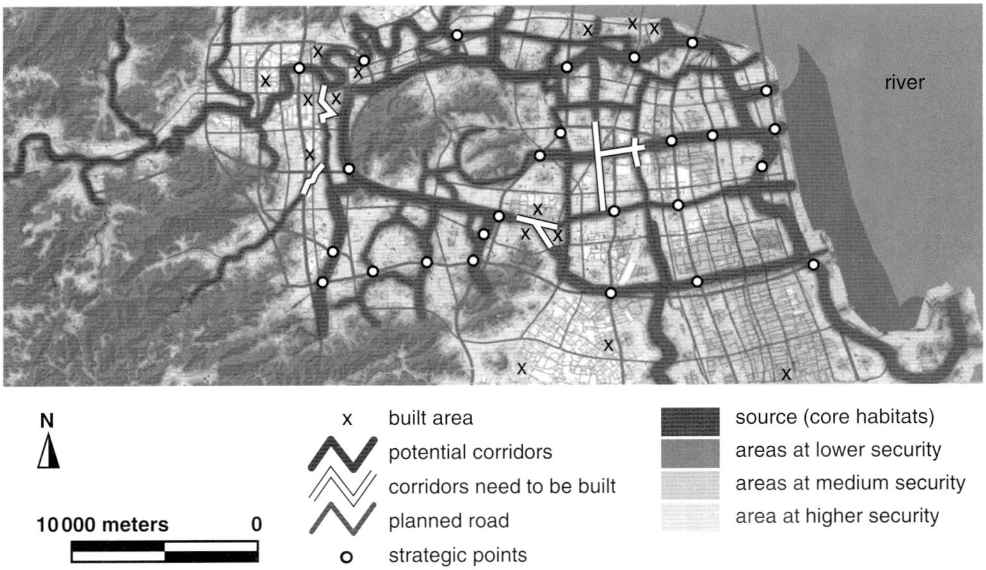

Figure 12.3 Large-scale: security patterns for biodiversity conservation.

one of the important bird areas, both nationally and globally. Feeding, nesting, and immigration processes are the main focus for the purpose of the protection of habitats and biodiversity (Figure 12.3).

3 Cultural processes: the targeted cultural processes include historical processes of cultural heritage sites and the cultural landscape and the process of recreational use of the landscape.

12.3.2 Defining landscape security patterns for the targeted processes

Borrowing the security pattern model from biological conservation, the general spatial model for the SP for all the horizontal processes across the landscape are composed of the following elements (Yu 1996).

- Sources: the core area and the origin of the of the target process, such as core habitats for the targeted species, heritage sites for preservation and fishing ponds for recreation.
- Buffer zones: the areas around the sources, which are potentially important for protecting the sources.
- Linkages: the connecting linear elements that link two or more of the sources.
- Radiating routes (for biological processes): from the sources, which is the potential network for species to take control of the landscape. Here, the target species are taken as active and initiating forces of control over the landscape.
- Strategic points: the spatial location that potentially controls the movement and connectivity of the target process.

SPs for any individual process are composed of more than one of the elements, but not necessary all of them: Three security levels – low, medium and high – are used to define the quality of the SP in safeguarding each of the targeted processes. The security levels are classified according to the area, number and quality of the landscape elements that make up the SP (Figure 12.4).

1 SP for floods: Various hydrological models are used to simulate floods based on rainfalls, tides, terrain and wetlands. Floodable areas are calculated for 10 years, 20 years and 100 year frequencies, which are used as the criteria for the definition of security level of floods. Flood SP include the existing water channel network, wetland, and potential wetlands and flood vulnerable areas.

2 SP for biodiversity conservation: The focal species approach for biodiversity conservation is widely accepted and used to identify critical landscape elements and locations of biodiversity conservation (Caro 2000, Opdam *et al.* 2003, Brooks and Kennedy 2004, Lambeck 1997, Eycott *et al.* 2007). Birds are the main concern for this area due to the rich fluvial

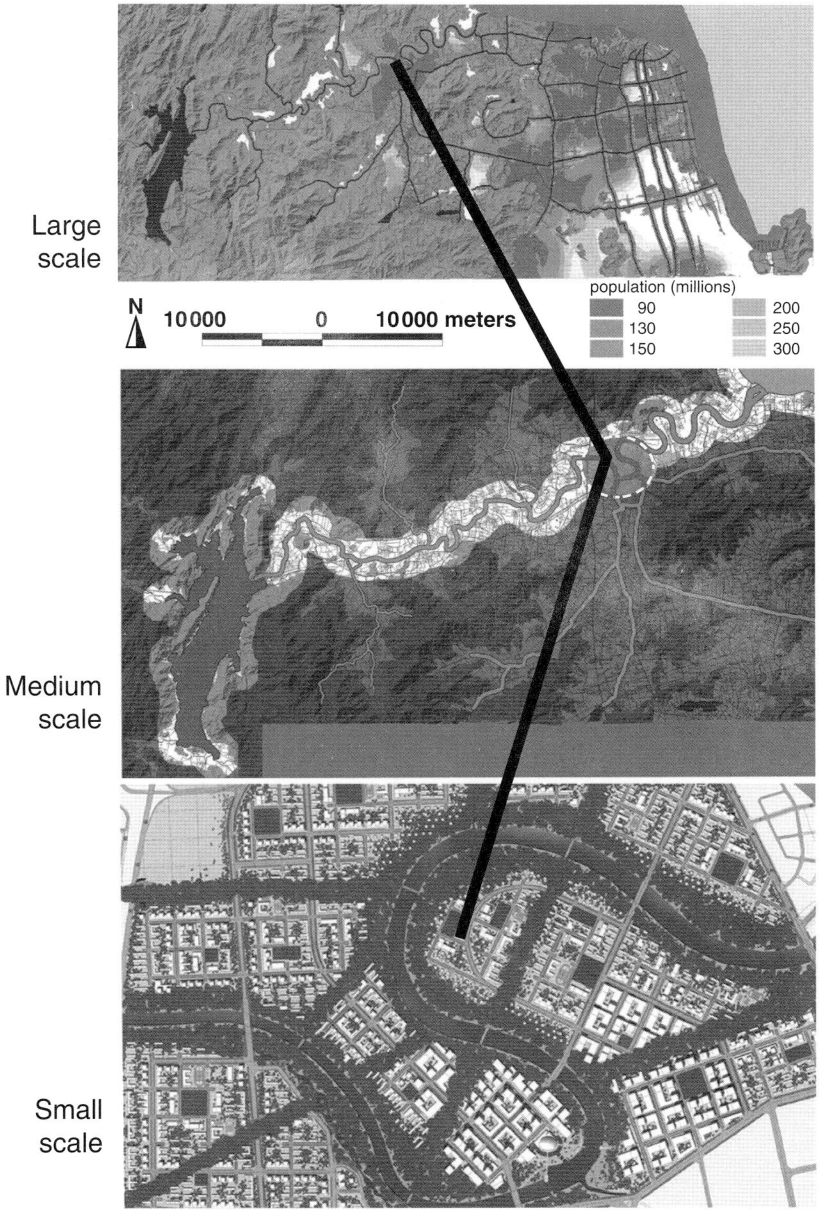

Figure 12.4 Large scale: security pattern for recreation. Landscape elements including wetland, forests, water features, and cultural heritage sites were identified as sources for recreation. The security pattern for recreation landscape was identified according to the recreational values of these resources and their accessibility. Based on their spatial relationships, a recreational landscape network was designed.

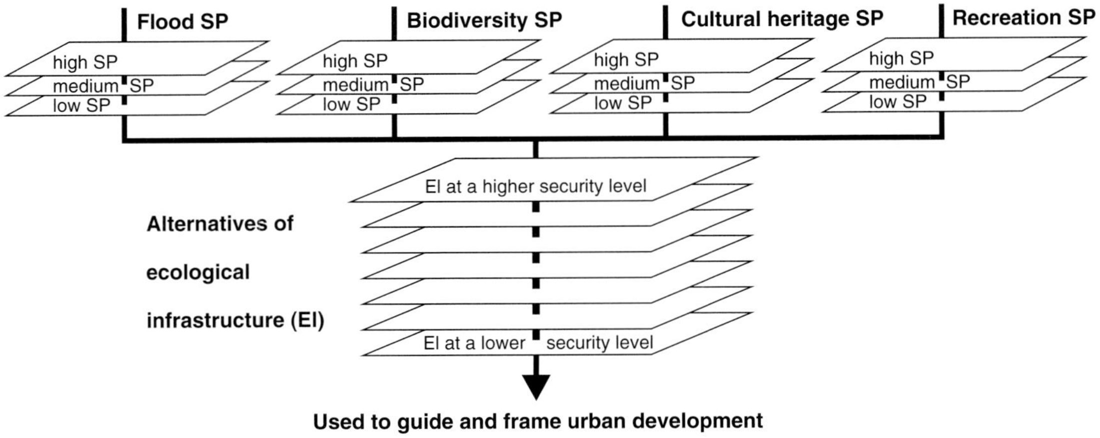

Figure 12.5 Large scale: the overall regional ecological infrastructure (EI) was an integration of security patterns and plans for flood control, biodiversity conservation, cultural heritage protection and recreation. Three alternatives of EI were developed corresponding to high, medium and low security levels. They will be used to guide and frame regional urban development pattern.

and inter-tide wetland ecosystems. Nationally ranked endangered birds (in the red list) are selected as the focal species to identify native habitats and biodiversity protection.

The Security Patterns are identified based on two overlaid analysis (Yu 1996): the suitability analysis of the habitats (sources) for the focal species, and the potential surface analysis based on the least-cost analysis and surface models (Knaapen *et al.* 1992, Yu 1995, 1996, 1998, Ferreras 2001, Graham 2001, Michels *et al.* 2001, Schadt *et al.* 2002, Adriaensen 2003). On the potential surface, buffer zones, potential linkages, radiating routes and strategic points were able to be identified (see Figure 12.3).

3 SP for the cultural processes: Both heritage sites and linkages that connect, or potentially connect, these sites are included in the construction of SPs for heritage protection. Recreational sources are then referred to (e.g., water body for fishing, forest for hiking, and historical site for sightseeing) and the linkages between these resources are based on the least-distance model result.

12.3.3 Defining ecological infrastructure

The overlaying technique is used to integrate the SP for individual processes. Alternatives of EI are developed at various quality levels: high, medium and low.

The EI is planned at three scales (Figure 12.5):

1 The regional scale (hundreds to thousands of square kilometers): at the regional scale the structural elements are defined as corridors and as restricted areas for construction. Alternatives of regional EI are developed at diverse quality levels: high, medium and low.

2 The intermediate scale (tens of square kilometers): at this scale, the overall design and management guidelines are developed for EI, and especially for the green corridors that work as critical EI elements in water management, biodiversity conservation, heritage protection and recreation.

3 The small scale (less than 10 square kilometers): at a specific site an urban green network is designed to allow ecosystem services to be delivered into the urban fabric.

12.3.4 Scenarios of urban growth pattern based on the regional ecological infrastructure

Using the three regional EI alternatives as a framework structure, scenarios of regional urban growth patterns were simulated using GIS: the Adjusted Sprawl Scenario, the Aggregated Scenario (Figure 12.4), and the Scattered Scenario.

Comparative impact evaluations were made for these scenarios by a planning committee composed of decision makers of the city, planning experts from all over the country, and stakeholders who are represented by officials from various functional departments of the Taizhou City government (including the departments of agriculture, water management, forestry, industry, tourism, finance, transportation, public affairs, security, culture education, tax, etc.). It also includes representatives of individual villages who originally owned the land as well as representatives of developers and representatives of investors who are eager to get the right to develop the land.

Decision makers finally selected one of the three urban growth scenarios as the most feasible, after multiple brainstorming sessions among the planning committee. As expected, the Aggregated Scenario, which is based on the medium quality EI, was considered the more balanced and less difficult to be implemented.

12.3.5 Shaping urban form at the intermediate scale

In shaping the urban form and structure at the city scale, an urban open space system is planned connecting the regional EI with the inner city's landscape elements, related to ecological, cultural and recreational values, and also integrated with the commuter routes, cycling network and pedestrian network.

12.3.6 Shaping urban land development at the small scale

Using a selected site ($10\,km^2$ in size) as a demonstration, alternative urban development models were designed to test the possibility of building an EI based city. In these alternatives, ecosystem services safeguarded by EI are delivered into the urban fabric so that the conventional urban development model can be avoided (Figure 12.6).

These new urban land development alternatives were presented to the developers and investors, as well as to the city's decision makers, to let them know that the business-as-usual models of land development can be avoided. The new way of development, by building the EI into their land use scheme, will not only help the whole city, but will also benefit site development ecologically and economically. These alternative development schemes show how the regional, large and small scales of EI can be realized in land development to handle the problem of urban growth.

12.4 CONCLUSION

It is argued that the current urban growth model in China is unsustainable. Recognizing this fact the Chinese leadership is now calling for ecological civilization (Hu 2007), a totally new concept proposed in the Chinese language and especially worded by the top Chinese leaders. It reflects an important change in the top Chinese leadership's understanding of development. Rather than emphasizing economic construction as the core of development as it did in the past, the Chinese leadership has come to realize that development, if sustainable, must entail a list of elements including the right relationship between man and nature. The ecological civilization concept is proposed at a time when ecological and environmental issues are at a very serious stage. Facing such a reality, the construction of ecological civilization was definitely not rhetoric for self adulation by officials in their speeches. It needs to be transformed into tangible measures that will change the way our economy develops and reshape the landscape to be able to meet the serious challenges of sustainable development.

Accordingly, it is important to recognize that the conventional approach to urban development planning, which is based on population projection and then built infrastructure, is unable to meet the challenges and needs of ecological and sustainable urban development, and is certainly unable to meet the goal of ecological civilization. it is in this situation that the negative approach is proposed. using the analogy of photography in describing the film and picture, the term "negative" is used to describe the urban development model being negatively enframed by EI, not the other way around. To put it another way, the EI is positively defining the urban form and growth pattern. Conventionally, landscape and green elements such as greenbelt and green heart are usually negatively defined by architectural and built infrastructure. By positively defining the EI for the sake of natural capital and cultural integrity of the land, the urban growth pattern and urban form are negatively defined. The negative approach, and especially the concept of EI, build a bridge between landscape urbanism, the disciplines of ecology and especially landscape ecology, the notion of natural capital and ecosystems services, and

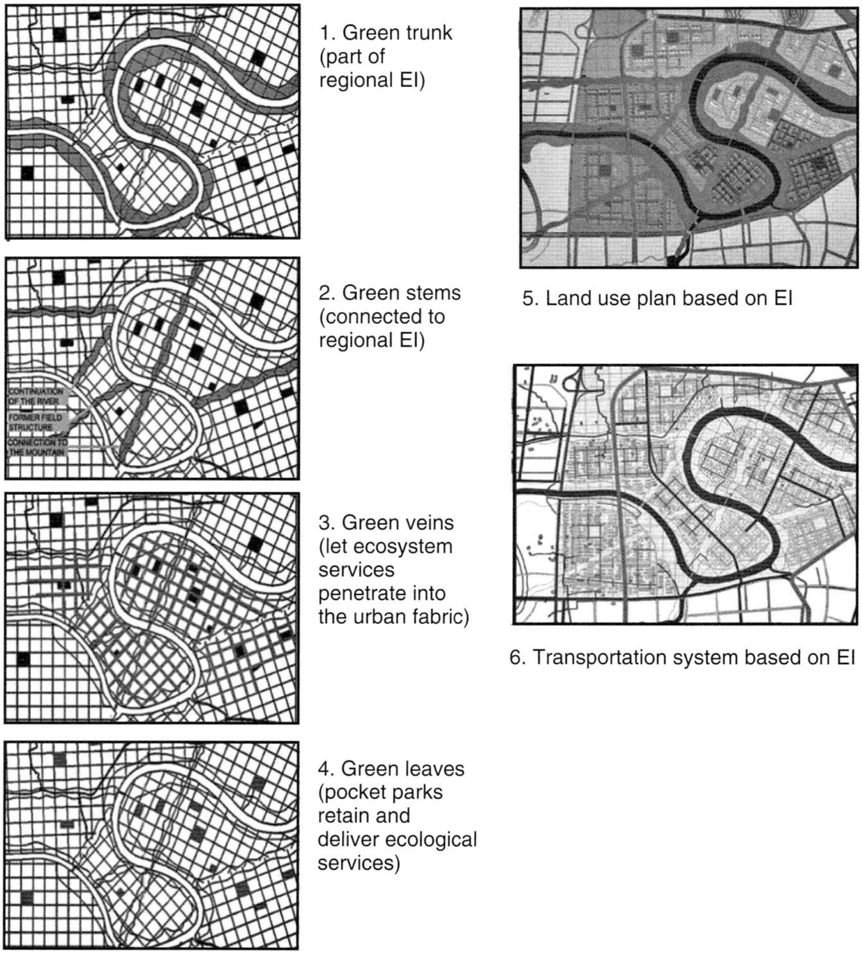

Figure 12.6 Small scale: the grid alternative. A grid system of green in stead of roads to deliver the ecological services into the urban fabric.

sustainable development. It is a bridge between smart growth and smart conservation.

REFERENCES

Adriaensen, F., Chardon, J. P., De Blust, G., *et al.* (2003) The application of "least-cost" modelling as a functional landscape model. *Landscape and Urban Planning* 64, 233–247.

Ahern, J. (1991) Planning and design for an extensive open space system: linking landscape structure to function. *Landscape and Urban Planning* 21, 131–145.

Ahern, J. (1995) Greenways as a planning strategy. *Landscape and Urban Planning* 33, 131–155.

Ahern, J. (2002) *Greenways as Strategic Landscape Planning: Theory and Application*. Wageningen University, The Netherlands.

Ahern, J. (2007) Green infrastructure for cities: The spatial dimension. In: Novotny, V., Brown P. (eds) *Cities of the Future. Towards Integrated Sustainable Water and Landscape Management*. IWA Publishing, London, pp. 267–283.

Amati, M., Yokohari, M. (2006) Temporal changes and local variations in the functions of London's green belt. *Landscape and Urban Planning* 75, 125–142.

ASLA (2005) *The Announcement of 2005 ASLA Professional Awards*. http://www.asla.org (accessed 22 March 2011).

Benedict, M. (2001) *Green Infrastructure: A Strategic Approach for Land Conservation Planning. National Conservation*

Training Center Conference, May 1–3, Shepherdstown, Virginia, quote on pp. 7–8.
Benedict, M., McMahon, E. (2006) *Green Infrastructure: Linking Landscapes and Communities*. Island Press. Washington, USA.
Bishoff, N. T., Jongman, R. H. G. (1993) *Development of Rural Landscape*. USDI National Park Service. Washington DC.
Blumenfeld, H. (1949) Theory of city form, past and present. *Journal of the Society of Architectural Historians* 8, 7–16.
Bohemen, H. (2002) Infrastructure, ecology and art. *Landscape and Urban Planning* 59, 187–201.
Brooks, T., Kennedy, E. (2004) Conservation biology: Biodiversity barometers. *Nature* 431, 1046–1047.
Buuren, M. van, Kerkstra, K. (1993) The framework concept and the hydrological landscape structure: a new perspective in the design of multifunctional landscape. In: Vos C. C., Opdam P. (eds) *Landscape Ecology of a Stressed Environment*. Chapman & Hall, London, pp. 219–243.
Caro, T. (2000) Focal species. *Conservation Biology* 14(6), 1569–1570.
Chen, K. L., Lü, Y., Zhang, X. H. (2004) No water without wetland. *China Environment and Development Review*, 296–309.
Cheng, J., Masser, I. (2003) Urban growth pattern modeling: a case study of Wuhan city, PR China. *Landscape and Urban Planning* 62, 199–217.
Corner, J. (1999) Recovering landscape as a critical cultural practice. In: Corner, J. (ed.) *Recovering Landscape: Essays in Contemporary Landscape Architecture*. Princeton Architectural Press, New York.
Corner, J. (2006) Terra fluxus. In: Waldheim, C. (ed.) *Landscape Urbanism*. Princeton Architectural Press, New York, pp. 21–33.
Costanza, R., Daily, H. E. (1992) Natural capital and sustainable development. *Conservation Biology* 6, 37–46.
Costanza, R., d'Arge, R., de Groot, R., *et al.* (1997) The value of the world's ecosystem services and natural capital. *Nature* 387, 253–259.
Daily, G. C. (1997) *Nature's Services: Society Dependence on Natural Ecosystems*. Island Press, Washington, DC.
Daily, G. C. (2000) Management objectives for the protection of ecosystem services. *Environmental Science and Policy* 3, 333–339.
Daily, G. C., Söderqvist, T., Aniyar, S., *et al.* (2000) The value of nature and the nature of value. *Science* 289, 395–396.
De Groot, R. S. (2006) Function analysis and valuation as a tool to assess land use conflicts in planning for sustainable, multi-functional landscapes. *Landscape and Urban Planning* 75(3–4), 175–186.
De Groot, R. S., Wilson, M. A., Boumans, R. J. (2002) A typology for description, classification and valuation of ecosystem functions, goods and services. *Ecological Economics* 43(3), 393–401.
Eycott, A., Watts, K., Moseley, D., Ray, D. (2007) *Evaluating Biodiversity in Fragmented Landscapes: The Use of Focal Species*. Forestry Commission, October, 2007, Edinburgh, UK.
Fabos, J. G. (1995) Introduction and overview: the greenway movement, uses and potential of greenways. *Landscape and Urban Planning* 33, 1–13.
Fabos, J. G. (2004) Greenway planning in the United States: its origins and recent case studies. *Landscape and Urban Planning* 68, 321–342.
Fedorowick, J. M. (1993) A landscape restoration framework for wildlife and agriculture in the rural landscape. *Landscape and Urban Planning* 27, 7–17.
Ferreras, P. (2001) Landscape structure and asymmetrical inter-patch connectivity in a metapopulation of the endangered Iberian lynx. *Biological Conservation* 100, 125–136.
Forman, R. (1995) *Land Mosaics: The Ecology of Landscapes and Regions*. Cambridge University Press, Cambridge, UK.
Forman, R., Collinge, S. (1997) Nature conserved in changing landscapes with and without spatial planning. *Landscape and Urban Planning* 37, 129–135.
Forman, R., Godron, M. (1986) *Landscape Ecology*. John Wiley & Sons, Inc., New York.
Frey, H. W. (2000) Not green belts but green wedges: the precarious relationship between city and country. *Urban Design International* 5(1), 13–25.
Gaubatz, P. (1999) China's urban transformation: patterns and process of morphological change in Beijing, Shanghai and Guangzhou. *Urban Studies* 36(9), 1495–1521.
Graham, C. H. (2001) Factors influencing movement patterns of keel-billed toucans in a fragmented tropical landscape in southern Mexico. *Conservation Biology* 15, 1789–1798.
Guo, M., Li, X., Xiao, D. N. (2005) Ecological security pattern analysis of Jiuquan oasis using remote sensing and GIS. *Geoscience and Remote Sensing Symposium. IGARSS '05. Proceedings*. 2005 IEEE International, 25–29 July 2005, pp. 1867–1870.
Guo, X. M. (2004) The economic cost of environmental pollution and ecological deterioration. *China Environment and Development Review*, 53–71.
Haines-Young, R. (2000) Sustainable development and sustainable landscapes: defining a new paradigm for landscape ecology. *Fennia* 178(1), 7–14.
Howard, E. (1946) *Garden Cities of Tomorrow*. Faber & Faber, London (first published 1898).
Hu, J. T. (2007) *Report to the Seventeenth National Congress of the Communist Party of China* on October 15, 2007. http://news.xinhuanet.com/newscenter/2007-10/24/content_6938568.htm (accessed 24 March 2011; in Chinese).
Huang, J. F., Wang, R. H., Zhang, H. Z. (2007) Analysis of patterns and ecological security trend of modern oasis landscapes in Xinjiang, China. *Environmental Monitoring and Assessment* 134(1–3), 411–419.
Jiang, G. M., Liu, M. Z. (2004) Sand storm. *China Environment and Development Review* 2, 310–322.
Jim, C. Y., Chen, S. (2003) Comprehensive greenspace planning based on landscape ecology principles in compact

Nanjing city, China. *Landscape and Urban Planning* 65, 95–116.

Jongman, R. H. G. (1995) Nature conservation planning in Europe: developing ecological networks. *Landscape and Urban Planning* 32, 169–183.

Jongman, R. H. G. (2001) The context and concept of ecological networks. In: Jongman, R. H. G., Pungetti, G. (eds) *Ecological Networks and Greenways: Concept, Design, Implementation*. Cambridge University Press, Cambridge, pp. 7–33.

Jongman, R., Pungetti, G. (eds) (2004) *Ecological Networks and Greenways: Concept, Design, Implementation*. Cambridge University Press, xxi + 345 p.

Kerkastra, K., Vrijlandt, P. (1990) Landscape planning for industrial agriculture: a proposed framework for rural area. *Landscape and Urban Planning* 18(3–4), 275–287.

Knaapen, J. P., Scheffer, M., Harms, B. (1992) Estimating habitat isolation in landscape planning. *Landscape and Urban Planning* 23, 1–16.

Lambeck, R. J. (1997) Focal species: A multi-species umbrella for nature conservation. *Conservation Biology* 11(4), 849–856.

Kühn, M. (2003) Greenbelt and Green Heart: separating and integrating landscapes in European city regions. *Landscape and Urban Planning* 64, 19–27.

Lip, E. (1979) *Chinese Geomancy*. Times Book International, Singapore.

Little, C. (1990) *Greenways for America*. John Hopkins University Press, Baltimore, MD.

Mander, U. E., Jagonaegi, J., Kuelvik, M. (1988) Network of compensative areas as an ecological infrastructure of territories. In: Schreiber, K-F. (ed.) *Connectivity in Landscape Ecology*. Proceedings of the 2nd International Seminar of the International Association for Landscape Ecology. Ferdinand Schöningh, Paderborn, pp. 35–38.

Mao, Q. Z. (2003) Green space of Beijing city: History and prospect. *Beijing City Planning & Construction Review*, supplementary issue, 137–138.

McHarg, I. L. (1969) (1992 edition) *Design with Nature*. John Wiley & Sons, Inc., New York

Michels, E., Cottenie, K., Neys, L., De Gelas, K., Coppin, P., De Meester, L. (2001) Geographical and genetic distances among zooplankton populations in a set of interconnected ponds: a plea for using GIS modelling of the effective geographical distance. *Molecular Ecology* 10, 1929–1938.

Millennium Ecosystem Assessment (MEA) (2005) *Ecosystems and Human Well-Being: Synthesis*. Island Press, Washington.

Min, X. Y., Yang, B. J. (2003) The Second Green Belt and Beijing Spatial Distribution. *City Planning Review* 9, 17–21. (In Chinese)

Moughtin, C. (1996) *Urban Design: Green Dimensions*. Butterworth Architecture, Oxford.

Naveh, Z., Lieberman, A. (1984) *Landscape Ecology: Theory and Application*. Springer, New York.

Noss, R. (1993) Wildlife corridors. In: Smith, D. S., Hellmund, P. C. (eds) *Ecology of Greenways: Design and Function of Linear Conservation Areas*. University of Minnesota Press, Minneapolis, pp. 43–68.

Noss, R., Harris, L. D. (1986). Nodes, networks, and MUMs: Preserving diversity at all scales. *Environmental Management* 10(3), 299–309.

Opdam, P., Verboom, J., Pouwels, R. (2003) Landscape cohesion: an index for the conservation potential of landscapes for biodiversity. *Landscape Ecology* 18(2), 113–126.

Ouyang, Z. Y. (2003) The ecological planning of Beijing green belt. *Beijing City Planning & Construction Review*, supplementary issue, 202–205. (In Chinese)

Potschin, M., Haines-Young, R. (2006) "Rio+10", sustainability science and landscape ecology. *Landscape and Urban Planning* 75, 162–174.

Rossbach, S. (1983) *Feng Shui: The Chinese Art of Placement*. E.P. Dutton, Inc., New York.

Schadt, S., Knauer, F., Kaczensky, P., Revilla, E., Wiegand, T., Trepl, L. (2002) Rule-based assessment of suitable habitat and patch connectivity for Eurasian Lynx in Germany. *Ecol. Appl.* 12, 1469–1483.

Searns, R. M. (1995) The evolution of greenways as an adaptive urban landscape form. *Landscape and Urban Planning* 33, 65–80.

Seiler, A., Eriksson, I.-M. (1997) New approaches for ecological consideration in Swedish road planning. In: Canters, K., Piepers, A., Hendriks-Heersma, A. (eds) *Proceedings of the International Conference on Habitat Fragmentation, Infrastructure and the Role of Ecological Engineering*. 17–21 September 1995, Maastricht and The Hague. Ministry of Transport, Public Works and Water Management, Road and Hydraulic Engineering division, Delft, The Netherlands, pp. 253–264.

Selm, A.J. van (1988) Ecological infrastructure: a conceptual framework for designing habitat networks. In: Schreiber, K.-F. (ed.) *Connectivity in Landscape Ecology*. Proceedings of the 2nd International Seminar of the International Association for Landscape Ecology. Ferdinand Schöningh, Paderborn, pp. 63–66.

Skinner, S. (1982) *The Living Earth Manual of Feng-shui*. Routledge & Kegan Paul, London.

Swenson, J. J., Franklin, J. (2000) The effects of future urban development on habitat fragmentation in the Santa Monica Mountains. *Landscape Ecology* 15(8), 713–730.

Taylor, L. (1978) *Urban Open Spaces*. Academy Editions, London.

Taylor, J., Paine, C., FitzGibbon, J. (1995) From greenbelt to greenways: four Canadian case studies. *Landscape and Urban Planning* 33, 47–64.

Toft, D. (1995) Green belt and the urban fringe. *Built Environment* 21(1), 54–59.

Troll, C. (1939) Luftbildplan und ökologische Bodenforschung. *Zeitschrift der Gesellschaft für Erdkunde zu Berlin* 7/8, 241–298.

Turner, M. G. (1989) Landscape ecology: the effect of pattern on process. *Annual Review of Ecology and Systematics* 20, 171–197.

Turner, M. G., Gardner, R. H., O'Neill, R. V. (2001) *Landscape Ecology in Theory and Practice*. Springer, New York.

United Nations (UN) (2007) *State of World Population 2007: Unleashing the Potential of Urban Growth*. Online-Report. United Nations Population Fund. http://www.unfpa.org/swp/2007/english/introduction.html (accessed 24 March 2011).

Waldheim, C. (ed.) (2006) *Landscape Urbanism*. Princeton Architectural Press, New York.

Walmsley, A. (1995) Greenways and the making of urban form. *Landscape and Urban Planning* 33, 81–27.

Walmsley, A. (2006) Greenways: multiplying and diversifying in the 21st century. *Landscape and Urban Planning* 76, 252–290.

Wang, J. (2003) *Cheng Ji*. Sanlian Shudian, Beijing. (In Chinese)

Wilson, E. O. (1992) *The Diversity of Life*. The Belknap Press of Harvard University Press, Cambridge, MA.

Wu, L. Y., Wu, T. H. (2003) From strategic plan to action plan. *City Planning Review* 12, 13–17. (In Chinese with English abstract)

Yang, B. J. (2003) An approach to reform confronting the reality: on the theoretical and practical implication of immediate planning. *City Planning Review* 3, 5–9. (In Chinese)

Yeh, A. G., Wu, F. (1999) The transformation of the urban planning system in China from a centrally-planned to transitional economy. *Progress in Planning* 51(3), 167–249.

Yokohari, M., Takeuchi, K., Watanabe, T., Yokota, S. (2000) Beyond greenbelts and zoning: A new planning concept for the environment of Asian mega-cities. *Landscape and Urban Planning* 47, 159–171.

Yu, K. J. (1994) Landscape into places: Feng-shui model of place making and some cross-cultural comparison. In: Clark, J. D. (ed.) *History and Culture*. Mississipi State University, USA. pp. 320–340.

Yu, K. J. (1995) *Security Patterns in Landscape Planning with a Case Study in South China*. Doctorial Thesis, Graduate School of Design, Harvard University, MA.

Yu, K. J. (1996) Security patterns and surface model in landscape planning. *Landscape and Urban Planning* 36(5), 1–17.

Yu, K. J. (1998) Landscape ecological security patterns in biological conservation. *Acta Ecologica Sinica* 19(1), 8–15. (In Chinese with English abstract)

Yu, K., Padua, M. (2006) *The Art of Survival*. The Images Publishing Group, Victoria, Australia.

Yu, K. J., Li, D. H., Liu, H. L., Han, X. L. (2005a) On the "Negative Planning" *City Planning Review* 9, 64–69. (In Chinese with English abstract)

Yu, K. J., Li, D. H., Liu, H. L., Han, X. L. (2005b) Growth pattern of Taizhou City based on ecological infrastructure – A Negative Approach physical urban planning. *City Planning Review* 9, 76–80. (In Chinese with English abstract)

Zhang, B. (2003) My view on the immediate plan. *City Planning Review* 6, 64–65. (In Chinese)

Zhao, Y. Q. (2004) Rapid development and spatial extension: a case study of Shenzhen. *City Planning Review* 6, 32–42. (In Chinese)

Zhao, J. X., Huang, P., Yang, C. F., Guo, X. M. (2004) Situation of China's environment and development. *China Environment and Development Review* 2, 23–50. Social Sciences Documentation Publishing House.

Zube, E. H. (1986) The advance of ecology. *Landscape Architecture* 76(2), 58–67.

Zube, E. H. (1995) Greenways and the US National park system. *Landscape and Urban Planning* 33, 17–25.

CHAPTER 13

Integrating science and creativity for landscape planning and design of urban areas

Antje Stokman[1] and Christina von Haaren[2]

[1] *Institute for Landscape Planning and Ecology, University of Stuttgart, Stuttgart, Germany*
[2] *Institute of Environmental Planning, Gottfried Wilhelm Leibniz University of Hannover, Hanover, Germany*

13.1 INTRODUCTION

Linking scientific ecological knowledge to the practice of deliberately shaping urban ecosystems is a great challenge. Since the 1970s urban ecologists and landscape planners have been addressing urban ecosystems. Moreover, urban ecology has delivered many important insights into urban ecological processes, and urban landscape planning has transformed this knowledge into restrictions and proposals for urban development (Forman and Godron 1986, Sukopp *et al.* 1995, Selman 2000, Sukopp 2002, 2008, Marzluff *et al.* 2008). Methodologies and processes have been developed that introduce ecological knowledge and integrate public information and participation into the urban planning process (Ermer *et al.* 1996, Council of Europe 2000, Forman 2008, von Haaren *et al.* 2008, Willemen *et al.* 2008). However, urban landscape planning as well as landscape planning in general still lack the critical quality to develop engaging ideas and future-orientated visions that politicians and citizens can identify with. Consequently, these groups frequently perceive landscape planning as a restrictive instrument of nature conservation, which is to be avoided whenever possible. This weakness of implementation may be (partly) compensated for by integrating landscape design approaches, which are better suited to create visions of spatial qualities. Such visions which integrate concepts of ecological performance and beauty have the potential to gain broad public support.

Both landscape planning[1] and landscape design[2] are summarized in many countries under the term landscape architecture. Landscape architecture looks at the urban ecosystem not only in terms of "environment" (focussing on the biophysical dimension of ecosystems) but also in terms of "landscape" (including the cultural

[1] In this chapter the term "landscape planning" is used as a synonym for the also common term "environmental planning". In the US context, the term "environmental planning" is connected to a legislative planning framework including a full range of environmental regulations administered federally by the Environmental Protection Agency. In the German and European context the term "landscape planning" is used in a very similar way and connected legally to the legislative federal planning frameworks. Landscape planning is a comprehensive environmental planning and can contribute to the assessment of the value and resilience of an area and to the evaluation of impacts and resulting measures (von Haaren *et al.* 2008). Instead of using the term environmental planning we prefer using "landscape planning" as well as "landscape design" as the word "landscape" comprises both the physical characteristics and the perceived features of the "environment", which means the totality of circumstances surrounding an organism or group of organisms. Therefore the word "landscape" in contrast to "environment" integrates humans' relationship with the land and combines both the scientific and the esthetic characteristics of the "environment" (see also Council of Europe 2000).

[2] In this chapter the term "landscape design" is tied closely to the disciplines of applied arts and engineering. In the context of art, design is viewed as a more rigorous form of art with a clearly defined purpose, creating something that is both functional and esthetically pleasing. In the context of engineering, design is viewed as an iterative decision-making process of producing plans by which resources are converted into products or systems that meet human needs and solve problems.

Applied Urban Ecology: A Global Framework, First Edition. Edited by Matthias Richter and Ulrike Weiland.

dimension of ecosystems). In this respect the discipline of landscape architecture is probably the most important partner for landscape ecologists, as it involves the investigation, planning and designed response to human-influenced ecosystems, focussing on their spatial configurations and management options at many scales and to varying degrees (Corner 1999, Johnson and Hill 2001, Meyer 2008). In that sense urban ecology and landscape architecture share a strong common goal which provides unique opportunities to cooperate in defining new directions for urban environments.

The discipline of landscape architecture represents a profession as well as an academic education. Despite this common ground, landscape planning and landscape design differ greatly in their tasks and approaches. For many years there has been an ongoing debate about how to overcome this divide between the rational, analytic instrumentality of the "planners" and the creative, cultural action of the "designers".

In view of these theoretical and practical discrepancies, this chapter attempts to:

analyse the different approaches of planning and design, and

show ways to reinforce the strengths of both strands of the profession by finding new ways of interaction.

The methodological approach to this investigation includes: an analysis of the relevant literature, a systematic comparison of the objectives and methodologies of landscape planning and design and an analysis of case studies. The two authors represent the two different approaches of landscape planning and design. Their underlying intent to understand and communicate about the respective approaches is documented in this chapter. Furthermore the chapter is based on the hypothesis that the convergence of the two approaches leads to a structured, iterative process of sustainable planning and design of urban areas.

In order to establish common knowledge between the disciplines of urban ecology and landscape architecture, the chapter starts with an introduction to the different objectives and methodological approaches of both landscape planning and design. Based on this information, the authors identify reasons for conflicts as well as interfaces and potential synergies between the two strands of the profession. Finally, drawing on the analysis of constraining and supporting factors as well as existing ecological concepts, initial steps are proposed to reinforce the cooperation between landscape planning and design.

13.2 LANDSCAPE PLANNING AS A LEGALLY BASED CONTRIBUTION TO SUSTAINABLE DEVELOPMENT IN GERMANY

Since the 1970s there has been growing international acceptance that the integration of environmental and nature conservation goals into urban planning is necessary in order to ensure sustainable development (Selman 2000, Sukopp 2002). The German Federal Nature Conservation Act (Nature Conservation Act 2002) assigns this task to landscape planning, which in Germany is carried out with comprehensive coverage in both rural and urban areas. In some German federal states, however, urban areas are excluded and planning is restricted to the regional level, where it contributes to defining restrictions for urban development, which are constituted in regional planning.

13.2.1 Tasks of landscape planning

The European Landscape Convention (Council of Europe 2000) defines the tasks of landscape planning in Europe and in Germany. More specific tasks are defined in the Nature Conservation Act (1976, last changes 2008). The general goal is to support environmentally friendly and resource-efficient development and to improve the ecosystems in terms of their functionalities for human requirements and biodiversity.

This encompasses a coordinated inventory and evaluation of all natural resources. Furthermore the general goals of the law are given concrete form and translated into objectives and measures for the real landscapes. Thus, landscape planning strives to develop coordinated measures that benefit different natural resources (soil, water, climate, plants and animals, or natural scenery and recreation). The measures are coordinated with each other in order to promote multifunctional effects. More specifically, the planning tasks are described in federal law and in different guidelines of the German federal states: For example, landscape planning should provide the necessary information for integrating environmental objectives into spatial planning and contributing to environmental impact assessment (Figure 13.1). It provides nature conservation officials and non-governmental organizations with a basis for initiating proactive landscape development projects and for making quick comments on potentially harmful proposed development. Last but not least,

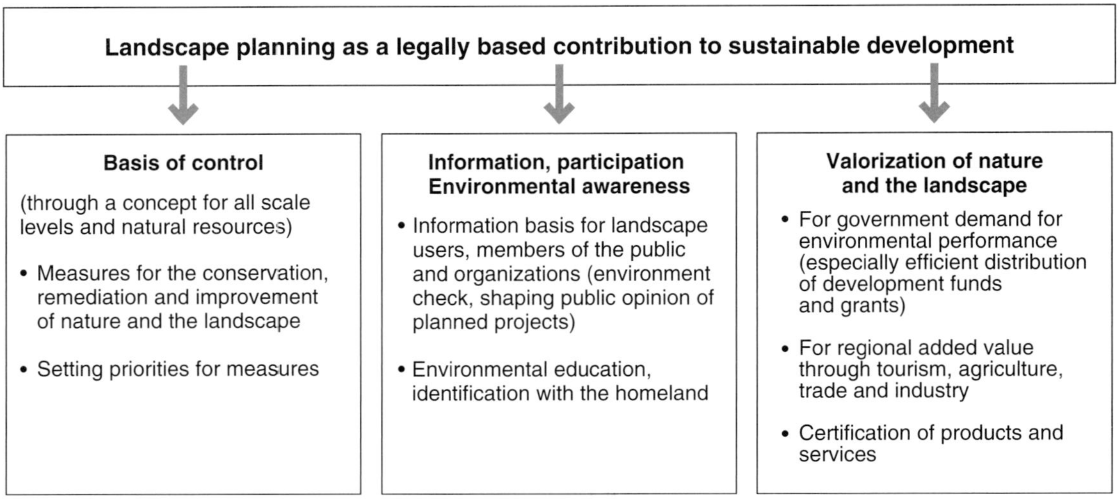

Figure 13.1 Tasks of landscape planning in the context of sustainable development (modified from von Haaren *et al.* 2008).

landscape planning, especially on the local scale, is a major instrument for providing citizens with information about the state of the environment. The regional landscape plan for the county of Verden (Germany), which was visited on the internet 200–500 times per month after its completion in 2008, illustrates how much citizens rely on this function.

In the special context of urban landscape planning the following topics and questions are especially important to consider:
• curtailing land consumption (von Haaren 2004, Heiland *et al.* 2006)
• integration of compensation measures into housing development, dealing with contaminated soils, developing urban greenways (von Haaren and Reich 2006)
• fostering good climatic conditions in towns and urban habitats, dealing with invasive species and protecting endangered species that are typical for urban areas
• organizing social and ecological multifunctionality (Schiller 2006).
In recent times, addressing climate change by promoting adaptation and mitigation measures has become an important task in urban ecosystems.

13.2.2 Methodologies of landscape planning

Landscape planning is basically an analytical process. Landscape objectives are drawn from both scientifically based landscape analysis and from normative democratically legitimized goals (as stipulated in laws). Ecosystems with their different goods and services (which are referred to in Germany as ecosystem's capability or landscape functions in the Nature Conservation Act) are the subject of landscape planning. Landscape planning uses methods derived from biology, ecology, soil science, and hydrology. To some extent, the methodologies for the evaluation of landscape functions and landscape inventory (habitats/biotopes, landscape esthetics, recreation, soil potential for biotope development) have been developed specifically for landscape planning. Some of these methods have been adapted to the urban context (e.g., Müller 1984: urban habitat inventory, Wilmers *et al.* 1987: urban climate and planning).

Based on the inventory and assessment of all natural resources, environmental risk analysis (originally Bachfischer *et al.* 1977) is used to estimate the interactions and cumulative effects of proposed projects and to forecast the expected environmental changes. The environmental risk analysis is based on the principles of benefit–cost analysis and resembles the Driving Forces–Pressures–State–Impacts–Responses (DPSIR) concept of the European Environment Agency (OECD 1993, EEA 1999). The effects of specific land use changes on nature and the environment (e.g., human settlement development, production of renewable energy) can be illustrated in scenario form. Prioritizing the measures and objectives (von Haaren 2004) helps preserving the most valuable and fragile landscape components. Distinguishing between priority objectives and objectives of minor importance also helps to focus public participation on the relevant topics (von Haaren 2002). Special adaptations of the landscape plan for urban areas are discussed by Küpfer *et al.* (2007).

Landscape planners use facilitation and moderation methods during public participation events, which accompany the landscape planning process. Techniques that promote creative approaches and solutions are used. However, such methods in the planning process are usually restricted to brain storming and design methods are rarely used. In the United States a landscape planning process has evolved which includes an explicitly iterative procedure as well as alternative futures, representing stakeholder perspectives (Steinitz 2002). This may be looked upon as a bridge to a more design oriented approach.

13.3 LANDSCAPE DESIGN AS A CREATIVE CULTURAL ACTION

The task of landscape design is the purposeful manipulation of space and its organization in terms of information, function and systems with regard to infrastructural performance, esthetic qualities and social affects. Landscape designers consider landscape "less a quantifiable object than an idea, a cultural way of seeing, and as such it remains open to interpretation, design and transformation" (Corner 1999). Von Seggern *et al.* describe spatial design in the context of designing urban landscapes as "searching for interpretations, for development possibilities and a spatial *Gestalt* which is able to unite multiple requirements – from function, to process, esthetics, construction, material, symbolism and aging – in a convincing whole" (von Seggern *et al.* 2008, p.37).

13.3.1 Tasks of landscape design

Landscape design is carried out by landscape architects within the scope of the profession, which is defined by the professional institutes of architecture and landscape architecture. The general goal of the profession is to investigate and design in response to the urban environment. The scope of landscape design includes architectural and urban design, site planning, environmental restoration, town and urban planning, parks, streetscape, waterscape, civil infrastructure, and recreation planning. It is concerned with the composition of space through built structures and plantings, topography and landform, water, paving, and other structures (Figure 13.2).

The history of landscape design, with which people adapt and craft their environment to suit their needs,

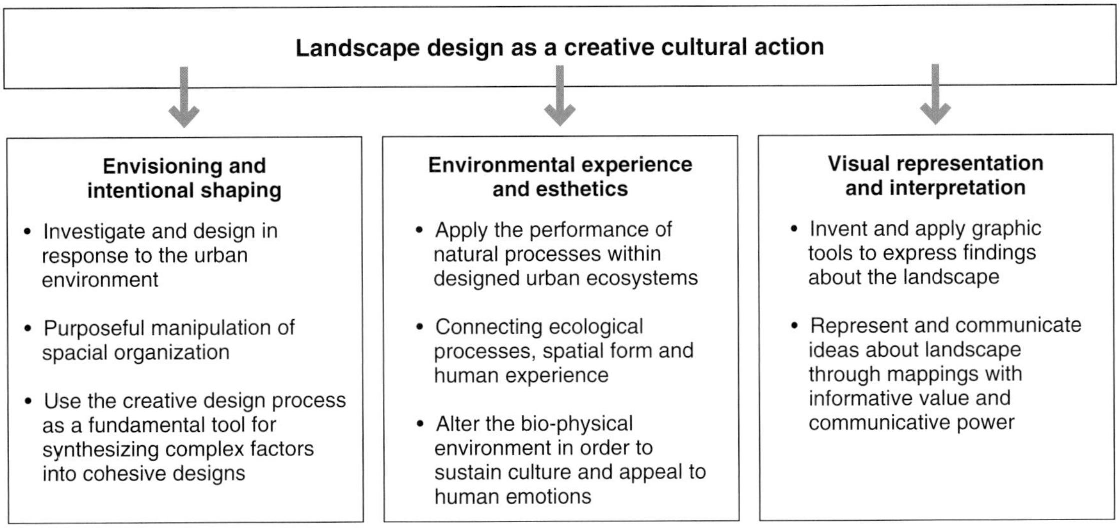

Figure 13.2 Tasks of landscape design as a creative cultural action.

both practically and esthetically, dates back to the early beginnings of human culture (Jellicoe and Jellicoe 1995). The roots of the profession are commonly seen within the art of gardening, with the Egyptian and Persian gardens around 1500 BCE being some of the earliest physical evidence of ornamental landscape design and the formal Renaissance gardens in the 15th and 16th century, the Baroque gardens in the 17th century and the English landscape gardens in the18th century being among the most prominent garden styles. However, other landscape architects argue that the roots of the profession should not only be seen in the design of the privileged landscapes of parks and gardens, but also in the skilfull design of the agricultural and infrastructural landscapes. These designed landscapes of survival structure the productive use and urbanization patterns of ecosystems on regional scales based on both pragmatic and symbolic principles (Yu 2006, Shannon 2007).

However, landscape architecture as a profession, for which the term is used to describe the whole professional task of designing a composition of planting, landform, water, paving and other structures, only developed in the 19th century. The term was first used by Frederick Law Olmsted and Calvert Vaux in 1858, who then designed and constructed New York's Central Park and used the term "landscape architecture" to describe their work for the planning of urban park systems. Olmsted intended to design the urban landscape in order to recapture the transformative power of esthetic experience in nature: "A park is a work of art, designed to produce certain affects upon the mind of men" (Olmsted cited in Meyer 2008, p.7). Therefore, landscape design is closely linked with art as a way of altering the biophysical environment in order to sustain culture and appeal to human emotions.

13.3.2 Methodologies of landscape design

Designing is an ancient, cultural technology of inventing and developing an idea/a plan and/or product with intention. Designers use the creative design process as a fundamental tool for synthesizing complex factors into cohesive designs (Felson and Pickett 2005). This means that design is based on a fundamental human subconscious activity that humans use to make decisions and survive within a complex environment that cannot be captured and analyzed only by analytical thinking. The philosopher of science Hans Poser describes "design as a way of life" and characterizes "design [as] the thinking up of a promising and feasible possibility [...] aimed at bringing about something new" (Poser in: von Seggern *et al.* 2008, p.35).

Designing requires the designers to work with their heads by using their imagination and with their bodies to use their drawing and modeling skills at the same time. An idea derived from a combination of knowledge and intuition is developed and reconsidered by exploring its possibilities and constraints, gradually making use of the potential of the idea on the way to realization. This process can be characterized by a series of typical steps within the design process which form an iterative, non-linear process: discovery (hypothesis of design ideas) – research (investigation of missing information), analysis (reviewing and combining information and ideas), problem-solving (conceptualizing design solutions), testing (checking design solutions), presentation (presenting design solutions), implementation (transferring design solution to reality), evaluation (inspecting and surveying realizes design solutions): all connected in an ongoing process of alternating application and adaptation. There are different temporal conditions and rhythms that combine in the process of design, ranging from the super-fast transmission speed of the nervous system within human imagination to the super-slow transmission speed from the world of thought to the material world of reality within landscapes and cities.

Although there are many theories on guiding the design processes and developing methods for designing in the professional context of landscape architecture, the process of design cannot be fully described in rational terms and the methods by which designs are realized vary from person to person (Prominski 2004). It is especially the sense of "landing" that Christophe Girot describes as the initial act of site acknowledgement and discovery that escapes clear scientific methodology and is almost always the result of chance and personal initial intuition (Girot 1999).

To express findings and ideas about the landscape, the visual representation tools of landscape design are of particular importance as they do not only represent an abstract system of color-codes but mainly operate as a mechanism for the new interpretation of landscape. It is only through representation and picturing that landscapes become culturally visible and meaningful – "the world is visually prefabricated through its potential for being seen" (Waldheim 1999, p.127). Using a variety of tools from pencil to computer, designs can be represented and communicated in the form of drawings, photomontages, plans, diagrams, models, films, texts, and calculations.

13.4 LINKING LANDSCAPE PLANNING AND DESIGN: DIFFERENCES, INTERFACES AND POTENTIAL SYNERGIES

The comparison of tasks and methodologies of landscape planning and design shows some crucial incongruities but also potentials for improvement by combining the two approaches. The differences in the applied methodology, work process and results can be seen in:

- Timeline and scale: how to deal with the contrast of the complexities of large-scale and long-term landscape development versus small-scale and short-term project implementation?
- Ecological and cultural perception: are environmental objectives and man-made works of infrastructure and art antagonistic or synergistic parts of shaping landscapes?
- Process and learning: can the deductive approach of planning be reconciled with the experimental approach of design in face of dynamic landscape changes and uncertainties?
- Involvement and experience: how to combine a comprehensive information system and convincing vision about the landscape in order to gain public and political attention and participation?

The differences in dealing with these aspects will be described in the following paragraphs. They are represented by the different procedures of landscape planning and design – albeit their common landscape architecture education. Seldom do they work together; perhaps the differences of methodologies indirectly correspond to different types of professionals who may have problems cooperating: The designer as the creative, artistic, intuitive type suspects bureaucracy and boredom in the work of the planner, and the landscape planner as the more analytical, systematic type suspects superficiality and piecemeal in the work of the designer.

However, the differences are not necessarily an obstacle for the integration of the approaches. To the contrary: the combination of landscape planning and design approaches holds the opportunity to broaden viewpoints and enlarge the methodological scope. In the following section the differences are discussed in more detail. For each point, possible solutions and their role in improving the communication of landscape objectives are illuminated using a case study.

13.4.1 A matter of timeline and scale: linking multidimensional perspectives on strategic landscape development

With respect to urban landscapes, there are dynamic land use types which are constantly changing in rapid adjustment to changing market conditions, sociocultural development and technological innovation (e.g., agriculture, urbanization, transportation). However, these dynamic land uses may endanger the functional capabilities of natural ecosystems that provide benefits to human societies. Clean drinking water, soil stabilization by plants, green spaces for leisure and health, wind movement and clean air are ecosystem services that are in most cases irreplaceable or very expensive to replace using technology. The protection of these ecosystem services requires long-term planning strategies and a critical amount of stability in space and time, which is largely dependent on government policies.

Landscape planning aims to define a sustainable landscape framework in order to support long-term ecological processes occurring in landscapes (e.g., groundwater and nutrient flow, wildlife corridors, air movement, etc.). Therefore it requires long-term planning information and objectives for municipal, regional and state governments. It also usually covers large-scale areas, defined by political boundaries ranging from cities, to regions and countries.

Landscape planning has to be integrated into the legislative planning framework of an administration working on statutory land use planning and certain legal instruments as well as the needs of the population for information. Due to the comprehensive process of inventory and assessment of all natural resources, the preparation of a landscape plan takes quite a long time, and once finished, it gives general and long-term guidance on landscape development. Although landscape plans may contain many concrete and smaller scale project proposals with short-term realization periods, they usually do not elaborate on them. Landscape plans are difficult for the general public to grasp due to their complexity, large-scale and long-term implementation periods. Therefore, the mobilizing effect of a prompt success to focus community involvement is usually not achieved.

Landscape design on the other hand is challenged by the task of accommodating for contemporary needs, being commissioned to design specific sites and tasks according to the requirements of specific clients. The designs are driven by short-term economic or politically driven goals and objectives, significant time pressure and often the necessary information for design decisions is incomplete. Because landscape designers are involved in the active development of dynamic urban land use in close cooperation with their clients, the requirements of their small-scale projects are not always in line with the requirements of the large-scale landscape framework and may result in conflicts with landscape planners. The challenges for the profession of landscape architecture are: to skillfully accommodate and balance the dynamic land uses within a more stable framework, to link the implementation of large-scale plans to the implementation of small-scale projects, to think strategically and make intelligent compromises between complementary and competing land uses.

By combining different time frames and scales for both long-term and short-term development, the strengths of both landscape planning and design can be used to develop a new way of strategic planning and design. It is necessary for landscape designers to get involved in the strategic large-scale and long-term landscape visions from their perspective of design – and it is necessary for landscape planners to link their long-term landscape strategies to short-term measures of implementation.

13.4.2 A matter of perception and meaning: linking environmental goals and cultural concepts

Landscape has not only a natural, but also a cultural dimension that results from the interaction of natural and human factors. Urban landscapes are cultural products with distinct forms. The experience of the

Case study "Plan Ooievaar", The Netherlands

The "Plan Ooievaar" (Plan Stork) in The Netherlands illustrates the strength of such an approach for implementation at a national scale (Sijmons 1998). It was the winning contribution to a national contest in 1986 called "Netherlands – Riverland". In their plan the three authors, who were landscape planners within the Dutch Ministry of Agriculture, Nature and Fishery, proposed a coherent national scale vision for the management of rivers, nature development and landscape architecture. This vision was presented by a very simple drawing and some schematic sections, which explained how to separate conflicting water functions such as nature development and agriculture and how to integrate water functions that would reinforce each other. Rather than claiming that agriculture was damaging ecosystems, they designed a synergetic spatial strategy that concentrated both natural development and exceptional forms of recreation in the river floodplains while managing the basin areas by agriculture, so that the existing farms could continue operation in the future.

While at the national level the minister embraced the plan due the debate about the upcoming dike enhancements, at the micro-level a number of experiments to combine "new nature" development, recreation and agriculture based on the "Ooievaar principles" started in different regions, for example, the Duursche Waarden (north of Deventer), in Rhenen and the Gelderse Poort (south of Arnhem). These successful strategies on the national level and projects on the micro-level demonstrated a new relationship between spatial planning, water management and nature conservation. Their success in dealing with the floods of 1993 and 1995 strongly influenced the discussion about the future of flood management in The Netherlands. Rather than only controlling water by raising dikes, the explicit integration of water issues into spatial planning was put forward in the "Space for River" national policy in 1995.

The combination of the different scales, the strategic large-scale level with the level of local implementation can provide new ways for the cooperation between landscape planners and designers.

cultural perception of the environment influences human behavior towards the landscape.

In landscape planning landscape is perceived mainly in terms of its ecological and landscape esthetic inventory. This necessarily incorporates the cultural aspects that shape the landscape. However, in landscape planning culture is understood as a descriptive term, which specifies those components of the world that are created by humans. The cultural dimensions are not usually interpreted as a proactive part of planning, nor are the creation of valuable works of culture part of the planning task. Landscape planners do not believe in ecological determinism, but the landscape inventory is used as a diagnostic tool in order to protect and develop landscape capacities and functionalities (ecosystem services) against harmful impacts. Landscape planning concentrates on administrative, regulative, institutional instruments and financial incentives for the implementation of guidelines and policies. As a consequence, there is a tendency to focus on a protective and defensive strategy more than on proactive development. Having to deal with the complexities of large-scale conditions of entire cities, regions and countries, it traditionally relies on rational–analytic, rather than abstract and schematic planning methodology and neglects design approaches that employ visual means or intentional shaping. In order to organize, evaluate and display complex layers of information linked to a specific location, Geographical Information Systems (GIS) are the main tools of landscape planners.

Landscape design, in contrast, is much more oriented towards the existing as well as the potential cultural dimensions of a landscape. Humans are seen as creative agents in a process of environmental discovery and intentional shaping of the environment (Stokman *et al.* 2008). With ecological realities being acknowledged more on the selective base of learned experience than comprehensive scientific completeness, the process of landscape change is not seen in terms of implementing a determined "solution" but rather in terms of making choices and trade-offs between environmental and human factors on the basis of culturally constructed landscape values and requirements (Nassauer 2002, p.217).

This means that the ecological factors of a landscape are not mainly seen as constraints to development, but rather as systems with special processes leading to spatial characteristics that can inspire and guide future landscape form and function. Nature is not something to be preserved but to be constructed. Due to this understanding, man-made systems like buildings, infrastructure systems, and parks are interpreted as designed parts of the urban ecosystem, accommodating man-made flows of water, waste, traffic and materials. Design goes beyond restoring a system to its natural state by inventing and creating new types of coupled natural–human systems. The inventory of landscape functions of landscape plans may offer a useful contribution and basis for this process. By combining the environmental knowledge and administrative instruments of landscape planning with the skills of inventing

Case study "Red Ribbon" project, Qinhuangdao, China

The power of such an approach is demonstrated by the "Red Ribbon" project, a new river park design on the banks of Tanghe River at the city of Qinhuangdao, China (Stokman and Ruff 2008). The project's challenge was to find a new way to deal with the natural habitats along the river without the existence of formal instruments for protection and at the same time allowing for urban development, recreational uses and outdoor activities. In order to inhabit the space perceived as wild and inhospitable by people, the landscape architecture firm Turenscape invented the spatial concept of the "Red Ribbon". It is a fiberglass structure, stretching over a length of 500 m (Figure **13.3**). Accompanied by a boardwalk, it allows for seating and at night is lit from inside. As such, it leads through the existing landscape as a central theme, providing for orientation and recognition. The riverbank is left in its natural state and changes its character depending on the current water level. In the river itself and along the transition zone between water and land aquatic and semiaquatic plants form dense patches that allow for flooding.

Figure 13.3 Case study "Red Ribbon" project, Qinhuangdao, China: finding a new way to deal with natural habitats along a river without the existence of formal instruments for protection and at the same time allowing for urban development, recreational uses and outdoor activities. Photos courtesy of Turenscape, Beijing.

In this case a landscape design approach contributed the courage to use a cultural anthropogenic solution for the ostensibly contradictory aim of nature protection. Instead of rigging up, the project technically withdraws and gives more space to nature and its forces. By doing so, the "Red Ribbon" project shows in an outstanding way, how a small intervention can have a great effect on the improvement of a site and its utilization. The project's approach of minimal intervention, however, not only applies to ecology, but also to form. As such it makes a strong statement against the usual practice of beautification in China, which is characterized by lavish flowerpot arrangements, monumental statues, well-groomed, but inaccessible lawns, coming along with elaborate water and light features and topped by acoustic installations. Both in its formal vocabulary and in the way it integrates design and ecological processes, the "Red Ribbon" project adopts a pioneer role in China.

powerful spatial design concepts, new ways can be developed to give visible forms to landscape functions.

13.4.3 A matter of process and learning: linking management and experimentation to achieve adaptive landscape development

Understanding landscape is about understanding change and processes. Any design that ignores these dynamics and refers to only one future state of landscape form and process cannot be sustainable. Landscape change as a dynamic process demands an adaptive approach to planning and design.

The process of landscape planning is largely based upon a deductive method: it follows a linear and straight path, leading from analysis to objectives and measures. The methodologies that are applied in Germany are also standardized to a large extent by always using the same factors to assess and evaluate within a systematic comprehensiveness. The reason for this is that the provided information and objectives should be transparent for affected stakeholders as well as the public. Comprehensible and comparable results are also required in landscape planning because of the necessity to set priorities for administrative action concerning the protection or rehabilitation of certain parts of the landscape. These preconditions represent constraining factors for the transition to a more creative deduction of planning objectives.

Nowadays landscape planning is trying to overcome the attitude "that good design as a rule should follow the dictates of nature's form and process, often at the expense of creativity and originality" as Nina-Marie Lister points out in reference to the "famous father of ecological design" Ian McHarg (Lister 2007, p.40). In general, landscape planning methodology has overcome such static paradigms or *Leitbilder* (von Haaren 1991, von Haaren 2002) in favor of an approach based on the principles of sustainability. Recently, driven by the possibilities of new technologies (von Haaren and Warren-Kretzschmar 2006), approaches to landscape planning have introduced more dynamic concepts: A landscape plan as a result of planning processes is no longer a static plan, but a system of modules that can be changed each at their own pace. Consequently, it may be looked upon as a system, in which the parts are constantly or periodically changing. Also the linear process of planning has been unhinged in the past: intermediate results are used for certain applications already before the plan as a whole has been passed. For example, projects are already implemented during the first steps of the planning process in order to stimulate citizens' interests and demonstrate the relevance of the plan.

Such an approach to landscape planning can be further developed by drawing on the concept of "adaptive environmental assessment and management", which was developed by the ecologists Crawford S. Holling and Carl J. Walters in the 1970s (Holling 1978, Walters and Holling 1990). It has been widely discussed and applied in the field of ecological restoration in order to deal with gaps in understanding and changes in the ecosystems regarding the effectiveness of human actions in achieving desired management objectives. Drawing on these pioneers' works, the ecologists Carol Murray and David R. Marmorek have further refined the methodology of what "adaptive management" is and how it can be systematically applied. They define adaptive management as "a rigorous approach for learning through deliberately designing and applying management actions as experiments" (Murray and Marmorek 2003, p.417). This approach explicitly shows a way to make management decisions and carry out actions despite inevitable uncertainty about how to achieve the desired outcomes by providing a science-based learning process characterized by using outcomes for evaluation and adjustment – the "Adaptive Management Cycle" (Murray and Marmorek 2004, p.2, von Haaren *et al.* 2008, p.41). Some landscape planning projects have elaborated such a "learning" step-by-step process (Tesch 2001).

Landscape design, on the other hand, needs to accept its role to be one of the drivers of change by making use of its positive potentials rather than opposing it. In contrast to the deductive process of landscape planning, the landscape design process starts with one or several ideas and is adapting these by an iterative process to the conditions of the reality through evaluation and objectification. The essence of these ideas is reduced in complexity by making use of the human capability to reduce complex information into coherent designs at the interface of empirical and intuitive knowledge. Both analysis and development ideas are presented in appealing mappings, making use of manual sketching and/or computer visualization techniques. It is hypothesized here that the advantage of landscape design may be that it produces more consistent ideas about many projects than the deductive process of landscape planning. At the same time, this procedure also enables design to meet the wishes and ideas of the addressees, be it a politician, members of local authorities or a private client. The disadvantage of the design process is that facts and framework conditions may be overlooked, which can cause the whole idea to be questioned at a later point in time. Also the solution of a multitude of problems related to the planning site that are outside the design focus, may be neglected. This risk increases with the size of the planning area and the complexity of the task. The concept of "designed experiments" may offer a strategy to address this dilemma. The ecologists Alexander J. Felson and Steward T.A. Pickett (Felson and Pickett 2005, Pickett and Cadenasso 2008) propose that ecologists and urban/landscape designers should strategically team up by architecturally designing and scientifically inserting experiments into the urban fabric. This concept allows the infusion of experimental goals and monitoring approaches into the projects' development over time while already being implemented. Such experiments deliberately take advantage of designed urban components including parks, streets and infrastructure systems. The results gained from the experiments

Case study "Schöneberg Nature Park" project, Berlin, Germany

The "Schöneberg Nature Park" project in Berlin (Figure **13.4**) is a good example for showing the benefit of a landscape strategy that integrates, accompanies and manipulates dynamic natural–human development (Grosse-Bächle 2003). The site is located within an overgrown railway wasteland and was designed with the aim to create an attractive park that combines the interests of nature development and recreation. The project was developed as a cooperative effort of landscape planners and designers with ecologists making long-term observations of the ecological development (Kowarik and Langer 2005). By defining which areas to leave for free successions and which areas to manage by different maintenance interventions, they allowed the spatial coexistence of various types of landscape dynamics. Interferences of visitors who use the territory were seen as a development impulse and integrated into the management strategy. By heightening the boardwalks, taking them around special vegetation in specific areas and bundling functions by setting up resting areas and viewpoints in special settings, large areas are kept free from intervention which ensures space for undisturbed development of vegetation.

Figure 13.4 Case study "Schöneberg Nature Park" project, Berlin, Germany: integrating, accompanying and manipulating dynamic natural–human development. Photo: Antje Stokman.

Thus the park is developed in an iterative process, including designers, ecologists and visitors. The challenge of generating ecological processes while considering functions for human use and esthetic experience is becoming particularly important in response to esthetic theories, suggesting that humans will value these processes more if they could see and understand them.

contribute to new knowledge that can be fed back into both experimental and practical design proposals.

Combining the new approaches of "adaptive management" linked to landscape planning and "designed experiments" linked to landscape design can lead to a new understanding which is gained through the adaptive process of design, implementation, monitoring and redesign of urban landscapes.

13.4.4 A matter of involvement and experience: linking information and participation

Building new frameworks for future urban landscape development needs the mobilizing action of people who help the process along. This requires an action and project-oriented approach based on sharing the necessary information. Landscape planning and design need to be a participatory process. Local people should collectively decide about the many options for their possible futures, in which landscape planners and designers having the role, on the one hand, of a "wise facilitator" (Lister 2007, p.51) and on the other hand, they also point out where the limits of local competencies lie.

The traditional task of landscape planning was to guide land uses and the development of landscape. However, due to the need to support implementation by acceptability of the objectives an important focus of landscape planning is nowadays to create a generally accessible and easy to update information base for participation and action (von Haaren and Warren-Kretzschmar 2006). Also the requirements of the EU Directive about strategic impact assessment foster the inclusion of public participation (Scholles and von Haaren 2004). Consequently, landscape planning increasingly supports the task of providing members of the public with environmental information and thereby enables their participation in developing sustainable cities. However, frequently only land users such as farmers, who are or may be directly affected by landscape planning proposals, take part in the participation process. A reason for this abstinence of "ordinary" citizens may be the large planning area, the resulting high number of objectives and ideas and complexity of information. Usually the representation of landscape plans focuses more on usability and GIS-visualization possibilities than on esthetic principles. Because of the large quantities of data, landscape planners still struggle with technical and methodological challenges of attractive plans and three-dimensional (3D) visualizations (Bishop and Lange 2005). Landscape plans contain many color-coded maps which address different topics, with each color representing specific information that is only understandable in conjunction with the maps' legend. Owing to their lack of sensuous–esthetic accessibility and their high degree of abstraction, formal landscape plans have difficulty reaching and appealing to the general public, the politicians and the authorities. Even a comprehensive 3D-visualization, which undoubtedly supports the understanding of the plan (von Haaren and Warren-Kretzschmar 2006, Sheppard *et al.* 2008), cannot force people to take part in the process. Visualizations cannot compensate for the absence of ideas that are simple enough to be well understood by citizens, which promise politicians to distinguish themselves as sponsors of a highly regarded change to the city and which enhance the beauty of their home town.

The aim of landscape design is to provide a vision for the future in order to convince the client as well as to engage the public in the realization of the design idea. Therefore, the complexity of the design needs to be condensed a strong design idea that is easy to communicate and appeals to the client and public. Comprehensive information that might have been accumulated within the design process recedes into the background and is aggregated into a comprehensive and convincing idea that signifies the future and gains support. This idea is communicated through a spatial concept that is expressed in drawings and texts that convey its essence and form a clear basis for more specific design decisions. Often carefully selected, meaningful metaphors and titles are used to explain spatial concepts, like the "Red Ribbon" described above. However, spatial concepts and metaphors tend to be become detached from the ecological landscape context, whereby the formal idea and symbolic meaning of specific forms and materials become more important than environmental concerns and awareness.

The task to create acceptability and engagement may be generally easier in landscape design, as the clients usually commission and want the project for their own interests. In landscape planning, in contrast, the project may not always be welcomed by all addressees. Thus the more important it should be to use all available means to strengthen acceptability.

Case study "Masterplan emscher:zukunft", Ruhr region, Germany

The "Masterplan emscher:zukunft" defines a vision and guiding idea to develop the valley of the river Emscher into an "Emscher Park" (Stokman 2006). This vision for the new Emscher Valley as a regional park is the expression of a new planning culture that follows a strategic and project-orientated planning approach to seize opportunities to add to the quality of the landscape. The need to transform the 85 km long Emscher River from an open sewage canal back to a river is considered a chance to restructure the urban regional landscape in order to become more ecological and more functional – but above all, more beautiful and meaningful. Therefore the concept of the "Emscher Park" involves all stakeholders from the region: to create an added value in their collective interest, to mobilize and bundle financial resources as well as combining ecological, economical and social interests. The process of designing and implementing the masterplan was set up as an adaptable and responsive public planning platform and process, providing a reliable consensus between all regional protagonists. To continually adjust the masterplan as a framework and impulse for new projects runs parallel with the active implementation of local pilot projects that immediately let people emotionally experience the change.

13.5 CONCLUSION

Being part of the institutionalized planning system, landscape planning in Germany is rooted in a science-based tradition. In general, the role of intuition, creativity and emotion remains largely unaddressed. Design in contrast is understood by many designers as a way of reinterpreting the human relationship with nature and to manage human interactions with nature by giving them new form and meaning. The better integration of a landscape design approach may be advantageous and efficient for landscape planning in strengthening the quality of landscape visions, their procedural implementation and in offering the opportunity for a better communication of landscape development objectives to the public and decision makers.

The findings of the preceding analysis show that for the most part landscape planning and design follow potentially complementary and seldom contradictory principles and methodologies. The differences between the two approaches refer to timeline and scale, the perception of cultural concepts, the integration of process and implementation as well as stakeholder involvement. All of these differences may be turned into impulses for an improvement of cooperation between the two approaches to enhance the quality and implementation of both landscape planning and design.

The following features should be considered in an integrated urban landscape planning and design process (Figure 13.5):

- the combination of large-scale planning with small-scale projects, which demonstrate feasible and convincing implementation solutions
- a step-by-step deduction and adaptive implementation of objectives and measures by integrating management, monitoring and evaluation processes
- the continuous reformulation of design and management tasks according to updated results and their integration within both strategic vision and small-scale implementation schemes
- the invention and development of new representation techniques to reduce complex environmental factors into cohesive ideas, and
- the creation of an added value by multifunctional landscapes in the collective interest of diverse stakeholders.

Albeit the advantages promised by such innovations of integration, both approaches should keep their autonomy and special strengths for the respective other approach to benefit from: For the tasks of landscape planning the comprehensive analysis and assessment of landscape functions is indispensible, as landscape planning has to create a legally compliant ground for argumentation and actions of the competent authorities. For the task of landscape design the creative skill to synthesize complex factors into cohesive spatial ideas is essential, as it needs to respond quickly to the requirements of clients for short-term project realization. However to combine these different strengths requires knowledge, appreciation and ability to develop

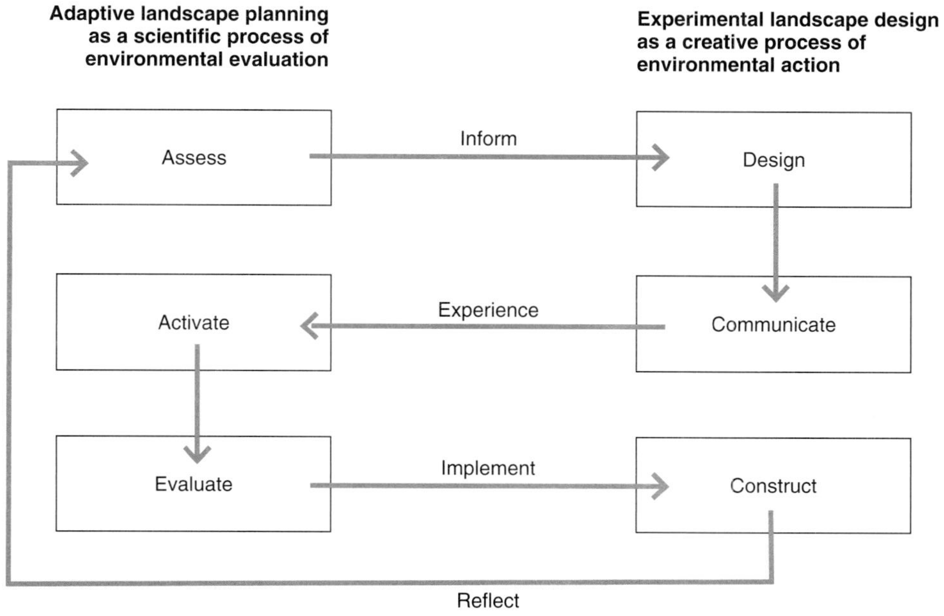

Figure 13.5 Linking adaptive landscape management and experimental landscape design towards process-driven project implementation.

cross-disciplinary research and projects. Ecologists and planners need to become more interested in design and designers need to become more interested in ecology – in order to employ ecology as an "agent of creativity" (Corner 1997).

These recommendations may suffice for further theory building as well as for using and testing new approaches in landscape planning and design in general. However, a strong need for research can be identified in several fields. Questions remain whether creative development of ideas is hampered by the necessary inventory and assessment aiming to be comprehensive and cover all relevant information. Also it may be a task of future research to determine how far the methodological approach of landscape design is able to cope with the complexity of landscape planning tasks and the ability to produce solutions, which are able to prioritize the right issues and comply with the requirements of transparency and traceability. In addition, the knowledge about the process and preconditions of creativity is still insufficient. Especially the role of intuition and whether information or experience plays an important part in intuitively choosing the most relevant factors in the creative process must be examined more closely. In order to better choose the right approach for different planning tasks and settings, the implementation success rates of the different methodologies should also be examined under comparable framework conditions.

Sustainable urban landscape cannot be created through greater technical or scientific knowledge alone: methods to link scientific and creative approaches are needed to put the meaning of ecological theory into practice. This demands new ways of teaching and thinking.

ACKNOWLEDGMENT

The authors would like to thank the editors for giving us opportunity and freedom to take up a long-time neglected disciplinary topic as well as Barty Warren-Kretzschmar for her revisions and encouraging advice.

REFERENCES

Bachfischer, R., David J., Kiemstedt, H., Aulig, G. (1977) Die ökologische Risikoanalyse als regionalplanerisches Entscheidungsinstrument in der Industrieregion Mittelfranken. *Landschaft und Stadt* 9, 145–162.

Bishop, I.D., Lange, E. (2005) *Visualization in Landscape and Environmental Planning*. Taylor & Francis, London.

Corner, J. (1997) Ecology and landscape as agents of creativity. In: Thompson, G., Steiner, F.J. (eds) *Ecological Design and Planning*. John Wiley & Sons, New York, pp. 80–108.

Corner, J. (1999) Recovering Landscape: Essays in Contemporary Landscape Architecture. Princeton Architectural Press, New York.

Council of Europe (2000) *European Landscape Convention*. Florence, Italy. http://conventions.coe.int/Treaty/EN/Treaties/Html/176.htm (accessed 24 March 2011).

Ermer K., Hoff, R., Mohrmann, R. (1996) *Landschaftsplanung in der Stadt*. Ulmer, Stuttgart.

European Environment Agency (EEA) (1999) *Environmental Indicators: Typology and Overview*. European Environment Agency, Copenhagen.

Felson, A., Pickett, S.T.A. (2005) Designed experiments: new approaches to studying urban ecosystems. *Frontiers in Ecology and the Environment* 3(10), 549–556.

Forman, R.T.T. (2008) *Urban Regions: Ecology and Planning beyond the City*. Cambridge University Press, Cambridge.

Forman, R.T.T., Godron, M. (1986) *Landscape Ecology*. John Wiley & Sons, New York.

Girot, C. (1999) Four trace concepts in landscape architecture. In: Corner, J. (ed.) *Recovering Landscape: Essays in Contemporary Landscape Architecture*. Princeton Architectural Press, New York, pp. 59–67.

Grosse-Bächle, L. (2003) *Eine Pflanze ist kein Stein. Strategien für die Gestaltung mit der Dynamik von Pflanzen*. Beiträge zur räumlichen Planung, Band 72, Schriftenreihe des Fachbereichs Landschaftsarchitektur und Umweltentwicklung der Universität Hannover, Hanover.

von Haaren, C. (1991) Leitbilder oder Leitprinzipien ? *Garten und Landschaft* 101(2), 29–34.

von Haaren, C. (2002) Landscape planning facing the challenge of the development of cultural landscapes. *Landscape and Urban Planning* 60, 73–80.

von Haaren, C. (2004) The role of the English and German planning system in controlling the change of the countryside to urban use. In: Kanning, H., Lang, A., Scholles, F. (eds) *Nachhaltigkeit als Herausforderung der Raum- und Umweltplanung in Europa – Sustainable Development as a Challenge for European Spatial Planning*. Beiträge zur räumlichen Planung, Band 75. Institut für Landesplanung und Raumforschung, Universität Hannover, Hanover.

von Haaren, C., Reich, M. (2006) The German way to greenways and habitat networks. *Landscape and Urban Planning* 76, 7–22.

von Haaren, C., Warren-Kretzschmar, B. (2006) The interactive landscape plan: use and benefits of new technologies in landscape planning, including initial results of the interactive landscape plan Koenigslutter am Elm, Germany. *Landscape Research* 31(1), 83–105.

von Haaren, C., Galler, C., Ott, S. (2008) *Landscape Planning: the Basis of Sustainable Landscape Development*. Federal Agency for Nature Conservation Germany, Bonn. http://www.bfn.de/fileadmin/MDB/documents/themen/landschaftsplanung/landscape_planning_basis.pdf (accessed 24 March 2011).

Heiland, S., Reinke, M., Siedentop, S. (2006) *Beitrag naturschutzpolitischer Instrumente zur Steuerung der Flächeninanspruchnahme*. BfN-Skripten Band 176, Bundesamt für Naturschutz, Bonn.

Holling, C.S. (ed.) (1978) *Adaptive Environmental Assessment and Management*. John Wiley and Sons, New York.

Jellicoe, G., Jellicoe, S. (1995) *The Landscape of Man. Shaping the Environment from Prehistory to the Present Day*. Thames & Hudson, London.

Johnson, B., Hill, K. (2001) *Ecology and Design: Frameworks for Learning*. Island Press, Washington, DC.

Kowarik, I., Langer, A. (2005) Natur-Park Südgelände: Linking conservation and recreation in an abandoned railyard in Berlin. In: Kowarik, I., Körner, S. (eds) *Wild Urban Woodlands*. Springer, Berlin, pp. 287–299.

Küpfer, C., Arnold, S., Deuschle, J., Müller-Pfannenstiel, K. (2007) Landschaftsplanung und Eingriffsregelung im Innenbereich: Plädoyer für einen "Landschaftsplan für den Innenbereich". *Naturschutz und Landschaftsplanung* 39, 107–113.

Lister, N.-M. (2007) Sustainable large parks: ecological design or designer ecology? In: Czerniak, J., Hargreaves, G. (eds): *Large Parks*. Princeton Architectural Press, New York.

Marzluff, J.M., Shulenberger, E.H., Endlicher, W., *et al.* (eds) (2008) *Urban Ecology: An International Perspective on the Interactions between Humans and Nature*. Springer, New York.

Meyer, E.K. (2008) Sustaining Beauty: The Performance of Appearance. *Landscape Architecture Magazine* 98(10), 92–131.

Müller, N. (1984) Grundlage und Arbeitsmethoden der Kartierung der Flora im besiedelten Bereich von Augsburg. *Ber. Naturwiss. Vereins f. Schwaben* 88, 33–38.

Murray, C., Marmorek, D. (2003) Adaptive management and ecological restoration. In: Friederici, P. (ed.) *Ecological Restoration of Southwestern Ponderosa Pine Forests*. Island Press, Washington, DC, pp. 417–428.

Murray, C., Marmorek, D. (2004) *Adaptive Management: A Spoonful of Rigour Helps the Uncertainty Go Down*. Proceedings of the 16th International Conference of the Society for Ecological Restoration, August 24–26, 2004, Victoria, Canada. http://www.essa.com/documents/Murray_Marmorek_adaptive_management_SER_conference.pdf (accessed 24 March 2011).

Nassauer, J. (2002) Ecological science and landscape design: A necessary relationship in changing landscapes. In: Johnson, B., Hill, K. (eds) *Ecology and Design: Frameworks for Learning*. Island Press, Washington, DC, pp. 217–230.

Nature Conservation Act/Gesetz über Naturschutz und Landschaftspflege (Bundesnaturschutzgesetz – BNatSchG) vom

25. März 2002, BGBl. I S. 1193, zuletzt geändert am 22. Dezember 2008, BGBl. I, p. 2986.

Organisation for Economic Co-operation and Development (OECD) (1993) *Core Set of Indicators for Environmental Performance Reviews*. Paris, France.

Pickett, S.T.A., Cadenasso, M.L. (2008) Linking ecological and built components of urban mosaics: an open cycle of ecological design. *Journal of Ecology* 96, 8–12.

Prominski, M. (2004) *Landschaft entwerfen*. Reimer, Berlin.

Schiller, J. (2006) *Expertenworkshop Landschaftsplanung: Planung für mehr Naturschutz in der Stadt*. November 2006, Vilm, Germany. http://www.bfn.de/fileadmin/MDB/documents/themen/siedlung/lp_stadtnatur.pdf (accessed 24 March 2011).

Scholles, F., von Haaren, C. (2004) Co-ordination of SEA and landscape planning. In: Schmidt, M., João, E., Albrecht, E. (eds) *Implementing Strategic Environmental Assessment*. Environmental Protection in the EU, Vol. 2. Springer, Berlin, pp. 557–570.

von Seggern, H., Werner, J., Grosse-Bächle, L. (2008) *Creating Knowledge: Innovation Strategies in Designing Urban Landscapes*. Jovis, Berlin.

Selman, P.H. (2000) *Environmental Planning: The Conservation and Development of Biophysical Resources*. SAGE Publications Ltd., London.

Shannon, K. (2007) Indigenous landscape urbanism: Sri Lanka's reservoir and tank system. *Journal of Landscape Architecture*, autumn 2007, 6–17.

Sheppard, S.R.J., Shaw, A., Flanders, D., Burch, S. (2008) Can visualization save the world? Lessons for landscape architects from visualising local climate change. In: Buhmann, E., Pietsch, M., Heins, M. (eds) *Digital Design in Landscape Architecture 2008*. Wichmann, Heidelberg, pp. 2–21.

Sijmons, D. (1998) *Landscape*. Architectura & Natura Press, Amsterdam.

Steinitz, C. (2002) On teaching ecological principles to designers. In: Johnson, B.R., Hill, K. (eds) *Ecology and Design: Frameworks for Learning*. Island Press, Washington, D.C., pp. 231–244.

Stokman, A. (2006) Wasserlandschaft neue Emscher im Ruhrgebiet – Impuls für den Wandel einer Industrieregion. In: Stadtkultur International e.V. (ed.) *Dialoge über das Bauen am Wasser – Dokumentation des 9. Deutsch-Chinesischen Symposiums zu Architektur und Stadtentwicklung in Ningbo im Oktober 2005*. http://www.stadtkultur-international.de/pubningbo/Stokman_Antje.pdf (accessed 24 March 2011).

Stokman, A., Ruff, S. (2008) The Red Ribbon Tanghe River Park – reconciling water management, landscape design and ecology. *Topos* 63, 29–35.

Stokman, A., Rabe, S., Ruff, S. (2008) Beijing's new urban countryside – designing with complexity. *Journal of Landscape Architecture (JoLA)* autumn 2008, 30–45.

Sukopp, H. (2002) On the early history of urban ecology in Europe. *Preslia* 74, 373–393.

Sukopp, H. (2008) On the early history of urban ecology in Europe. In: Marzluff, J.M., Shulenberger, E.H., Endlicher, W., *et al.* (eds) *Urban Ecology: An International Perspective on the Interaction between Humans and Nature*. Springer, New York, pp. 79–97.

Sukopp, H., Numata, M., Huber, A. (1995) *Urban Ecology as the Basis of Urban Planning*. SPB Academic Publishing, The Hague.

Tesch, A. (2001) *Ökologische Wirkungskontrollen und ihr Beitrag zur naturschutzrechtlichen Eingriffsregelung*. Beiträge zur räumlichen Planung Band 60. Institut für Landschaftspflege und Naturschutz, Universität Hannover, Hanover.

Waldheim, C. (1999) Aerial representation and the recovery of landscape. In: Corner, J. (ed.) *Recovering Landscape: Essays in Contemporary Landscape Architecture*. Princeton Architectural Press, New York, pp. 121–139.

Walters, C.J., Holling, C.S. (1990) Large-scale management experiments and learning by doing. *Ecology* 71(6), 2060–2068.

Willemen, L., Verburg, P.H., Hein, L., van Mensvoort, M.E.F. (2008) Spatial characterization of landscape functions. *Landscape and Urban Planning* 88, 34–43.

Wilmers, F., Scholz, K.-D., Katzschke, D. (1987) Stadtklima und räumliche Planung: Klimaökologische Funktion und Beurteilung der Freiräume der Kernrandzone des Großraums Hannover. Beiträge zur regionalen Entwicklung Heft 17. Hanover.

Yu, K. (2006) *Position Landscape Architecture. The Art of Survival*. China Architecture and Building Press, Beijing.

CHAPTER 14

Landscape as a living system: Shanghai 2010 Expo Houtan Park

Kongjian Yu

Graduate School of Landscape Architecture, Peking University, Beijing, China

14.1 INTRODUCTION

As one of the central green spaces of the Shanghai 2010 Expo, Houtan Park is not just designed as a revived ecological waterfront for the exhibition but due to the flexibility integrated into the design has been transformed into a public waterfront open space for Shanghai afterwards. The 14 ha (34.6 acre) site is located on the southern boundary of the Expo between the east bank of the Huangpu River waterfront and Puming Road. A former steel manufacturer and boat repair facility remained on site when the design was initiated in early 2007; construction was completed in October 2009 and has been open to the public since May 2010. Table 14.1 provides basic information about the project.

14.2 OBJECTIVE

The objective of the park design was: to create a green Expo, accommodate a large influx of visitors during the exposition from May to October, demonstrate green technologies, transform a unique space to make the Expo an unforgettable event, and to enable its transition into a permanent public waterfront park after the Expo.

14.3 CHALLENGES

Integrating public space into such a degraded riverfront presented several design challenges:

14.3.1 Pollution

This brownfield, previously owned by a steel factory and a shipyard was largely used as a landfill and storage yard for industrial materials (Figure 14.1). The Huangpu River is severely polluted and has been designated as a Grade V water body, which is considered unsafe for swimming and recreation and devoid of aquatic life. The foremost design challenge was to remediate the site to create a safe and healthy public space in a limited timeframe.

14.3.2 Flooding

The site is relatively flat with an elevation ranging from between 4 and 7 m (13–23 feet) and the existing floodwall, designed to protect against a 1000-year flood event, has a top elevation of 6.7 m (22 feet). The average tide elevation of the Huangpu River is 2.24 m (7.3 feet), the average high tide elevation is 3.29 m (11 feet) and the average low tide elevation 1.19 m (3.9 feet). The disparity between the water elevation and levee elevation of 3.4–5.5 m (11–18 feet) has restricted public access to the waterfront. A sloped riverfront design would require more space than is available and possibly erode, whilst maintaining the existing floodwall would continue to prevent accessibility and will remain devoid of aquatic life and natural habitat. Another major challenge associated with the flood control was to design an ecologically sensitive

Applied Urban Ecology: A Global Framework, First Edition. Edited by Matthias Richter and Ulrike Weiland.

Table 14.1 Basic information about the Houtan Park project.

Location:	Pudong, Shanghai City
Size:	14 ha
Design:	January 2007–October 2009
Construction/ Completion:	October 2009, Opening: May 2010
Client:	Shanghai Expo
Design company:	Turenscape (Beijing Turen Design Institute) in cooperation with Peking University, Graduate School of Landscape Architecture
Design principal:	Kongjian Yu

For further project collaborators: see background research group[1] and design team[2].

riverfront and provide direct public access to the waterfront, whilst protecting the city from flooding.

14.3.3 Circulation

The Expo's west entrance is located on the southern boundary of Houtan Park and is the sole access point for the waterfront, which is the central green space for the Expo. Due to the park's location as one of the main entrances it was critical to address pedestrian circulation, safety, and evacuation routes during the design phase. The creation of organized well-developed pathways and meeting spaces was crucial to bring about a comfortable and safe pedestrian environment.

14.3.4 Transformation

During the Expo, the site demonstrated an ecological city ideology – a productive and healthy urban public green space that focussed on simultaneously addressing pedestrian flows, safety and evacuation routes. However, after the Expo, the site functions as an openly accessible public riverfront. One of the major challenges was to provide flexible strategies to address the practical aspects of the Expo and also allow for a smooth and economical transformation into an ecologically sensitive and accessible riverfront park afterwards.

14.3.5 Identity

The site is positioned between the World Expo Exhibit and the Huangpu River. Houtan Park needed a distinct character to stand out amongst the dozens of bold exhibitions that complement modern Shanghai. The site also needed to provide a memorable outdoor experience while realizing the more conventional ecological, circulatory, educational and exhibition functions. The competing needs of the site were a great inspiration for the design.

[1]Background research group: Wanli Fang, Cheng Cheng, Xuesong Xi.

[2]Design team: Shihong Ling, Wei Hong, Tianyuan Yuan, Yuanyuan Jin, Hongqian Yu, Xiang Long, Xiangbin Kong, Dongli Ren, Yuan Zhang, Haibo Tang, Weirong Jiang, Yu Qiu, Shaohui Bai, Yuan Fang, Meicai He, Fan Yang, Junying Zhang, Yang Pan, Jing Zhang, Xiangjun Liu, Hongxia Ding, Jing Niu, Yuan Zhang, Shaohua Lin, Yanan Zang, Zongbo Shao, Dehua Liu, Fumin Yu, Rao Chen, Wei Zhang, Jiwei Chen, Xiangrong Wan.

(a) (b)

Figure 14.1 (a, b) The former site: a brownfield with a degraded environment. Photos: K. Yu.

14.3.6 Form

The site is narrow, locked between the Huangpu River and a main city thoroughfare. This thin band of land challenged the ability to effectively organize public spaces over a long distance with a water frontage over 1700 m (1 mile) in length, while averaging only 50–80 m (164–260 feet) in width, with the narrowest area being approximately 30 m (100 feet) wide.

14.4 DESIGN CONCEPT AND STRATEGY: A LIVING SYSTEM

A living organism has the ability to adapt, change, and protect itself. Using this design concept, the Houtan Park was designed as a living system with ecological services including: provisioning services (including food, water and energy), regulating services (such as purification of water, carbon sequestration and climate regulation, waste decomposition and detoxification, crop pollination, pest and disease control), supporting services (such as nutrient dispersal and nutrient cycling, seed dispersal and primary production), cultural services (including cultural, intellectual and spiritual inspiration, recreational experiences, ecotourism and scientific discovery), especially by incorporating Chinese culture and the site's industrial heritage. The resulting design is a contemporary high performance and low maintenance landscape (Figure 14.2).

A layered approach was used to organize the space, to integrate multiple functions (and ecosystem services)

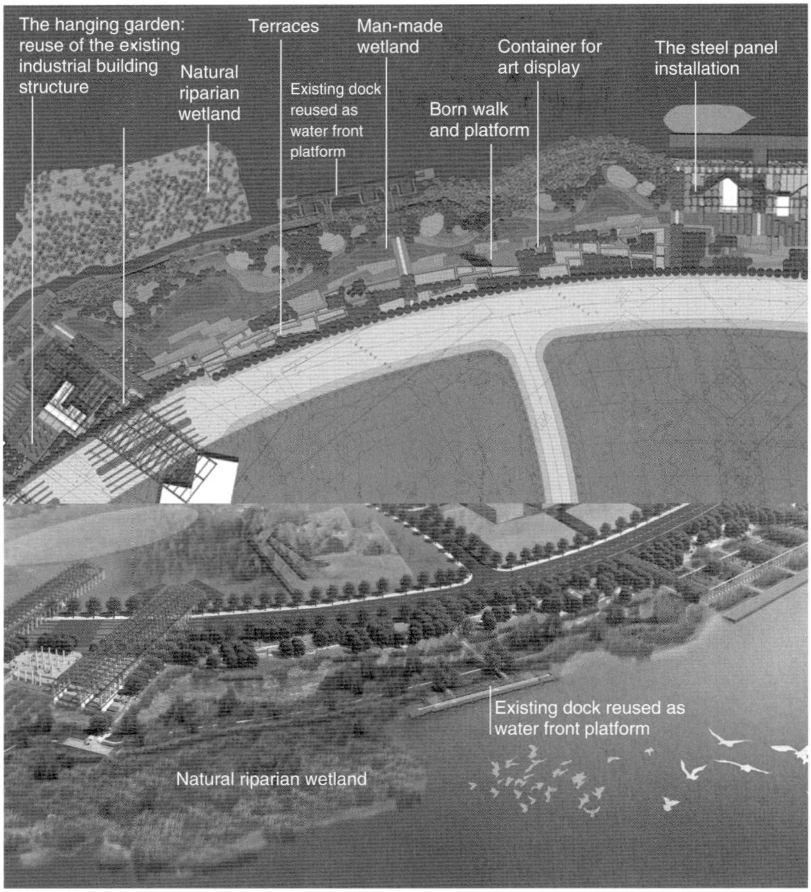

Figure 14.2 Site plan (the central part) for Houtan Park.

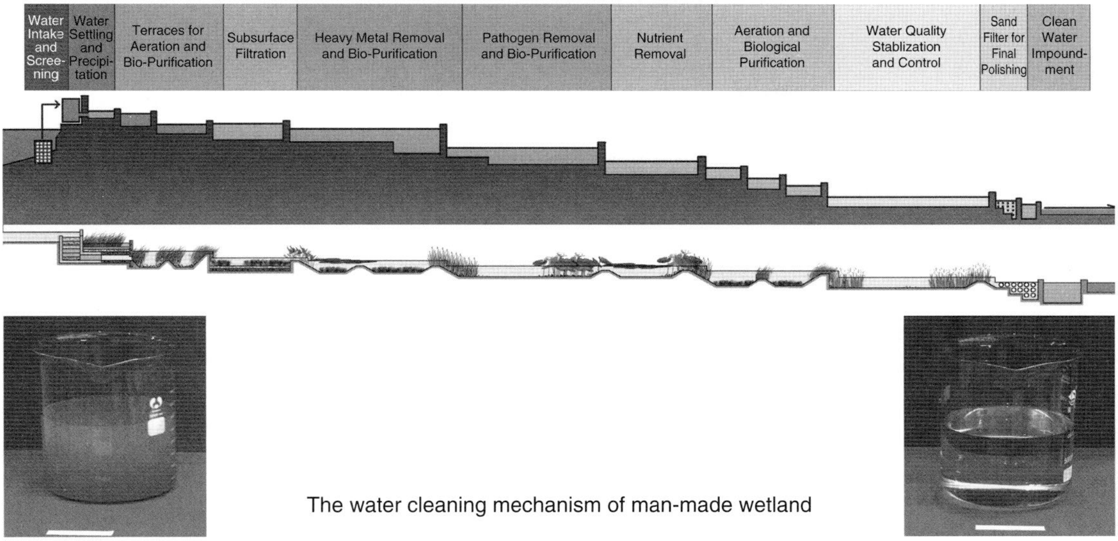

Figure 14.3 The water-cleaning mechanism of man-made wetland.

of the park, to interpret the site's history and future potential, to express the unique character of the park, and to address access and circulation challenges. The timeline, spatial context and amenities of the landscape were used to build upon these layers. The layering of the ecological landscape combined with the regional agricultural and industrial heritage create an environmentally sensitive postindustrial landscape that speaks specifically about the past, present, and future of Shanghai. Weaving these layers of landscape into an integrated system is a network of paths and places that create an esthetically pleasing experience for visitors during the Expo and for the future.

14.4.1 Ecological landscape

The design preserves a rare remnant 4 ha (10 acre) wetland located along the Huangpu riverfront. The existing and proposed vegetation community, composed of lush willows and reeds removes pollutants, prevents riverbank erosion, and creates avian and aquatic habitats. An existing concrete levee was reconfigured as a tidal river wetland and was planted with native wetland species to ecologically improve flood control and is reinforced with permeable riprap. River water is pumped from Huangpu River to the treatment wetland located on the upper bank in the riparian zone above the tidal levee to improve its quality during the Expo and more importantly for the demonstration of the ecosystem services of the designed landscape.

Purification wetland

The Huangpu River water has been designated as Lower Grade V, the poorest quality on a scale of I to V. Water from the Huangpu River is diverted to the constructed wetland where it is biologically treated to an improved Grade III designation via a series of wetland cells to facilitate settling, aeration, and vegetative and microbial processes (Figure 14.3). The treated river water was then used safely for landscape irrigation and other non-potable uses throughout the Expo. Full-scale pilot testing results indicate that the Houtan Park treatment wetland has the ability to treat over 2400 t of water per day (500 000 gallons per day).

In addition to its water treatment capabilities, the wetland created new waterfront avian and aquatic habitats and offers great recreational and educational opportunities for adults and children (Figures 14.4 and 14.5).

Resilient flood control

The constructed wetland acts as an additional buffer between the Huangpu River and the city to absorb

Figure 14.4 The linear wetland in the middle of the park with terraces covered by a variety of native grasses and wetland plants (view facing north). Photo: K. Yu.

Figure 14.5 The banded pattern of the wetland plant community demonstrates the filtration process of the wetland at the park's entrance. Photo: K. Yu.

and retain flood water between the 20-year and 1000-year flood event levees. The terraced design of the constructed wetland reduces the relative elevation change between the city and the river and subsequently the strength requirements for the levee. The replacement of the existing concrete retaining wall by a more ecologically sensitive riprap system encourages native vegetation and new habitats to emerge.

Valley landscape

The valley terrain formed by the constructed wetland has been planted with native trees to form a serene pastoral landscape, a quiet refuge within a bustling metropolis. The valley meanders down through a series of stairways, boardwalks, and platforms with large swathes of native wetland plants surrounding the visitor in what has become a relict landscape.

Native vegetation

The recreation of a native landscape using a myriad of wetland plants and wildflowers has formed a lush and biologically productive wetland riverfront that will become a carbon sink in the future.

14.4.2 Three dimensions of meanings

Three cultural meanings are conveyed in the creation of the park.

Agricultural heritage

From the Tang dynasty to Shanghai's opening as a port city in 1843, Houtan witnessed 1200 years of the rise and fall of the agricultural economy along the Huangpu River – making its agricultural heritage an important design consideration.

The terraced belt forms a transition zone between the constructed wetland and the Puming Road and is inspired by China's famous terraced agricultural fields. The terraced fields address the 3–5 m (10–17 foot) elevation drop and also aid in aerating, settling, and purifying the river water. Agricultural crops from Shanghai's past and wetland plants with water purification capacities were selected to create an urban agricultural garden rich with seasonal changes. The memories of agrarian culture are evoked through the scene of the golden blossoms of oil seed rape flowers *(Brassica capestris)* in spring, the splendid sunflowers in summer, the fragrance of rice in fall and green clover (Chinese milk vetch, *Astragalus sinicus*) in winter (Figure 14.6). The design showcases the nostalgia of the productive earth, reconnects the urban dweller to the land, and provides an educational venue for the children of Shanghai to learn about agriculture and ecosystems.

The terraced structure increases the diversity of the landscape and the vegetation reinvigorates the waterfront for a richly textured pedestrian experience. People enter the wetland through the paths traversing the fields and experience the agricultural landscape nestled around an ecologically sensitive constructed wetland.

Figure 14.6 Crops (sunflowers and rice in summer) are grown on the terraces throughout the seasons to filter excess nutrients from the Huangpu River water and to make the landscape productive and educational. Photo: K. Yu.

The paths, like capillaries of a sponge, greatly improve the pedestrian flow into and through the park.

Industrial heritage

Shanghai is the birthplace of industrialization in China and the city has witnessed the nation's modern industrial emergence and growth. Factories (only a few structures are located within the site) and cargo piers are the most important industrial structures that remain on site. The design transformed the existing infrastructure using extraction, infill, and interspersion methods to preserve, reclaim and recycle them to celebrate the site's industrial past.

• The "hanging garden": an industrial transformation. The factory building structure located to the south of the Houtan Wetland Park and to the north of the temporary car park have been reclaimed as a multiservice center – a "hanging garden" hosting a mix of bars and cafes. There used to be two building structures from the former steel factory running across the planned Puming Road, and in the original design, both structures were reused to become a unique entrance to the Expo and to Houtan Park. However approval was only given for the retention of one half of these to be integrated into the "hanging garden".

• The floating garden: a terminal transformation. A cargo pier is located in the middle section of Houtan Park near the preserved existing wetland. Since the pier is no longer operational, it has been reclaimed as a floating garden known as the "reed platform". Today the pier is an observation platform where visitors are immersed among the wetland with views of Shanghai's skyline.

• Steel panels: an artistic installation. An industrial-style installation was constructed using steel panels reclaimed from the site. This is made of a series of folded panels parallel to the stream valley thus casting shadows and framing views. The panels are also used on the boardwalk as a paving material. These modern industrial art pieces merge with the wetland vegetation and landscape pattern creating a design juxtaposition of the site's past, present and future.

Prophecy of the ecological civilization

As a vision and experiment of the postindustrial era, the design promotes a low carbon landscape ideology, emphasizing low cost sustainable construction and maintenance processes. The design embodies an ecological framework by employing a constructed wetland where biological processes showcase a new ecological culture emerging from the urban environment.

Unlike more conventional engineered designs, the Huangpu riverbank is designed to be environmentally sensitive; clay is used as the lake bottom as a naturally impermeable material to aid in establishing a self-sustaining wetland and riparian habitat. A wide variety of wetland species and crops are used for water purification to increase biomass production and in turn operate a low cost and low maintenance water treatment system. The site design uses recycled materials such as old bricks, tiles and biodegradable bamboo flooring to minimize project costs and to integrate innovative energy efficient technologies into the architecture.

14.4.3 Experience network

Houtan Park is designed as a living system offering multiple ecoservices to restore riverbank habitats and establish a self-sustaining treatment wetland while retaining symbols of the agricultural and industrial past. This site serves as a messenger of postindustrial civilization by incorporating ecosystem services into an urban environment, enhancing the understanding of nature and stewardship in a reclaimed public space whilst hoping to extend this new conscience throughout China and the world. The design integrates agrarian, riparian, and biological systems within a pedestrian network forming a "linkage network + node" enabling a rich experience through a textured space.

Pedestrian system

With safety, convenience and flexible capacity as the principle circulation strategy, the design organizes the pedestrian system into a main loop, six "roads" bisecting the wetland and multiple footpaths diverging off the main route. This system ensures seamless connections from the city to the wetland park and not only successfully managed the massive pedestrian influx during the Expo within the site, but also ensures a pleasant and accessible urban park afterwards. The loop is the main path of the park surrounding the wetland that connects to the footpaths and secondary roads. So visitors have options when exploring the space, but remain closely connected to the main route.

Links and nodes

The links and nodes of Houtan Park can be divided into three categories:

• Platforms. Platforms are gathering spaces including the "hanging garden" plaza transformed from the former industrial structure that acts as the central piece of the park's service center, the floating garden plaza transformed from the cargo pier, and the water gate pier plaza that acts as the only entrance from the river.

• Masses. These are groups of trees and bamboos massed throughout the park; they allow unique experiences of penetrating and passing through large vegetation.

• Containers. Containers are spaces enclosed by groves that are intended for exhibitions and resting; modern art and reclaimed industrial elements are located here.

14.5 CONCLUSIONS

Houtan Park demonstrates new ecological water treatment and flood control methods and demonstrates landscape as a living system that has the ability to adapt, change, and protect itself and that can provide multiple ecosystem services for society and nature. The postindustrial design demonstrates a unique productive landscape evoking memories of the past and the future of the ecological civilization, paying homage to a new esthetics based on the qualities of low maintenance and high performance.

F. ENVIRONMENTAL URBAN POLITICS

CHAPTER 15

Geographical perspectives on a radical political ecology of water

Alex Loftus

*Department of Geography, Royal Holloway, University of London, Egham, United Kingdom**

> "The history of cities can be read as a history of water"
>
> Gandy 2002, page 22

15.1 INTRODUCTION

The distribution of water in global cities provides a particularly powerful lens through which to make sense of our contemporary world. Not only does it demonstrate profound global injustices – over half the world's population, for example, is estimated to be confronted with some form of water scarcity – but through analyzing the processes through which potable water circulates through our cities it suggests ways in which injustices are reproduced and, ultimately, might be contested. On the simplest level, it takes very little time before one realizes that there are sufficient water resources and easily enough technical know-how in our world to provide a guaranteed, safe supply of water for all in the world, even with a rapidly expanding urban population. And yet, unless something dramatic happens, and in spite of high-profile declarations to the contrary, the lot of many of the poorest in the world is unlikely to improve when it comes to accessing water. Indeed, things may well get worse. Although often hidden behind a cloak of technocratic jargon, these simple facts are repugnant and yet strangely accepted.

The starting point for a range of approaches that have flourished in recent years within the disparate field of urban political ecology is a scream at these injustices. Captured in this scream is an effort to show the world is open to change and a rejection of the complacent acceptance that that some will remain poor and thirsty while others will remain rich and sated. Beyond this scream it is an effort to capture the radical potentials through which cities might be remade in fairer, more democratic ways. Thus, for many geographers, such work has provided both a means of radical critique and a method for beginning to consider a political approach through which change might be achieved. As Robbins' (2005) definition of political ecology neatly puts it, political ecologists are against apolitical ecologies. They are against Malthusian readings of resource wars that naturalize the scarcity of resources, arguing instead for recognition of the power relations through which resources are both produced and distributed. Moreover, a political ecology approach contains, as Robbins would have it, both a hatchet and a seed. It employs critical insights to chop its way through the acquiescent acceptance that the world is unchangeable: it then employs

*This chapter is a substantial rewrite of *Loftus, A. (2009) Rethinking political ecologies of water. Third World Quarterly* 30(5), 953–968. It includes very few sections from the original.

Applied Urban Ecology: A Global Framework, First Edition. Edited by Matthias Richter and Ulrike Weiland.

this to develop normative claims about how the world should be. Firmly opposed to environmental injustice, a political ecology approach commits us to helping to bring about a better world through contesting the reproduction of socionatural inequalities.

Robbins' catholic conception of political ecology is helpful. The approach has developed through many different influences and, in the contemporary moment, political ecology adopts several varying guises. Perhaps the only unifying idea is the desire to politicize environments as a way of changing them. Because of this, much work on the political ecology of water has emerged as something of a distinct subfield within broader urban political ecological approaches. A loosely based Oxford school of urban political ecology emerged in the late 1990s and early years of the 21st century, which took as its starting point David Harvey's suggestive claim that the city should be viewed as a created ecosystem. Some of the most productive and exciting insights within urban political ecology have come from work with an empirical focus on water struggles. Often more avowedly historical materialist, work on the urban political ecology of water has been at the forefront of developing new approaches to human–environment relations. In recent years, there has been a noticeable move between the study of how the distribution of water has been shaped by relations of power and an analysis of how water itself shapes those relations. As water comes to embody and express social relations, so it becomes crucial to the production and reproduction of power in the world. Thus, not only does work on the political ecology of water have something to say about the production of the world water crisis, it also opens up new understandings of political ecologies of power.

I begin the chapter with a review of work on the urbanization of nature. This provides a vehicle for presenting some of the analytical tools mobilized by geographers who have sought to develop a radical political ecology of water. In the next section I take this forward through a more detailed survey of this work. Then I move on to apply some of these insights to an analysis of water privatization, noting that debates around water privatization have slowed somewhat in recent years. The latter part of the chapter focusses on the questions left hanging within such work and I offer some highly selective thoughts on the questions that might motivate future research. This may appear something of a partisan review, neglecting as it does many other fruitful approaches and for this I apologize. My aim is to present work that is the most likely to shatter the illusion that the distribution of potable water cannot change and, pursuing this, I choose issues from the more radical aspects of urban political ecology.

15.2 THE URBANIZATION OF NATURE

If urban political ecologists seek to confront the passive acceptance of the world as it is, one of the ways in which they have sought to do this is through demonstrating the processes and relationships out of which cities are constituted in historically and geographically specific ways. Understanding these relationships suggests that if they were constituted differently, urban form would differ in fundamental ways. Thus, if we want to make our cities better places in which to live, we must first grasp the relationships out of which they have come to entrench injustice and seek to change these. This, of course, builds on a much longer tradition within urban theory and a much more extensive set of debates in which the entire concept of the urban has come to be questioned. David Harvey poses the problem neatly, writing that: "The thing we call a 'city' is the outcome of a process we call 'urbanization'. But in examining the relationship between processes and things, there is a prior epistemological and ontological problem of whether we prioritize the process or the thing and whether or not it is possible to separate the process from the thing embodied in it" (Harvey 1996a, p.50).

In a classic statement of his overall approach to urbanization, Harvey goes on to call for an approach that regards processes as more fundamental, whilst recognizing that all processes are mediated through the things they "produce, sustain and dissolve" (Harvey 1996a). Urban geographers – and by implication, urban ecologists – should focus not on the city as their primary object of analysis but on the process of urbanization as it produces, sustains, and dissolves individual cities in historically and geographically specific ways. A central aim of work in urban political ecology has been to take forward this approach, but it has done so by emphasizing not only relations amongst people, but also relationships with a range of non-human others. Thus, picking up on Harvey's analytical shift from "the city to urbanization", Swyngedouw and Kaika (2000) write that "the environment of the city (both social and physical) is the result of a historical geographical process of the urbanization of nature" (p.569). Here they not only draw out a process-based understanding of the city but collapse the long-standing dualism between the city and the country. The city as an object of analysis is thereby displaced in favor of a historical geographical materialism that highlights the processes and relationships out of which it is made and remade. They conclude by arguing that:

> viewing the city as a process of continuous, but contested, socioecological change, which can be understood through the analysis of the circulation of socially and physically metabolized 'nature', unlocks new arenas for thinking and acting on the city; arenas that are neither local nor global, but weave networks that are almost simultaneously deeply localized and extend their reach over a certain scale, a certain spatial surface. (Swyngedouw and Kaika 2000, p.577).

The call for such work has become a central plank in the theoretical framework within urban political ecological approaches. Although sharing many affinities with the broader field of political ecology described by Robbins (2005), the historical materialist roots are often far more evident in some of this work. Perhaps its clearest expression is found in the edited collection by Heynen *et al.* (2006), which takes as one of its starting points Harvey's oft-cited expression that the city is a created ecosystem and that "in a fundamental sense, there is nothing unnatural about New York City" (Harvey 1996b, p.186). For Heynen *et al.* (2006):

> Questions of socioenvironmental sustainability are fundamentally political questions. Political ecology attempts to tease out who (or what) gains from and who pays for, who benefits from and who suffers (and in what ways) from particular processes of metabolic circulatory change. (Heynen *et al.* 2006, p.12)

To define urban political ecology as a coherent body of thought (or even as a loosely defined "school") would be to over-systematize what is a remarkably diverse set of approaches to considering the politicized environment of cities. Influences have been similarly broad, with writers drawing variously on Cronon's (1992) magisterial account of the development of Chicago in *Nature's Metropolis*, to environmental justice debates and a focus on the development of environmental social movements within the city (see Bullard 1990, Gottlieb 1993). Nevertheless, there has been a concerted effort to read the city as one moment in the ongoing socionatural processes of transformation. Smith's (1984) quixotic claim that nature is materially produced in ways that vary according to the relations through which that production is organized has been another clear influence. Empirically, work has ranged across cities in both the global North and South with foci on anything from urban forests (Heynen 2006) to fat in sewerage infrastructure (Marvin and Medd 2006).

15.3 URBAN POLITICAL ECOLOGIES OF WATER

Some of the key arguments of Heynen *et al.* (2006) have been best captured in work on water (Swyngedouw 1997, 1999, 2004). Drawing on both actor–network approaches but, above all, historical materialist accounts of the production of nature, Swyngedouw has sought to demonstrate the manner in which water and social power come to be fundamentally intertwined. As a socionatural hybrid, water embodies and expresses the relations through which it is reproduced. The flow of water through a city makes life possible and, at the same time, helps to structure the ways in which that life is lived. It produces and reproduces hierarchies, naturalizing those relations which are socially produced and socializing the dependence of the city on the environments out of which it is produced. For Swyngedouw (2006), key tropes for interpreting the political ecology of cities are those of circulation and metabolism. As water and capital become fundamentally intertwined through a metabolic process, so circulation becomes a key element in the production, reproduction and survival of capitalist relations of production.

Swyngedouw's approach is often a historical one, best exemplified in his work on Guayaquil and on the engineering of the Spanish waterscape: these two studies are amongst the foremost influences on the urban political ecology of water in recent years. In the former, Swyngedouw disentangles the fundamental relationships between water and social power that shape the Ecuadorean city. Urbanization is shown to be a simultaneously political economic and political ecological process. This historical understanding of the political ecological constitution of the city is taken as an entry point to the intense social struggles over access to water that have erupted in the city in recent years, where over a third of the population still remains excluded from supplies of potable water. Perhaps more so in his writings on the Spanish waterscape, there is a noticeable shift of emphasis from an analysis of the social production of an uneven waterscape to the role that water plays in shaping relations of power. Water and social power are shown to be mutually constitutive. Thus, in the Spanish context, he shows how the scalar politics developed within Franco's fascist project for Spain was produced and reproduced through transforming the hydraulic engineering of the country. Embarking on a huge dam-building project, Franco sought to unite the various territorial interests within the country in a simultaneously material and discursive project of re-engineering the waterscape. While many of Swyngedouw's concerns appear to echo those of Wittfogel (1957) or Worster (1992), whose work on large-scale hydraulic infrastructure seeks to unravel the relations between dam-building and social power, his work seeks to counter a latent or overt environmental determinism in both these works by focusing on the dialectical coevolution of water and social power. Again, this is the case within Gandy's research on New York City as well as Kaika's (2005) on Athens. Within the latter (see especially Kaika 2004), the movement between the intimate politics of the home and the large symbolically charged engineering projects of the nation-state helps to further embed such understandings within a critique of everyday life.

The twin focus on circulation and metabolism has helped to transform research on water politics and opened up numerous new avenues for radical scholarship. Kaika's (2005) research on Athens manages to traverse the material and symbolic, showing how large scale acts of water infrastructure engineering serve powerful ideological functions. Similarly, Gandy's (2002) study of New York City demonstrates how the urbanization of nature not only shapes the material form of the city but also the way in which power circulates and is stabilized. This opens up a new agenda for understanding the role of produced natures within the stabilization of specific social formations and the making of specific subject identities. Picking up on these themes, Ekers and Loftus (2008) have argued that much of the recent research on the political ecology of water provides a key entry point into a second set of debates. Through engaging with the work of Antonio Gramsci and Michel Foucault, they suggest that new insights might be possible into the consolidation of power within liberal capitalist societies and attendant processes of subjectification. To date, work on neoliberalism and neoliberalization has touched on these subjects inadequately. Yet, in the work of Swyngedouw, Kaika and Gandy – though rarely confronting Gramsci and Foucault directly – there is a shared interest in the role that water plays in the production and reproduction of power. By connecting the individual with broader networks of power, there is the possibility of seeing the subject as both the iteration of practices of power and also a knowing, situated and active agent, contesting the consolidation of hegemony through the integral state. Potentially, this opens up new inroads into a politically emancipatory and radically democratic project. Through better understanding of how water serves to consolidate particular forms of power, these may, in turn, be undermined. Although such a shift is implicit in the literature and the entry points are numerous, it has to date been relatively ignored.

15.4 PRIVATIZATION QUESTIONS

Within several other studies, an urban political ecological approach has developed at a somewhat less abstract level and has focused on much more concrete questions such as debates over water privatization. If any debate has dominated the literature on water provision since the 1980s, it is surely this one around the correct role of the private sector. The divestiture of water provision in England and Wales in 1989 spearheaded a wave of pressure from private companies to free up further concessions for profitable investment. With the ascendancy of the Washington Consensus, privatization of water services rapidly acquired orthodoxy within policy spheres, even if the actual number of concession contracts has never affected more than 10% of the world's population. Through both loan conditionalities and through loans directly to the private sector, the World Bank assumed a role as willing handmaiden to water privatization schemes.

Importantly, water privatization can vary widely from outright concession contracts to far more limited service contracts in which the private sector may only take on a role around billing or construction. As Bayliss and Fine (2008) note, most of the more recent policy prescriptions by the international financial institutions have tended more toward commercialization rather than full divestiture. Nevertheless, the wave of anticorporate globalization protests found the perfect target as water privatization seemed to epitomize the freeing up of the development process to the profit motive. The dominance of large monopoly interests served to fuel this anger still further. Thus, with water provision having unfolded in France through a mixture of public and private provision, French water companies were the best placed to acquire a market-leading position within this sector and, by the end of the 1990s, the large French multinationals Suez and Vivendi (subsequently named Veolia), along with the German multinational RWE, were operating in over 100 different countries. The role of the European Union in pushing for privatization of water services came increasingly to the fore through the early years of the new millennium as it pushed for the inclusion of water service liberalization in the World Trade Organization's General Agreement on Trade in Services (GATS). This was finally blocked in 2006 when it was agreed that GATS would not include water for human use.

For David Harvey (2003), the freeing up of water service provision to the private sector and attendant efforts to ensure full cost recovery represent a new front in what he terms accumulation by dispossession. Arguing that the violent appropriation of common resources marks not a brief and passing phase in the history of capitalism but an ongoing process necessary for capitalism's survival, Harvey suggests that the privatization of water is a response to a more general crisis of overaccumulation. Having exhausted many of the opportunities for spatial, temporal and spatiotemporal fixes to crises of overaccumulation, capital seeks new terrains for profitable investment in what previously had existed outside of its orbit. Thus, water becomes a key battleground for this new round of accumulation. The argument is well put and has been taken forward within several papers that seek to make better sense of the role of water privatization within broader political economic processes.

Nevertheless, if used too crudely (and Harvey does not use the framework in the way), this risks overlooking the fact that corporations have actually found it remarkably difficult to profit from water privatization in recent years. Part of this lies in the fact that it is not water itself that is generally privatized. Rather it is the contract for the provision of water services. Some of the most high profile water privatizations in the global South – such as those in Buenos Aires, Guayaquil, Jakarta and Manila – have therefore been concession contracts lasting for up to 30 years, in which the private interests involved have agreed to expand the number of connections within a city and to improve the delivery of services. Such contracts commit such companies to high levels of long-term investment in fixed infrastructure. Given that many of these investments are to be made in areas where the majority of the population subsist on incredibly low incomes, these are exactly the kinds of investments that capital tends to dislike. Add to this, high profile struggles in such cities as Cochabamba and Dar es Salaam and the risks of investing act as a further disincentive for capital. Nevertheless, it has taken nearly 20 years from the signing of the first concession contracts in the global South to come to the realization that privatization does not provide a panacea to the global water crisis. Notwithstanding the efforts of the London-based Adam Smith Institute, using funding from the United Kingdom's Department for International Development, to promote privatization in various countries in the South, the fact remains that it is incredibly difficult to make the profits expected by private investors from the large, long-term needs of infrastructural development for poor people. For this reason, even after signing contracts binding private companies to investments in improving water provision, many have been reluctant to meet commitments to invest in poorer localities: contracts have been renegotiated soon after they have been signed, often under highly dubious circumstances and many of the high profile contracts of the 1990s have now been abandoned. The second World Water Development Report (2006) does, in many ways, capture a shifting mood within policy spheres as new mixes of public private partnerships or trisector partnerships (involving civil society too) have come to replace larger concession contracts. As several works have pointed out, processes of corporatization have taken precedence over formal privatization. The public sector is now expected to act far more like a private sector provider and the citizen–consumer nexus has shifted decisively in favor of the latter in many parts of the world. In short, two key themes appear to have dominated the debates on water provision over the last two decades. The first, overarching theme, concerns how to achieve water security for the many currently experiencing scarcities. The second concerns the role that the private sector might play in choreographing the governance arrangements to ensure adequate provision of water.

Studies by Bakker (2004), Budds and McGranahan (2003), Loftus and McDonald (2001), Page (2005), Ruiters (2002), Smith and Ruiters (2006) and Swyngedouw (2004) lay out some groundwork to a political ecological approach to water privatization. Bakker's (2004) work, in particular, although largely focused on a northern context, has gained widespread recognition for a nuanced approach to the complexities of privatization. She argues that the deregulation of water governance has given rise to a process of reregulation. Cautioning against crude assumptions that privatization will have necessarily negative environmental consequences, she emphasizes the important role of a new regulatory agency in the English and Welsh context – the Environment Agency – and the role this plays in seeking out ecological safeguards for local water supplies. Furthermore, her work goes on to chart the apparent shift back towards mutualization within the English and Welsh context. Her overarching claim, that water is an "uncooperative commodity", chimes with several of the themes developed later in this paper. When turning her attention to the global south, Bakker's (2003) work has mobilized the analogy of archipelagos to capture (more effectively than the claim that this is a form of splintering urbanism) some of the spatial implications of privatization strategies in southern cities. Later, in an important contribution to debates over the right to water, Bakker has drawn attention to the danger in eliding struggles against privatization with those seeking to achieve universal rights to water. She favors struggles that put access to the commons at the forefront on their efforts, for in these battles the right to water is not muddied by neoliberal efforts to instate a system of water rights. This has been one of the more recent vehicles, building on the Chilean experience (see Bauer 1997 and Budds 2008) through which property rights that favor elite groups have been introduced to the water sector. Seeking a right to water fails to specify the means through which that right might be achieved. As the South African example (see Loftus 2006) has shown only too well, even a constitutionally guaranteed right to water does not prevent routine disconnections of those who have exceeded their guaranteed free basic allowance. Indeed the corporatization of many municipal services in South Africa has opened up a maelstrom of contradictions within the rights-based framework on which much of the post-apartheid project has been constructed.

A paper by Budds and McGranahan (2003) develops a far more synthetic approach to privatization experiences around the world. Attempting to shift the debate from the polarizing positions that the profit motive is either antithetical to the public good or necessary in achieving it, the authors argue that neither large public infrastructural works nor private sector investments are likely to achieve the Millennium Development Goals at the moment. Neither public nor private entities are well-suited to the needs of the urban poor in the global South: neither of these are capable of addressing the current barriers to the development of water infrastructure in such areas. The paper makes a convincing argument against the Panglossian view that privatization will succeed where the public sector has failed and also the naïve counterpoint – that an unreformed public sector can achieve universal benefits. The last half decade of publicly operated water networks would appear to show that, in its current state, the public sector has not been up to the job. Budds and McGranahan (2003) weave several key claims from political ecological research into their more general argument.

Most other contributions from within the political ecological literature on the question of privatization have, not surprisingly, put a theoretically informed approach to work in analyzing concrete cases of water privatization in a range of different locations. Page's (2005) work in Cameroon takes an analysis of the process of commodification in the water sector further into the literature on commodities and shows how a rapprochement with certain psychoanalytical approaches might prove fruitful through a deeper consideration of the role of desire in our relationship with commodities. McDonald and Ruiters (2005) provide a series of essays on the process of corporatization in South Africa that draw more or less explicitly on political ecological frameworks. Loftus and McDonald's (2001) work in Buenos Aires develops an agent-centered approach to understanding the political ecology of water privatization that echoes the methodological framework developed by Bryant and Bailey. Here, the politicized environment is shaped out of the interactions of a number of key agents, from the state to business. Arguing that a coalition of international forces has created a myth of the Buenos Aires model and exported it to a range of different locations through Washington Consensus reforms, they seek to show how the model was shaped more by economic ideology than good sense, how it has opened the door to a new wave of corruption, and how, in the process, it has heightened both sociospatial and environmental inequalities within South American cities.

Overall, water privatization and commercialization has been one of the key terrains over which urban political ecological research has been taken forward. As a study such as Bayliss and Fine's (2008) makes so clear, this work can shine light on the perversities of current debates in water governance, which are all-too-often limited to exploring the appropriate (but always necessary) role of the private sector. This prevents far more fundamental questions from being posed that might have a far more dramatic impact on efforts to develop water service provision appropriate to the needs of the urban poor in the global south.

15.5 TAKING THE DEBATES FORWARD

If the preceding sections suggest some key areas being researched by urban political ecologists, the final part of the chapter is an effort to consider the ways in which such work might be taken forward. My focus here is much less on developing a normative framework for water governance, given, as I stated earlier, my concern to expose the fundamental limits to contemporary policy debates. Instead, my concern is to explore a more fundamentally *political* project. This, I would argue, is far truer to the principles on which urban political ecology has developed in its effort to undermine taken-for-granted assumptions and use research on water as a vehicle for getting to more radical political changes. I will frame this final part around two key moments whilst also showing how they shape and are shaped by other moments with an urban political ecological framework. First, I will focus on the theorization of water infrastructure, looking at how this opens up significant possibilities for rethinking: neoliberal rationalities; the engineering of scale; the reproduction of gendered subjectivities and ways of reworking hegemony. Concluding this section, I look at the potential for democratizing water technologies in ways that confront the violence of neoliberalism, the hierarchies through which the socionatural is structured, gendered inequalities and the maintenance of class power. Second, and more briefly, I will look at the ways in which urban political ecological research might permit the development of an immanent critique of everyday life. Few activities can be so mundane and yet so fundamental to human life as that of collecting and consuming water, and yet historical materialist theorizations of the everyday have only rarely found their way into work on urban political ecologies. This is to the detriment of the latter, for, through a more sophisticated understanding of the everyday, new entry points are opened up for political praxis. Concluding, I seek to bring these points together in rethinking the conditions of possibility for a radical urban political ecology of water.

15.6 INFRASTRUCTURES OF POWER: DEMOCRATIZING WATER TECHNOLOGIES

It is often within specific technologies that power is embodied and expressed within the waterscape. Thus, water meters might regulate the rhythms of a person's day-to-day life (Loftus 2006), dam-building projects might serve to foster a potent nationalism (Kaika 2006) and tubewells have been shown to play a role in the shaping of gendered subjectivities (Sultana 2009). Debates around the coevolution of water, power and infrastructure extend back some way. Thus, for Wittfogel (1957), the geography of Asia necessitated peculiarly centralized forms of irrigation. This, in turn, gave rise to particular forms of rule. Whilst not denying the role that geographical context makes, more recent work has sought to show how nature (and thereby scarcity too) is produced as an indivisibly socionatural phenomenon. This socionatural process can be located within specific technologies. Building on actor-network theory, many authors have sought to understand the multiplicity of actors involved in such processes (see Staddon 2009) and the role of non-human agency in animating such networks. Others have found inspiration in historical materialist understanding of the operation of power within urban form.

Emphasizing the key role played by water and water infrastructure in the consolidation of forms of rule, the Fordist social compromise has been shown to be part-produced and also expressed within the move from ostentatious displays of municipal engineering achievements to hidden networks that penetrate the intimacies of the modern home (Kaika and Swyngedouw 2000). These hidden infrastructures, in turn, became taken-for-granted facets of the social welfare provisioning of specific societies. In the contemporary era, underinvestment in, and the privatization of, networked infrastructure has led to a splintering urbanism, perhaps more representative of the post-Fordist or neoliberal moment (Graham and Marvin 2001). Turning to Foucault's later writings on governmentality, some have commented on the dispersed form of power that operates through the water network (Ekers and Loftus 2008). This shapes the field of action in which individuals are able to operate, defining what Foucault refers to as "the conduct of conduct". As Joyce shows, the provision of clean drinking water is part of a liberal governing rationality of "ruling from a distance" through ensuring the predictable circulation of people, pathogens and commodities. In Budds' (2008) work, the development of the Chilean Water Code is shown to be both a central plank in the constitution of authoritarian neoliberalism in the country, as well as a key expression of some of the contradictions within the emerging governing rationality. In the transition to democracy these historical contradictions within the law have deepened further finding a fundamentally socionatural expression (Budds 2009). More recent work on biopower picks this debate up directly through seeking to employ Foucauldian insights to the microlevel practices through which water is distributed and consumed. Within the South African case, work by Ruiters (2007) and by Hellberg (2009) has taken this forward in intriguing and suggestive ways. In short, processes of subjectification should be seen to flow through the water network, making and remaking citizens through the operation of biopower. Whilst performing a key role in developing neoliberal rationalities, water infrastructure also consolidates these rationalities at a variety of spatial scales. The ongoing debate over the production of scale has thereby been enlivened by recent contributions on water infrastructure. Here, scale has been shown to be produced socionaturally and not simply socially. Swyngedouw (2007) thereby demonstrates that Franco's authoritarian project for Spain rested in part on developing connections between different interest groups through the engineering of the water network. In turn, shifting relationships with the United States came to be embodied and expressed within the Francoist dam-building program. Water infrastructure is thereby crucial in the production, reproduction and networking of differing scalar formations from the home to the city, state and global. Here, further connections are possible with literature that emphasizes the importance of social reproduction to the production of scale. Thus scale is both a class and a gendered project. The day-to-day task of provisioning a household with water in the global south acts as a crucial determinant of these gendered spatial scales.

The production of these gendered relationships is absolutely fundamental to understanding the role that water plays within different societies. In the global South the act of provisioning the household with water (and often paying a bill, negotiating with water vendors and protesting over scarcities) is, almost invariably, assigned to women. Operating different water technologies is an everyday act through which gendered subjectivities come to be produced, embodied and performed. Nevertheless, much of the decision-making over *which* water technologies are to be developed within a particular social formation tends to be assigned to an elite of technocratic experts. More often than not, these are men. More often than not, their knowledge is based on a radical separation from the direct, sensuous experience of working specific technologies and infrastructures. As Sultana's (2009) work has shown, the implications of developing rudimentary technologies within the arsenic-contaminated groundwater supplies of Bangladesh has serious implications for women. In Loftus's (2007) work, gendered knowledge of processes of democratizing such technologies was explored.

Finally, water technologies have been shown to play a key role in consolidating the varying combinations of consent and coercion that Antonio Gramsci refers to as hegemony (Ekers and Loftus 2008). Hegemony, in short, must be understood as a socionatural, or technonatural phenomenon that operates in part through a reshaping of everyday experiences. One of the more subtle, but nonetheless crucial, ways in which this has taken place is through re-shaping the technological experience of accessing water. Swyngedouw and Kaika's (2000) earlier work thereby linked water infrastructure to the Fordist social compromise. More recently Loftus and Lumsden (2008) have sought to demonstrate the ways in which changes in the water technologies operating in informal settlements in South Africa have subtly shifted the citizen–consumer nexus in the post-apartheid country. Through attacking a culture of non-payment and rooting this attack in an army of new metering technologies, the ANC government has sought to foster a new atomized consumer of services amongst the very political activists who sought to overturn the apartheid regime. Paradoxically, within this they find grounds for hope. A new socionatural terrain for challenging hegemony is opened up within the waterscape and the possibility of democratizing technologies deepens.

What this work suggests is that there is the potential to begin to see infrastructure as both a tool through which forms of rule might be constituted but also as a potential weapon through which those forms of rule might be contested. With regard to the latter, debates within science and technology studies have, in recent years, begun to question the relationship of specific technologies to processes of democratization (Feenberg 1999). Within both science studies and the popular press, concerns over the development of new technologies, from genetically modified crops to nuclear energy, have raised questions over whether innovations foreclose debate because of the need for technically specific knowledge. In contrast, the huge expansion of the internet and various open access softwares have suggested other models of technological innovation that might open up rather than close down spaces for debate, participation, and democratic dissonance. The intellectual contributions of the Frankfurt School (Marcuse 1964) have, for many, been central in exploring some of the paradoxes of technical developments (Feenberg 1999). These debates are of fundamental concern to those seeking to understand water infrastructure in the contemporary moment. Conversely, as has been seen with other debates, the contributions of those researching water technologies provide insights that might further enliven such arguments. Rooted as it is in the daily practices of every citizen in the world, the debates that are spawned from the exploration of water service infrastructure have a particularly universal pertinence.

Taking this forward, through seeking to democratize water technologies, struggles against those very processes through which power is consolidated and through which gendered subjectivities, hierarchies and hegemony are reified have been crucial to the genesis of a radically democratic politics. Water struggles from Cochabamba in Bolivia to Durban in South Africa have shown the ways in which neoliberal rationalities might be challenged and through which an alternative geography of the city might be constructed. In the process, gendered subjectivities have formed both a starting point for such an alternative project and have been challenged by women seeking to subvert and transform unequal distributions of power. In this way, "situated knowledge" of water distribution and consumption provide a crucial route into the topographies of power that fix gender in certain ways at certain moments. Reconfiguring access to water at the same time restructures the very processes through which scale is produced and reproduced, whether through acts of producing potable water or through acts of social reproduction in ensuring a household has access to a safe, sufficient supply of water. Thus, through positioning water infrastructure as a particular terrain of struggle it becomes possible to envision and fight for a fairer world. This, in turn implies an immanent critique of everyday life – a critique that might emerge from the daily round of walking to a standpipe, arguing with a local councilor, reconnecting a household with water and forging new social collectivities out of such acts.

15.7 THE EVERYDAY

For Lefebvre, social subjects must be everyday or they must not be at all. In making this claim he seeks to develop a critique of everyday life that respects quotidian acts of production and reproduction whilst seeking to go beyond them. Such a critique, I would argue, needs to be placed at the center of a radical perspective on the political ecology of water. Although within geography there has been considerable interest in the writing of Michel de Certeau on the practice of everyday life (1984), there has been much less of a focus on the rich tradition of immanent critique within historical materialist debates. In the work of Trotsky (1994), Lukacs (1971), Lefebvre (1991), Gramsci (1971) and several others, there is something of a move against a vanguard politics that seeks to position critique and political practice within the quotidian realm. For Lukacs this emerges in his standpoint methodology in which a dialectical conception of the totality emerges not from the realm of abstract philosophical thought but from the sensuous acts of laboring and the attendant class struggle. For Gramsci, attention to subaltern knowledge and practices serves to counter the emerging scientific materialism of his day and thereby forge a new symbiosis between proletariat and Party in which the latter serves as an agent, rather than a vanguard, which seeks to organize the fragmented but potentially transformative common sense of the many. For Lefebvre, this means building on Marx as a key theorist of the everyday and one in which everyday life might be transformed into a work of art to be crafted by conscious subjects.

Given the historical materialist roots of much of the work in urban political ecology, it makes considerable sense to bring these debates about the everyday into much closer conversation with the effort to theorize the injustices produced on a daily basis through the provision of water. One way of doing this might be through questioning infrastructure in the manner suggested above. Here the manifestation of power in specific technologies comes to be questioned through the daily process of operating taps, water meters, pipes, and orifice valves. The reification described in such abstract fashion by Lukacs makes it appear as through the facts of water distribution are immutable. Nevertheless, through the laboring act of providing safe sufficient water for a household, process-based understandings are opened up that could be the starting point for a radical politics. This is one of the starting points of some feminist standpoint theorists whose work cries out to be put to work in making sense of struggles to ensure a more adequate distribution of water. Nevertheless, such moments of dialectical consciousness need situating within the common sense conceptions of the world that traverse the everyday. Here, Gramsci's insights might provide further tools for a critique that does justice to the lived realities of the urban poor.

15.8 CONCLUSIONS

In this chapter I have sought to demonstrate the ways in which studies of urban political ecology can form the basis for a radical critique of the contemporary world. The violence of water disconnections and the forms of consciousness that emerge from the day-to-day acts of supplying a household with water are some of the starting points for such an analysis. This, I argued, is the scream that provokes much of the work within this field. Within this loosely defined field of urban political ecology, approaches vary enormously. Nevertheless, it is possible to outline several themes that appear to unite some of these disparate frameworks. Urban political ecologists seek to understand the world as a means of changing it for the better. Whether focusing on renewed rounds of accumulation by dispossession through water privatization or the reproduction of gendered subjectivities through acts of social reproduction, urban political ecological approaches to water begin from the topsy-turvy lived reality of water distribution and consumption. They seek to politicize this and develop process based understandings of the city. Such understandings disrupt the sense in which the city appears as an immutable never-changing thing, one to which our theories must simply accommodate themselves. Instead they demonstrate the ongoing reciprocity between theorization and action, philosophizing about and changing the world. They seek ways in which the processes that make the city as a thing might be reworked in fairer, more socially and ecologically egalitarian ways and how understandings that might permit this are not the property of abstract theorists but emerge from the ongoing efforts of men and women to make sense of their own worlds through their own ability or inability to access water in different ways.

Taking some of these claims forward, in the final part of the chapter, I suggested a highly personalized set of questions that confront political ecological understandings of urban water. These concern the operation of power through infrastructural assemblages and the potential for building on an immanent critique of everyday life as a step towards reconfiguring our world. Within these lie conditions of possibility in which radical conceptions might germinate. In turn, they hold out hope for a better world, one in which all might have access to a fundamental resource such as water and one in which our cities might be remade in radically democratic ways.

REFERENCES

Bakker, K.J. (2003) Archipelagos and networks: urbanization and water privatization in the South. *Geographical Journal* 169(4), 328–341.

Bakker, K.J. (2004) *An Uncooperative Commodity*. Oxford University Press, Oxford.

Bauer, C.J. (1997) Bringing water markets down to earth: the political economy of water rights in Chile, 1976–1995. *World Development* 25(5), 639–656.

Bayliss, K., Fine, B. (2008) *Privatization and Alternative Public Sector Reform in Sub-Saharan Africa: Delivering on Electricity and Water*. Palgrave Macmillan, London.

Bryant, R., Bailey, S. (1997) *Third World Political Ecology*. Routledge, London.

Budds, J. (2008) The 1981 Water Code: the impacts of private tradable water rights on peasant and indigenous communities in Northern Chile. In: Alexander, W. (ed.) *Lost in the Long Transition: The Struggle for Social Justice in Neoliberal Chile*. Lexington Books, Lanham, pp. 35–55.

Budds, J. (2009) Contested H20: Science, policy and politics in water resources management in Chile. *Geoforum* 40(3), 418–430.

Budds, J., McGranahan, G. (2003) Are the debates on water privatization missing the point. *Environment and Urbanization* 15(2), 87–113.

Bullard, R. (1990) *Dumping in Dixie: Race, Class and Environmental Quality*. Westview Press, Boulder.

De Certeau, M. (1984) *The Practice of Everyday Life*. University of California Press, Berkeley.

Cronon, W.A. (1992) *Nature's Metropolis*. WW Norton and Co., New York.

Ekers, M., Loftus, A. (2008) The power of water: developing dialogues between Foucault and Gramsci. *Environment and Planning D: Society and Space* 26(4), 698–718.

Feenberg, A. (1999) *Questioning Technology*. Routledge, London.

Gandy, M. (2002) *Concrete and Clay: reworking nature in New York City*. MIT Press, Cambridge.

Gottlieb, R. (1993) *Forcing the Spring*. Island Press, Washington DC.

Graham, S., Marvin, S. (2001) *Splintering Urbanism: Networked Infrastructures, Technological Mobilities and the Urban Condition*. Routledge, London.

Gramsci, A. (1971) *Selections from the Prison Notebooks*. Lawrence and Wishart, London.

Harvey, D. (1996a) Cities or Urbanization? *City* 1, 38–61.
Harvey, D. (1996b) *Justice, Nature and the Geography of Difference*. Blackwell, Oxford.
Harvey, D. (2003) *The New Imperialism*. Oxford University Press, Oxford.
Hellberg, S. (2009) *Governing Lives through Hydropolitics in eThekwini*. Paper presented to the Center for Civil Society at the University of KwaZulu Natal on April 1, 2009.
Heynen, N. (2006) Justice of eating in the city: the political ecology of urban hunger. In: Heynen, N., Kaika, M., Swyngedouw, E. (eds) *In the Nature of Cities*. Routledge, London, pp. 124–136.
Heynen, N., Kaika, M., Swyngedouw, E. (eds) (2006) *In the Nature of Cities*. Routledge, London.
Joyce, P. (2003) *The Rule of Freedom*. Verso, London.
Kaika, M. (2004) Interrogating the geographies of the familiar: Domesticating nature and constructing the autonomy of the modern home. *International Journal of Urban and Regional Research* 28(2), 265–286.
Kaika, M. (2005) *City of Flows: Modernity, Nature, and the City*. Routledge, New York.
Kaika, M. (2006) Dams as symbols of modernisation: the urbanisation of nature between materiality and geographical representation. *Annals of the Association of American Geographers* 96(2), 276–301.
Kaika, M., Swyngedouw, E. (2000) Fetishising the modern city: the phantasmagoria of urban technological networks. *International Journal of Urban and Regional Research* 24(1), 120–138.
Lefebvre, H. (1991) *The Critique of Everyday Life: Volume 1*. Verso, London.
Loftus, A. (2006) The metabolic processes of capital accumulation in Durban's waterscape. In: Heynen, N., Kaika, M., Swyngedouw, E. (eds) *In the Nature of Cities*. Routledge, London, pp. 173–190.
Loftus, A. (2007) Working the socio-natural relations of the urban waterscape. *International Journal of Urban and Regional Research* 31(1), 41–59.
Loftus, A., Lumsden, F. (2008) Reworking hegemony in the urban waterscape. *Transactions of the IBG* 33(1), 109–126.
Loftus, A.J., McDonald, D. (2001) Of liquid dreams: a political ecology of water privatization in Buenos Aires. *Environment and Urbanization* 13(2), 179–199.
Lukacs, G. (1971) *History and Class Consciousness*. Merlin, London.
Marcuse, H. (1964) *One Dimensional Man*. Routledge and Kegan Paul, London.
Marvin, S., Medd, W. (2006) Metabolisms of obe-city: flows of fat through bodies, cities and sewers. In: Heynen, N., Kaika, M., Swyngedouw, E. (eds) *In the Nature of Cities*. Routledge, London, pp. 143–156.
McDonald, D., Ruiters, G. (2005) *The Age of Commodity: Water Privatization in Southern Africa*. Earthscan, London.
Page, B. (2005) Paying for water and the geography of commodities. *Transactions of the Institute of British Geographers* 30(3), 293–306.
Robbins, P. (2005) *Political Ecology*. Blackwell, Oxford.
Ruiters, G. (2002) Debt, disconnection and privatisation: the case of Fort Beaufort, Queenstown and Stutterheim. In: Pape, J., McDonald, D. (eds) *Cost Recovery and the Crisis of Service Delivery in South Africa*. HSRC Press, Pretoria, pp. 41–57.
Ruiters, G. (2007) Contradictions in municipal services in contemporary South Africa: disciplinary commodification and self-disconnections. *Critical Social Policy* 27(4), 487–508.
Smith, L., Ruiters, G. (2006) The public/private conundrum of urban water: a view from South Africa. In: Heynen, N., Kaika, M., Swyngedouw, E. (eds) *In the Nature of Cities*. Routledge, London, pp. 183–198.
Smith, N. (1984) *Uneven Development: Nature, Capital and the Production of Space*. Blackwell, Oxford.
Staddon, C. (2009) *The Social Life of a Water Meter*. Paper presented at the Annual Conference of the Association of American Geographers 23rd March, 2009.
Sultana, F. (2009) Community and participation in water resources management: gendering and naturing development debates from Bangladesh. *Transactions of the IBG* 34(3), 346–363.
Swyngedouw, E. (1997) Power, nature, and the city. The conquest of water and the political ecology of urbanization in Guayaquil, Ecuador: 1880–1990. *Environment and Planning A* 29, 311–332.
Swyngedouw, E. (1999) Modernity and hybridity: Nature, regeneracionismo, and the production of the Spanish Waterscape, 1890–1930. *Annals of the Association of American Geographers* 89(3), 443–465.
Swyngedouw, E. (2004) *Flows of Power – The Political Ecology of Water and Urbanisation in Ecuador*. Oxford University Press, Oxford.
Swyngedouw, E. (2006) Circulations and metabolisms: (hybrid) natures and (cyborg) cities. *Science as Culture* 15(2), 105–122.
Swyngedouw, E. (2007) Technonatural revolutions: The scalar politics of Franco's hydro-social dream for Spain, 1939–1975. *Transactions of the Institute of British Geographers* 32(1), 9–28.
Swyngedouw, E., Kaika, M. (2000) The environment of the city ... or the urbanization of nature. In: Bridge, G., Watson, S. (eds) *A Companion to the City*. Blackwell, Oxford, pp. 567–580.
Trotsky, L. (1994 [1923]) *Problems of Everyday Life*. Pathfinder, New York.
Wittfogel, K.A. (1957) *Oriental Despotism: A Comparative Study of Total Power*. Yale University Press, New Haven.
Worster, D. (1992) *Rivers of Empire: Water, Aridity and the Growth of the American West*. Oxford University Press, Oxford.
World Water Development Report 2 (2006) *Water: A Shared Responsibility*. Oxford University Press, Oxford.

PART IV

SYNTHESIS

CHAPTER 16

Synthesizing urban ecology research and topics for urban environmental management

Matthias Richter[1] and Ulrike Weiland[2]

[1]*Environmental Scientist, Publicist and University Lecturer, Germany**
[2]*Institute for Geography, University of Leipzig, Leipzig, Germany*

The subject of this last chapter is the question as to which urban ecological research results of the authors of this book contain starting points for urban environmental management and thus for an "applied urban ecology." For this evaluation the articles of Part III Selected Fields of Applied Urban Ecology will be used, in which research and action fields of urban ecology are prepared thematically.

"*Applied Urban Ecology – A Global Framework*" is a contribution to support the improvement of environmental quality in urban settings by mutually learning from other attempts. The purpose of this book is to provide urban ecological knowledge in a nutshell tailored for supporting environmentally sound urban development. As an applied scientifically based approach, urban ecology gives advice about where and how the urban environment may be protected from further harm, and how environmental quality and thus physical human living conditions may be improved.

If one subjects the articles of urban ecological research in general and the contributions of this book in particular to a comparative analysis one determines that not only the thematic areas and subjects of urban ecology boast a large range; the respective understanding of urban ecology is also different. Urban ecology can be conducted not only in an analytical–descriptive perspective ("how it is") but also in a normative perspective ("how it should be"). Whilst many urban ecologists take an analytical–descriptive perspective, in accordance with their view of themselves as natural or social scientists, others (who follow the goal of a change to and improvement of the urban environment) take a normative perspective. The editors of this book support the view, that a careful and "sober" normative perspective ("should-be perspective") must be added to the analytical–descriptive perspective ("is perspective") of urban ecological analysis out of which then impulses for the protection and the improvement of the urban environment could arise.

The interesting question is whether the different normative views about urban ecological topics in this book can be brought into line? Due to the variety of topic areas and subjects mentioned, the author's view is that this will only work under general terms such as "long-term environmentally sound" or – more broadly – "sustainable" urban development. Thus as soon as one is dealing with real decisions on the ground, priorities are to be set in favor of certain goals, which in turn more or less exclude the realization of other goals. In this way, for example, urban climate protection usually requires different measures from those of biodiversity protection and this also requires different ones from resource protection in the sense of a prioritization of renewable resources.

In Chapter 4 topics such as the urban heat island effect, biodiversity and the CO_2 concentration of the

*Contact information: http://www.Dr-Matthias-Richter.de

Applied Urban Ecology: A Global Framework, First Edition. Edited by Matthias Richter and Ulrike Weiland.

atmosphere are dealt with by Wu *et al.*, building on the LTER study in Phoenix, Arizona. A central concern of the chapter is to deliver a transdisciplinary view of urban ecosystems. The contribution from Wu *et al.* deals with an improved understanding of the driving forces in relation to patterns and processes in urban systems. The research on urban development is based on a landscape ecological approach, which is portrayed through computer-based modeling and landscape metrics. The effects of urbanization on different partial aspects are presented in particular, whereby to a great extent an analytical–descriptive perspective is taken. The normative perspective refers to the improvement of the linking of social, economic and ecological aspects within urban ecological research. The proclamation of sustainable urban development remains, however, very abstract:

> Urban ecology is expected to play an instrumental role in improving existing cities and developing new ones that are more sustainable ecologically, economically, and socially. To realize this goal, urban ecology needs to go beyond the city to consider broader landscapes and go beyond ecology to embrace design sciences. (Wu *et al.*, 2011, Chapter 4, pp. 48–49)

Material aspects and the question of sustainable urban development are topics of Chapter 5 written by Baccini. In this, existing analyses of the "material situation" of urban regions and social systems are portrayed in relation to the resources of water, biomass, construction materials and energy alongside the presentation of strategies and suggestions for sustainable urban development. Two main questions are posed in the contribution:

- What are the main characteristics of urban systems?
- In what way should they be reconstructed?

Thus not only the analytical–descriptive but also the normative perspectives are represented. The chapter is implicitly accompanied by the concern that with the current growth of the world population the limitation of resources could lead to a disaster. As the author himself explains, the integration of ecological principles, such as the reuse of building materials and the increase in energy efficiency, comprises approaches that can prevent the horror scenarios from becoming reality.

It must be commented here that humans have always been very resourceful at using alternative materials or products. It should also be asked whether the limitation of fossil fuels will not be compensated for in the near future by the intensive use of renewable energy, e.g., in the form of geothermal energy, solar energy, wind energy, and water power. Large investments in the further development of these resources are both in preparation and being undertaken. The question still remains as to whether their use will also lead to an equality of distribution. Especially here skepticism has its place and is thematized in relation to water resources in Chapter 15.

In Chapter 6 Aggarwal and Butsch look at the environmental problems in Indian mega-cities using the example of Delhi. These include serious air, water, and soil pollution, which lead to massive health problems of the urban inhabitants. In addition, the contribution reveals that the poor population classes are more strongly affected by these environmental problems than the middle and upper classes; the question of the social distribution of environmental pollution is made clearly. Much environmental data is presented in this chapter; however the normative perspective is taken in which conclusions for the regulation of urban development are made. The social–political dimension is rightly referred to, without the inclusion of which solutions to the problems are only possible to a limited extent. To quote:

> From a socioecological perspective societal reform is also urgently needed to overcome the vicious circle of environmental degradation and poverty. In brief, to produce environmentally sound, sustainable, inclusive and harmonious mega-cities, there is a need to reinvent urban planning, to bring about reforms in urban administration, and to have good governance, strong political will and a persistent civil society engagement. (Aggarwal and Butsch, 2011, Chapter 6, p. 80)

It will be difficult, and probably only possible in the long term, to bring about the required reforms and ways of thinking and to lead to practical improvements of the environmental situation. However, it must be remembered that the environmental conditions in European and North American cities were disastrous in the early days of industrialization, but were improved in the long term. The current mega-cities thus have an advantage over the early industrial cities: today there is much more knowledge, technology and many more methods for the protection of the environment than back then. Since the articulate and mobile middle classes develop quickly in upcoming nations such as India and China and (can) make higher demands with regards to their quality of life and on the environment than the industrial workers in the early days of industrialization in Europe and North America, there is a glimmer of hope for an improvement of the environmental situation in newly industrialized countries and "young industrial nations" in the near future.

In the contribution by Rink and Herbst in Chapter 7 the topic of "urban wilderness" is considered from various perspectives, including from the planning perspective and the social and ecological perspectives. The reference area for the literature study for this article extends from central, western and northern Europe to the United States. The article includes a portrayal of facts and definitions and thus takes an analytical–descriptive perspective. However, it also has an appealing character, in which the importance of areas for experiencing nature is emphasized and in which sites of urban wilderness are propagated as places for environmental education.

What is interesting is the consideration about how urban wilderness is viewed in an intercultural comparison. It seems to be that first a certain cultural shaping of the living conditions in the industrialized nations has led to the valuation of these areas, whereas in other regions of the world, such as in Africa or in the former Soviet Union, these areas are either ignored or seen to be of little value. As shown in this chapter, even in Europe these areas are not always valued. It can also be assumed that in the long term the relationship between indoor and outdoor experiences of urban inhabitants will continue to move in favor of the indoor experiences. Thus the skepticism remains valid as to whether areas for experiencing nature will continue to play a role on derelict land. If Konrad Lorenz is correct in his assumption in relation to nature, which is that we only protect that which we know, then this seems to indicate the necessity to think more deeply about several things: first, experiences of "nature" are becoming virtual more often, this means conveyed by media and, second, a filtering of urban nature experiences to more exotic and spectacular examples is gaining ground. Thus, the visitor numbers in zoos continue to be high and the keeping of exotic animals in urban apartments is on the increase. Thus the third point could be that areas for experiencing nature on derelict land are only selective and are important only for certain groups of the population and that the high point for their acceptance has already been reached.

The contribution from Kerle and Alkema in Chapter 8 deals with the risk of flooding and the prevention of floods using the example of Naga City in the Philippines and thus provides a contribution to the linking of urban ecology and risk research. The chapter addresses specific data and experiences with the authorities and the urban inhabitants. The normative perspective is implicitly included, in which the protection of human life and the entire values threatened by flooding are mentioned. According to the authors it is necessary to use various geoinformatic tools and techniques in order to optimize this protection. It is important to emphasize that not only disaster is perceived as an acute and immediate threat, which is promoted by reports in the media, but also that the creeping threat through, for instance, air and water pollution and long-term health effects are also heeded appropriately by the urban environmental policies.

In Chapter 9 Nikolopoulou presents the experiences from seven European cities from the EU-funded project RUROS. The legally protected good here is also human health, as in the previous article, whereby mainly the climatic quality of the surroundings in urban open spaces is dealt with. The chapter clearly portrays what the human adaptability mechanism looks like in unfavorable climates. The normative perspective becomes clear from the implicitly included demands for the, when possible, climatically acceptable urban open spaces. Solutions are also specifically mentioned, such as the examples in the part of this chapter named "design interventions," in which landscape architectural and material–technical suggestions are given for the design of urban open spaces.

In Chapter 10 plant diversity in gardens is investigated by Cilliers *et al.* In contrast to many studies in central Europe, not only the species diversity of spontaneous and subspontaneous flora is included, but also of planted species. In this chapter it is determined that the cultural, social, and economic make-up or the status of the garden owner is decisive for the design of the garden and its species diversity. Using the example of two South African settlements, Potchefstroom and Ganyesa, it is shown that apart from the "luxury effect", which says that the species diversity increases with socioeconomic status, further factors come into play with respect to the diversity of the plant species, such as previous use, or whether existing vegetation is deliberately included in the garden design.

The normative perspective remains more under-represented overall, even when, at the beginning, health aspects are briefly named as a service of urban green and towards the end a mixture of bottom-up and top-down approaches is requested regarding management for the provision of ecosystem services. Furthermore, conservation oriented planning is implied, which should use a participatory approach to include all parties on all levels.

In particular, at the transition from the analytical–descriptive to the normative perspective, the limited possibilities for controlling urban biodiversity are revealed in the case of gardens. The actors here are the garden owners, or those who are responsible for the design of the garden and the selection of the plant collection. For these people the decisive factor is whether the garden suits the needs of the user; this is, however, not dependent on a high or low biodiversity. Rather, the species diversity is a result of the needs as well as the social and economic possibilities and the level of education of the user, as is revealed. Thus, in the conclusion of the contribution the question of social justice is rightly posed with respect to the availability of resources and services by local governments, as well as the aspects of participation of the population in the relevant decisions.

In Chapter 11 Ignatieva shows which influences Anglo-American and European garden and landscape traditions have and had on garden design (with an emphasis on the British influence and, in particular, Victorian gardens). The homogenization and globalization that arise with the modern use of planting material are lamented. As an alternative the use of native species is shown in historical perspectives and at a country-specific level. Although it is mostly facts that are given, intrinsically, and in the layout of the article, the Euro-American consumer's vision as well as the globalization of garden styles and plant material is criticized and the use of native plant material is propagated as the normative perspective. As already revealed in Chapter 10, it becomes clear that the biodiversity of plants in cities is steered by cultural influences, which find their expression in the layout of open spaces, as well as in the use of species. One question is interesting: how the use of species is driven by the behavior of the markets and the attitude of the actors; inexpensive plant material becomes established because the buyer and the customer decide to purchase it. For the majority of those responsible for garden and open space management, a good esthetic effect and a low price are more important in the end than the knowledge that native material has been used.

Chapters 12, 13 and 14, which are placed in Section E of Part III (Environmental Urban Design), will be discussed together since they show some similarities in relation to the analytical–descriptive and normative perspectives. Section E deals with the connection between scientifically funded landscape and environmental planning on the one hand and landscape architecture oriented towards a design principle or rather landscape design on the other hand. The implementation of urban ecological principles is demonstrated on various scales. The practical examples come from China, The Netherlands, and Germany.

It becomes clear from Section E that landscape architecture and ecology have found mutual crystallization points through various practical projects in the last few decades. This means, for example, that concepts such as the creation of seminatural recreation possibilities are connected to the topic of biodiversity. Whilst the basic ideas leading design come from ecology, design techniques and techniques from landscaping are used for their implementation. This can be linked to the transformation of contaminated derelict areas to semi-natural, esthetically pleasing and useful open spaces.

The administrative preconditions for urban landscape planning are different in the three countries dealt with in Section E – Germany, The Netherlands, and China; this becomes clear in the individual chapters. What they have in common, however, is that the analytical–descriptive and normative perspectives are closely interlinked. The realization of ecological principles is dealt with under the explicit inclusion of the possibilities for use by users of green- and open space. Since the green and open space planners usually do not have any power of control over the planned areas, they are dependent on the acceptance of their ideas by the decision makers. In the meantime, ecological principles have blended into the general conscience so that also in landscape architectural competitions at least now and then success can be achieved in the form of winning prizes. It is obvious here to which degree the implementation of urban ecological principles is dependent on the consciousness of the decision maker (and the general public).

The identity of spaces is addressed in Chapter 14 in view of the redesign of an open space in Shanghai into Houtan Park; it is also mentioned at the beginning of Chapter 11, where, in view of the effects linked to globalization, "the loss of the uniqueness of the place" is lamented. The use of native plant material can, in contrast, contribute to the establishment of an identity of urban landscape areas.

It should be questioned critically, whether a uniform design (in contrast to "loss of uniqueness") cannot have an identity establishing role and whether non-native species could also have a potential for establishing identity? Uniformity of designer elements per se is not necessarily negative, but can also be something experienced positively by user groups or provide something meaningful in the sense of a clear use or mood association. This includes the increasing use of palms in central European roadside cafes. The fact that these are seen positively by visitors as "southern holiday flair" shows that identification and identity do not, or do not have to, have anything to do with native species. Identity is not static, and for many people globalization and some of its consequences, such as the partly connected cosmopolitanism, are something positive. In places where, however, due to globalization local and familiar things get lost, and where no urban spaces with a new positive identity arise in their place, there is a danger of the complete loss of identity of an area.

In Chapter 15 "urban political ecologies" are portrayed through the example of water as a resource and its daily availability using global practical examples. The focus of this portrayal lies in the approaches that can contribute to a democratization of water availability. The criticism of several aspects of the daily use of water has a target that neither the infrastructure linked to it, nor the exertion of power for water distribution are unchangeable. Rather, it should be understood that the availability and distribution of water can be changed.

The normative approach in this chapter is constituted from a plea for a "democratization of water supply", which is understood as a contribution to a better and more just life. In this respect a social utopia is implicitly defined on the one hand. On the other hand, the path shown is oriented towards existing practical examples and to existing political–ecological theoretical approaches, in which not only social–political concepts but also up-to-date discussions about "ecologies of water" are taken into account. Thus, it is shown that in many cases neither the privatization of the water sector, nor its exclusive retention as being the municipality's responsibility reveal the optimal way for a guaranteed daily water supply for everyone. In contrast, a corporatization of water supply (water distribution and consumption) and the associated infrastructure and services can be the best solution.

The generally formulated approach for urban political ecologies should be emphasized, which is shown using the example of the "lived reality of water distribution and consumption": urban political ecologists endeavor – building on a process- and actor-related understanding of the distribution of ecological resources – to reduce the existing unfairness by politicizing topics such as the "lived reality of water distribution and consumption." To quote from Loftus (Chapter 15):

> Such understandings disrupt the sense in which the city appears as an immutable never-changing thing, one to which our theories must simply accommodate themselves. Instead they demonstrate the ongoing reciprocity between theorization and action, philosophizing about and changing the world. (Loftus, 2011, Chapter 15, p. 202)

In this quotation the interaction between the analysis and description of the existing relationships ("is perspective") and the normative urban ecological positions ("should-be perspective") are adequately revealed.

To conclude, the chapters of this book convey an "applied urban ecology" even when they build on different areas and topics and even though the normative perspective is dealt with in different ways. Taking into account the different geographic and political contexts from which the contributions and examples come and on which they are based, no further generalizing statements about their implementation can be made, the national and regional differences are simply too big. Let this book serve as a stimulus to reflect on the approaches shown here in relation to their respective national and cultural backgrounds and to use them for one's own analyses with the goal of improving the environmental quality encountered!

Index

Note: Page numbers of illustrations are shown by *italic* type; references to tables by **bold** type.

Applied Urban Ecology: A Global Framework, First Edition. Edited by Matthias Richter and Ulrike Weiland.